Maritime Transportation

DEDICATION

To my mentor, Stian Erichsen.

Maritime Transportation

Safety Management and Risk Analysis

Svein Kristiansen

AMSTERDAM • BOSTON • HEIDELBERG • LONDON • NEW YORK • OXFORD
PARIS • SAN DIEGO • SAN FRANCISCO • SINGAPORE • SYDNEY • TOKYO

Elsevier Butterworth-Heinemann
Linacre House, Jordan Hill, Oxford OX2 8DP
30 Corporate Drive, Burlington, MA 01803

First published 2005

British Library Cataloguing in Publication Data
A catalogue record for this book is available from the British Library

Library of Congress Cataloguing in Publication Data
A catalogue record for this book is available from the Library of Congress

ISBN 07506 59998

Cover images: Berggren, B. et al. (eds.) 1989, Norwegian Shipping (In Norwegian: Norsk Sjøfart). Vol. 1, p. 135. Dreyers Forlag, Oslo.

For information on all Elsevier Butterworth-Heinemann publications visit our website at http://books.elsevier.com

Typeset by Keyword Typesetting Services Ltd.
Printed and bound in Great Britain by Biddles Ltd, Kings Lynn, Norfolk

CONTENTS

PREFACE

Shipping or commercial seaborne transport is largely an international activity by the very fact that ships are operating on the high seas between different countries and parts of the world. The international character is also emphasized by the disintegrated nature of shipping companies, where ownership, management, crewing and operations are located in different countries. Even the country of registration (the Flag State), which has the primary responsibility for safety, may not have any immediate link to the commercial activities. The international character eventually led to the establishment by the United Nations of the International Maritime Organization (IMO). IMO has the prime responsibility for adopting safety regulations but has no power to enforce them. It is a regrettable fact that certain Flags of Convenience (FOCs) show more interest in the fees they collect than in exercising safety control. Shipping is also tarnished by both unregulated and substandard employment practices, which have negative effects on safety. Lastly, it has been questioned whether the commercial orientation of ship classification is justifiable in all respects.

These weaknesses of the safety regime have become more visible in recent decades as a consequence of some large catastrophes to tankers and passenger vessels. Both the public and governments were aroused by the accidents to the tankers *Erika* and *Prestige* and not least to the passenger vessels *Herald of Free Enterprise* and *Estonia*. Today's society is less willing to accept environmental damage and fatalities. The conflict between the coastal state and shipping interests has therefore become more visible, and we also see that consumer groups are targeting ferry and cruise shipping. The authority of IMO has been somewhat reduced by the unilateral actions of certain states: the US has put tougher liability requirements on the shipowner (OPA'90), the European Union has speeded up the implementation of new safety regulations, and some coastal states have started to inspect vessels on their own initiative through so-called MOUs (Memorandums of Understanding).

The negative focus on shipping has had the effect that both the industry itself and the regulators have taken steps to heighten the safety level. IMO has during the last decade introduced both risk analysis (Formal Safety Assessment – FSA) and systematic safety management (International Safety Management Code – ISM). We now see that all stakeholders in the industry are striving for a more professional attitude towards controlling the risks. FSA is based on scientific methods supported by probability theory, reliability techniques and systems engineering. Likewise safety management finds its basis in organization and work psychology, quality management thinking and even anthropology. The striving for higher safety will therefore in the future be based on rational knowledge and not only the subjective experience of individuals.

Naval architects and marine engineers have key roles in the design and building of ships and thereby have considerable impact on safety. In many respects the engineering profession is focusing primarily on safety: hull strength, stability and vessel controllability. But the engineers are also working on the interface between systems and humans: navigating bridge, engine control room and related systems. However, the requirements of efficient building and maintenance processes are often given higher priority than ergonomic and human factors considerations.

The author of this book started some ten years ago to give a course on risk analysis for master's degree students in marine engineering. The modest ambition was to give the students a broader understanding of the safety aspects of the ship itself and as a transportation system. It was also important to address the fact that safety is not only about methods and techniques, but also about priorities and knowledge about safe behaviour. Engineers are also involved in operations and daily decision-making processes that influence risk. They may sometimes have the key responsibility for managing safety in competition with economic and time-pressure considerations. I have also given courses to personnel with a nautical background based on the material in this book. Present nautical education gives the necessary training in mathematics and statistics to follow the more technical aspects of risk monitoring and estimation. The book will hopefully therefore have a broad readership.

The book is organized in 4 parts or 15 chapters:

I. Background	1. Introduction
	2. Maritime risk picture
	3. Rules and regulations
II. Statistical methods	4. Statistical risk monitoring
	5. Decisions in operation
III. Risk analysis	6. Traffic-based models
	7. Damage estimation
	8. Risk analysis techniques
	9. Cost-benefit analysis
	10. Formal safety assessment
IV. Management and operations	11. Human factors
	12. Occupational safety
	13. Accident analysis
	14. Emergency preparedness
	15. Safety management

The parts can also be studied independently or in other sequences depending on whether the focus is on risk analysis that is mainly related to design and planning, or whether the the interest is in safety management.

The first part outlines the present situation with respect to safety in shipping in terms of risk level and dominating accident phenomena. It also focuses on some key problems relating to risk acceptance.

Part II gives a foundation for consistent application of statistical methods in risk monitoring and typical decision-making situations. It assumes that the reader has a basic knowledge of probability and statistics. Safety initiatives are often based on the assessment of the present risk. It is, however, a fact that the risk concept is often ill formulated and understood and, combined with limited data, this may lead to erroneous decisions with large consequences.

Risk analysis methods are outlined in Part III. It is covered on two levels: the ship as an element in a traffic scenario and as an entity in itself. Traffic-related accidents such as grounding and collision must be analysed in a wider context taking the environment, the fairway and maritime traffic into consideration. The consequences of an accident are dependent on the damage to the ship. This book focuses on impact-related damage to the hull. We give an outline of the general methods in risk analysis that have evolved during the last decades in land-based and process-oriented industries. These methods have also found wide application in the marine field. One of the fundamental paradoxes in risk analysis is the fact that we have 'hard' methods but 'weak' decision criteria: What is safe enough? An outline of different decision approaches is therefore given, together with the so-called cost-benefit method. Finally, it is demonstrated how the FSA method may be applied to concrete problems in ship design.

The final part of the book discusses a few aspects of systematic safety management. The first topic is human factors, which is important by the very fact that ships still are operated by humans. An overview is given of the limitations of human performance and how it is influenced by the typical conditions onboard. We choose also to focus on the ship as a workplace and have outlined some of the emerging knowledge with respect to occupational accidents. It should, however, be emphasized that no clear relation has been shown between ship and work accidents, although one may suspect there is one. In order to improve safety it is necessary to understand how and why things go wrong and lead to accidents. Without credible basic knowledge, risk analysis and decisions will be futile. Accident investigation and analysis have therefore been given considerable room in the book. In serious accidents the crew and passengers have to evacuate the vessel. The emergency situation is dramatically different from what one experiences under normal conditions. Design of escape routes and life-saving equipment are therefore critical and must be based on a realistic understanding of how people react in those circumstances.

Finally, it is necessary to admit that the scope of this book is perhaps too large for the number of pages in a typical textbook. This is for others to judge, my hope is only that it may inspire the reader to, further study of this large topic. The references in each chapter may also be of some help.

The manuscript of this book was originally written in Norwegian. In the rewriting process I have had vital help from doctoral student Torkel Soma, M.Sc., and Geir

Fuglerud, M.Sc. They have also given me constructive input and proposals for new text sections and useful examples. Without their help I am not sure the project would have been completed. But, as always, I take full responsibility for any errors and ignorant statements.

The inspiration to write the book came partly from my involvement in a number of EU-sponsored research projects. Directly and indirectly I am indebted to the following colleagues: Lars Egil Mathisen, Egil Rensvik, Odd T. Mørkved, Geir Langli, Martin Olofsson, Piero Caridis, Carlos G. Soares amd Mauro Pedrali.

I would like to dedicate the book to the numerous persons in the seafaring community who are the victims of bad ship design of engineers and incompatible orders from managers.

Svein Kristiansen
Trondheim, March 2004

PART I
BACKGROUND

1

INTRODUCTION

A catastrophe that pleases none is really bad.
(Danish proverb)

1.1 INTERNATIONAL TRADE AND SHIPPING

Waterborne transport of materials and goods has for centuries been the main prerequisite for trade between nations and regions, and has without doubt played an important role in creating economic development and prosperity. The cost of maritime transport is very competitive compared with land and airborne transport, and the increase to the total product cost incurred by shipping represents only a few percent. Negative aspects of waterborne transport include longer transport time as a result of relatively low ship speed, congestion in harbours resulting in time delays, as well as less efficient integration with other forms of transport and distribution.

Shipping has from time to time been under attack for unacceptable safety and environmental performance, and this will be discussed in the next chapter. At this point we only make the following remark: in view of the relatively low cost of transport, it is a paradox that some areas of shipping have a relatively low standard of safety. Efficient transport should be able to pay for acceptable safety.

It has been discussed for some time whether basic economic mechanisms could ensure safe shipping. In this context the following questions are relevant:

- Is there any economic motivation for high levels of safety?
- Who should pay for increased safety?
- Are there any trade-offs between safety and efficiency?

These questions will be addressed briefly in this chapter, as well as throughout this book.

1.2 THE ACTORS IN SHIPPING

In shipping there are a number of actors that have an influence on safety, and the most important of these are presented in Table 1.1.

Table 1.1. Actors in shipping that influence safety

Actor	Influence on safety
Shipbuilder	• Technical standard of vessel
Shipowner	• Decides whether technical standards will be above minimum requirements • Selects crew or management company for crew and operation • Make decisions regarding operational and organizational safety policies
Cargo owner	• Pays for the transport service and thereby also the quality and safety of the vessel operation • May undertake independent assessments of the quality of the shipper
Insurer	• Takes the main part of the risk on behalf of the shipper and cargo owner (i.e. vessel, cargo, third party – P&I) • May undertake independent assessment of the quality of the shipper
Management company	• Responsible for crewing, operation and upkeep (i.e. maintenance) of the vessel on behalf of the shipowner
Flag state	• Control of vessels, crew standards and management standards
Classification society	• Control of technical standards on behalf of insurer • Undertakes some control functions on behalf of the flag state
Port administration	• Responsible for safety in port and harbour approaches • May control safety standard of vessels, and in extreme cases deny access for substandard vessels

It should be evident that the different actors within the shipping domain to some degree have competing interests that may complicate the issue of safety, and this is a result of various factors such as the following:

- Who is controlling whom?
- Who sets the quality standards?
- What is the motivation for safe operation?
- Who is picking up the bill after an accident?

We will return to the questions of safety management and the regulation of shipping in later chapters.

1.3 THE SHIPOWNER

In the case of severe shipping accidents and losses, the shipowner and/or ship management company will be subjected to particular attention. This is natural given the fact that the shipowner owns the damaged/lost vessel, as well as manning, maintaining and operating it.

Questions that are always raised in the context of maritime accidents are whether the shipowner has demonstrated a genuine concern for safety, and whether the standards of the vessel and its crew have been sacrificed for profit. The shipowner may counter such questions by claiming that the standards will not be better than what the market is willing to pay for.

With regard to vessel safety standards there has recently been an increasing focus on the cargo owner, as this is the party that decides which ship to charter and at what price. The charter party (i.e. the contract) gives the cargo owner considerable authority to instruct the Master with respect to the operation of the vessel. Given this important role in terms of safety, it may be seen as a paradox that the cargo owner has minimal, if any, liability in the case of shipping accidents.

Shipowners take some key decisions that have profound consequence for safety. The choice of flag state for registration of vessels, choice of classification society and arrangements for insurance are some key decisions. There exist international markets for these services in which different standards and corresponding fees can be found. The safety standard will therefore to a large degree be a result of what the owner is willing to pay for these services. A much discussed and fairly controversial topic is the increasing practice of 'flagging out', in which the shipowner registers a vessel in a country other than where it operates. Flagging out is mainly done for economic reasons, as shown in Figure 1.1. Availability of cheap labour, and the costs and strictness of safety control, seem to be key concerns for the owner. Based on this it must be asked whether shipowners, through their choice of flag, sacrifice safety.

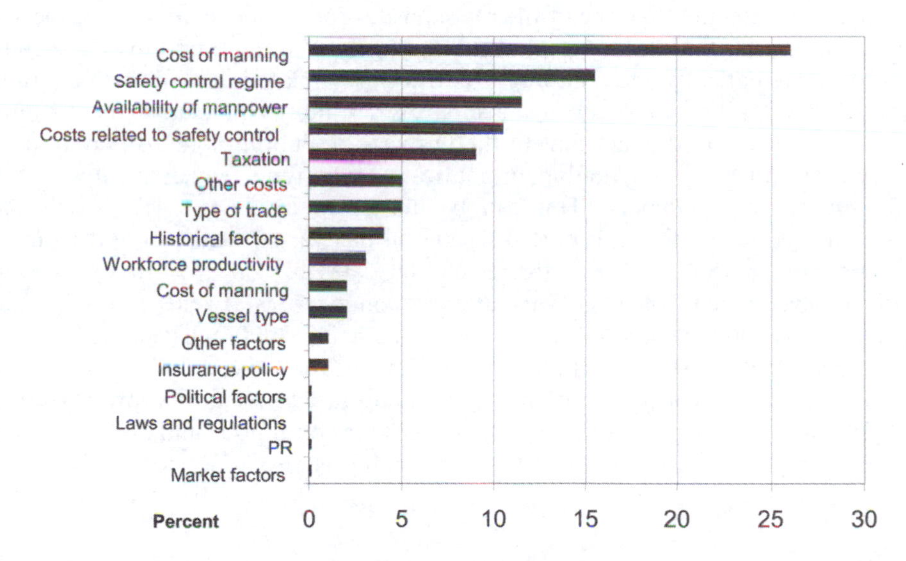

Figure 1.1. Reasons for flagging out: distribution of answers from questionnaire study. (Adapted from Bergantino and Marlow, 1998.)

The safety aspect of ship operation should also be seen in a wider context as shipowners have a number of different objectives that need to be balanced. These objectives include the following:

- Stay in business: return on investment.
- Marketing: win well-paying freight contracts.
- Service: minimize damage on cargo, keep on schedule.
- Efficiency: operate and maintain vessel.
- Employer: attract competent personnel.
- Subcontracting: select efficient service providers.
- Availability: minimize unplanned off-hire.

It is not necessarily obvious that these objectives and priorities for a given company are consistent with a high safety standard at all times. In this view it is of great importance that shipowners have clearly defined policies that never compromise on safety. An alternative view is that there is no conflict between cost, efficiency and safety. The main argument for this position is that in order to stay in business and thrive in the long term, shipowners have to operate safely and keep their fleets well maintained and up to standard. This view may, however, be a little naïve, as substandard shipping companies may not necessarily have a long-term perspective of their business. An OECD (Organization for Economic Cooperation and Development) study has shown that such substandard shipping companies are competitive on price and take their fair share of the market.

1.4 SAFETY AND ECONOMY

It is reasonable to assume that one of the prerequisites for achieving an acceptable safety standard for a vessel is that the company it belongs to has a sound economy and thereby is able to work systematically and continuously with safety-related matters such as training of personnel, developing better technical standards and improving management routines. This is made difficult by the fact that in the business of shipping one usually finds that income and revenues are fluctuating dramatically over time, which creates a rather uncertain business environment. This can be illustrated by Figure 1.2, which shows how the charter rate for the transport of crude oil has varied in the period from 1980 to 2002. As can be seen from this figure, charter rates for oil are heavily influenced by political development such as wars and economic crises. From a top quotation of approximately 80,000 USD/day in the autumn of 2000, charter rates fell to approximately 10,000 USD/day within a period of one year, i.e. a fall in charter rates by a factor of 8. In light of the fact that the minimum rate to result in a profit is around 22,150 USD/day[1], it is clear that the economic basis for continuous and systematic safety work is not the best. The volatility of tanker shipping is also reflected in tanker vessel prices, as shown in Figure 1.3.

---

[1]Front Line Ltd., Investor presentation, 3rd Quarter 2002.

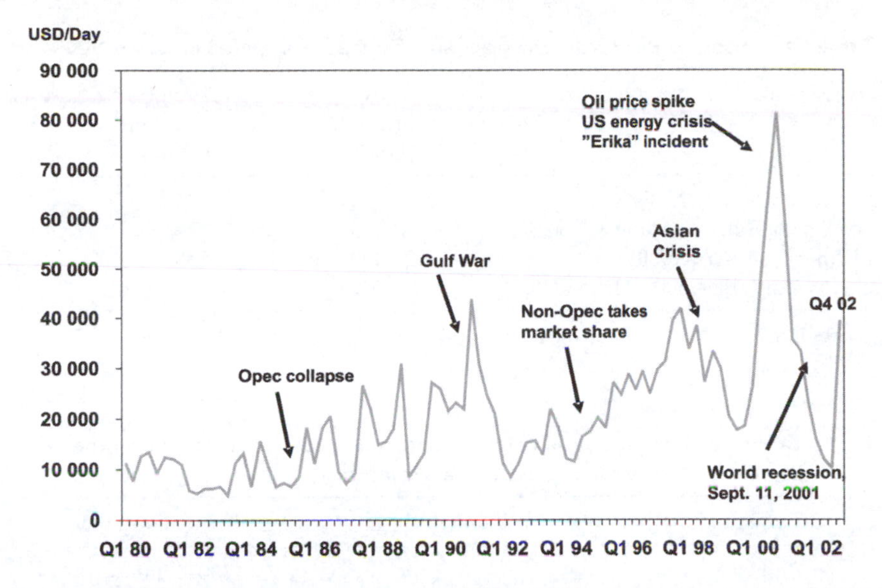

Figure 1.2. VLCC T/C equivalent rate for key routes, 1980 to 2002, given quarterly (Source: P. F. Bassøe AS & Co., *Tanker Fundamentals*, Nov. 2002.)

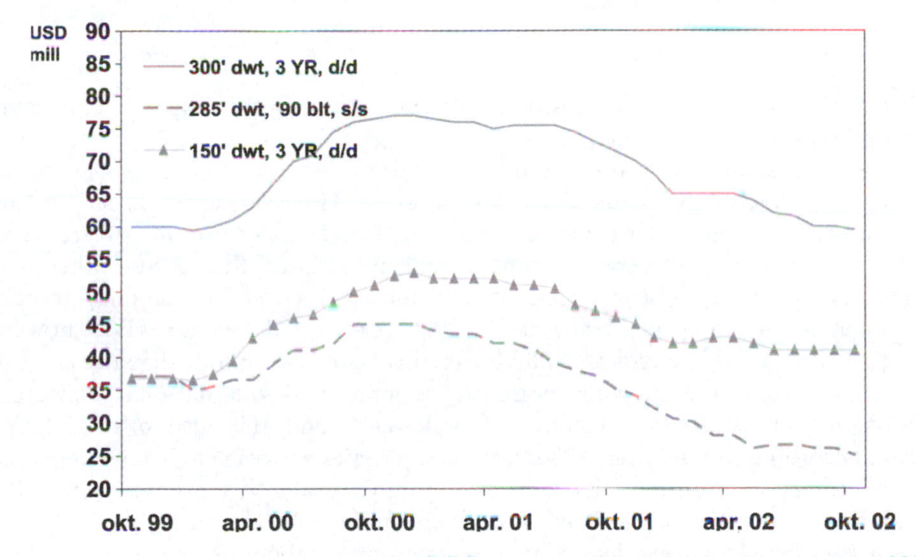

Figure 1.3. Crude oil tanker prices. (Source: P. F. Bassøe AS & Co., *Tanker Fundamentals*, Nov. 2002.)

It is also possible to argue for the opposite view, namely that the low cost of sea transport should put shipping in a favourable position compared with more expensive transport modes. As can be seen in Table 1.2, the relative cost of sea transport (i.e. freight cost in percent of sales price) is in the order of 2–6%. If we assume that the safety of

Table 1.2. Relative price of seaborne transport, Far East–Europe/US (distance: 9000 nm)[a]

Product/vessel/capacity utilization (%)	Sales price/unit (USD)	Freight cost/unit (USD)	Relative freight cost
1 barrel of crude oil/VLCC/50%	30	1.5	5.0%
1 tonne of wheat/Bulk 52,000 DW/100%	220	14	6.4%
1 car/Multipurpose Ro-Ro/50%	21,000	558	2.7%
1 refrigerator/Container 6600 TEU/80%	550	9	1.6%

[a]Market prices in fall 2002.

Table 1.3. Comparison of costs for different modes of transport on the Barcelona–Genoa route (costs expressed in Euros)

Transport mode	Trailer 16.6 m	40 feet container
Railway	—	1300–1500
Road	900	—
Short sea, Ro-Ro	1300	1200

Source: Abeille et al. (1999).

shipping could be improved significantly by a 50% increase in the freight rate, this would have resulted only in a 1–3% increase in the sales price.

An argument against sea transport in these kinds of discussions is the longer transport time due to relatively low speeds and delays in ports. However, an increasing competitiveness of shipping on shorter routes can now be seen. The increase in road transport is currently representing a great environmental problem in central Europe and other densely populated areas due to exhaust emissions, and the road system is becoming more and more congested. A study by Abeille et al. (1999), illustrated by Table 1.3, shows that seaborne transport may be economically competitive even on shorter distances.

There is currently a growing national, regional and international concern for the emissions related to the burning of fossil fuels, and this also affects shipping. However, according to Kristensen (2002), some ship types perform environmentally better than road transport. Some results from Kristensen's study of the environmental cost of road and sea transport can be found in Table 1.4. In studying Table 1.4 it must, however, be recognized that estimating environmental consequences and economic aspects of transport is a highly uncertain and controversial exercise. Nevertheless, Kristensen's study indicates that container and bulk carriers are far better than road transport. Ro-Ro vessels, on the other hand, are less favourable due to large motor installations (resulting in higher speed) and lower cargo capacity. The author also points out that these figures are subject to change due to the continuous toughening of emission standards.

Table 1.4. Environmental cost of road and sea transport, in euros per 1000 tonne-km

Transport mode/source of cost		Environmental cost, €/1000 tonne-km
Truck transport	Emissions	7–12
	Accidents	4–7
	Noise	5–15
	Congestion	5–12
	Total cost	21–46
Container vessel, 3000 TEU		6–8
Bulk carrier, 40,000 dwt		2–3
Ro-Ro cargo ship, 3000 lane-meters		33–48

Source: Kristensen (2002).

1.5 MARITIME SAFETY REGIME

Given the factors pointed out earlier in this chapter, it is not completely obvious that safety is an important issue to companies in the maritime transport domain. One may, however, argue that advanced modern ship design achieves high levels of safety, that training of crew is now of a fairly high standard, and that shipping companies are relatively advanced when compared with similar types of businesses. In addition to this, shipping is subject to rigorous control and continuously has the attention of both governments and the public. Table 1.5 shows that seaborne transport today is strictly regulated as a result of a series of internationally ratified safety conventions.

The average loss rate for the world fleet, measured in annual percent relative to the fleet at risk, has been reduced significantly during the same period studied in Table 1.5. In 1900 the average loss rate was 3%. This had been reduced to 0.5% in 1960, and further down to 0.25% in 2000.

It is too early at this stage of the book to discuss whether the safety level in maritime transport is acceptable. However, it can on the other hand be argued that there is a case for increased safety efforts unless it can be shown that this cannot be defended with regard to the resources spent. Another way of thinking is that safety should be on the agenda as long as accidents are rooted in trivial errors or failures (very often human errors). Thirdly, ship accidents should have our attention as long as they lead to fatal outcomes and the consequences for the environment are unknown. The examples given in Table 1.6 show that maritime safety still is on the agenda and will continue to be so in the foreseeable future.

1.6 WHY SAFETY IMPROVEMENT IS DIFFICULT

Despite the fact that safety is at the top of the agenda both in the shipping business itself and by regulators, it may appear that the pace of safety improvements is rather low. The degree to which this general observation is true will not be discussed in any depth here.

Table 1.5. Milestones in maritime safety[a]

Year	Initiative or regulation
1914	Safety of Life at Sea (SOLAS): Ship design and lifesaving equipment
1929	First international conference to consider hull subdivision regulations
1948	The International Maritime (Consultative) Organization (IMO) is set up as a United Nations agency
1966	Load Line Convention: Maximum loading and hull strength
	Rules of the road
	The International Association of Classification Societies (IACS): Harmonization of classification rules and regulations
1969	Tonnage Convention
1972	International Convention on the International Regulations for Preventing Collisions at Sea (COLREG)
1974	IMO resolution on probabilistic analysis of hull subdivision
1973	Marine Pollution Convention (MARPOL 73)
1978	International Convention on Standards for Training, Certification and Watchkeeping for Seafarers (STCW)
1979	International Convention on Maritime Search and Rescue (SAR)
1988	The Global Maritime Distress and Safety System (GMDSS)

[a]An excellent summary is given by Vassalos (1999).

However, some explanations for such a view are presented below:

- *Short memory*: When safety work is successful, few accidents tend to happen. This lack of feedback can make people believe, both on a conscious and unconscious level, that they are too cautious and therefore can relax on the strict requirements they normally adhere to. An even simpler explanation is complacency, i.e. that people tend to forget about the challenges related to safety if no accidents or incidents give them a 'wake-up call'. This weakness seems to degrade the safety work effectiveness of both companies and governments.
- *Focus on consequences*: People have a tendency to focus on the consequences of an accident rather than its root causes. There is, for instance, great uncertainty attached to whether oil pollution is reduced in the best way by double-hull tankers or heavy investment in containment and clean-up equipment. Doing something about consequences is generally much more expensive compared to averting, or reducing the probability and the initiating causes of an accident.
- *Complexity*: Safety involves technological, human and organizational factors, and it can be very difficult to identify the most cost-effective set of safety-enhancing measures across all potential alternatives. There is also a tendency among companies, organizations and governments to go for technical fixes, whereas the root causes in a majority of cases are related to human and organizational factors. It seems to be easier to upgrade vessels than to change people's behaviour.
- *Unwillingness to change*: Humans have a tendency to avoid changing their behaviour, also when it comes to safety critical tasks. People sometimes express their

Table 1.6. Recent maritime accidents and responses

Background	Response
Need to increase maritime safety, protection of the marine environment, and improve working conditions on board vessels. Flag state control is not regarded as efficient enough	Declaration adopted in 1980 by the Regional European Conference on Maritime Safety that introduced Port state control of vessels, known as the *Paris Memorandum of Understanding (MOU)*[a]
The loss of Ro-Ro passenger ferry *Herald of Free Enterprise* (Dover, 1987), and the loss of passenger ferry *Scandinavian Star* (Skagerak, 1990)	IMO adopts the *International Management Code for the Safe Operation of Ships and for Pollution Prevention (ISM Code)*: Ship operators shall apply quality management principles throughout their organization
Grounding of oil tanker *Exxon Valdez* in Alaska 1989, resulting in oil spill and considerable environmental damages	US Congress passes the *Oil Pollution Act (OPA '90)*: Ship operators have unlimited liability for the removal of spilled oil and compensation for damages[b]
The flooding, capsize and sinking of the Ro-Ro passenger vessel *Estonia*	*Stockholm agreement (1995)*: NW European countries agree to strengthen design requirements that account for water on deck
A need for greater consistency and cost-effectiveness in future revisions of safety regulations	*Interim Guidelines for the Application of Formal Safety Assessment (FSA) to the IMO Rule-Making Process*, 1997
Hull failure and sinking of the oil tanker *Erika* off the coast of France, 1999	European Commission approves a directive calling for tighter inspection of vessels, monitoring of classification societies, and elimination of single-hull tankers[c]
Oil tanker *Prestige* sinks off the coast of Spain, 2002	The European Commission speeds up the implementation of ERIKA packages 1 and 2
Spreading of exotic organisms through dumping of ballast water has resulted in widespread ecosystem changes	Increased focus on research on these issues, and introduction of new regulation and control measures[d]

[a]http://www.parismou.org/
[b]OPA (full text): http://www.epa.gov/region09/waste/sfund/oilpp/opa.html
[c]Erika Package 1: http://www.nee.gr/Files/erika1.pdf
[d]Australian initiative: http://www.ea.gov.au/coasts/pollution/

understanding of the need for change, but in practice use all means to sabotage new procedures. In some companies, 'cutting corners' is unfortunately a natural way of behaving.

- *Selective focus*: Formal safety assessment (i.e. a risk analysis and assessment methodology described in a later chapter) is in general seen as a promise for more efficient control of risk. However, such methods may be criticized in a number of ways: they oversimplify the systems studied, a number of failure combinations are overlooked due to the sheer magnitude of the problem, and operator omissions (e.g. forgetting or overlooking something) are not addressed in such models.

1.7 THE RISK CONCEPT

The concept of risk stands central in any discussion of safety. With reference to a given system or activity, the term 'safety' is normally used to describe the degree of freedom from danger, and the risk concept is a way of evaluating this. The term 'risk' is, however, not only used in relation to evaluating the degree of safety and, as outlined in Table 1.7, the risk concept can be viewed differently depending on the context.

Engineers tend to view risk in an objective way in relation to safety, and as such use the concept of risk as an objective safety criteria. Among engineers the following definition of risk is normally applied:

$$R = P \cdot C \tag{1.1}$$

where $P =$ the probability of occurrence of an undesired event (e.g. a ship collision) and $C =$ the expected consequence in terms of human, economic and/or environmental loss.

Equation (1.1) shows that objective risk has two equally important components, one of probability and one of consequence. Risk is often calculated for all relevant hazards, hazards being the possible events and conditions that may result in severity. For example, a hazard with a high probability of occurrence and a high consequence has a high level of risk, and a high level of risk corresponds to a low level safety for the system under consideration. The opposite will be the case for a hazard with a low probability and a low consequence. Safety is evaluated by summing up all the relevant risks for a specific system. This objective risk concept will be studied in much greater detail in later chapters.

An important question is how people relate to and understand the concept of risk. Table 1.8 gives a brief overview of some of the factors that determine the subjectively

Table 1.7. Different aspects of the risk concept

Aspect	Comments
Psychological	People often relate to risk in a subjective and sometimes irrational way. Some people are even attracted to risk
Values/ethics	Risk can be perceived in the light of fundamental human values: life is sacred, one should not experiment with nature, and every individual has a responsibility for ensuring safety
Legal	Risks and safety are to a large degree controlled by laws and regulations, and people might therefore be liable for accidents they cause
Complexity	The nature of accidents is difficult to understand because so many different influencing factors and elements are involved: machines, people, environment, physical processes and organizations
Randomness	There is often a fine line between safe and unsafe operation. A lack of system understanding may lead to a feeling that accidents happen at random
Delayed feedback	It is difficult to see the cause and effect mechanisms and thereby whether introduced safety measures have a positive effect on safety. Some measures even have to be applied for a considerable period of time before they have a real effect on system safety

Table 1.8. Different perception of risk

Factor	Negative → higher perceived risk	Positive → lower perceived risk
Is the hazard confronted as a result of a personal choice or decision?	Involuntary	Voluntary
Is the consequence (effect) of the act evident?	Immediate	Delayed
Is the cause and effect mechanism clear to the decision-maker?	Uncertain	Certain
Is the individual in a 'pressed' situation that leaves no alternatives?	No alternatives	Alternatives
Does the decision-maker experience some degree of control?	No control	Have control
Risk at work is not the same as risk in your spare time	Occupational	Hobby, sport
Is the risk unknown in the sense that it is seldom or still not experienced?	'Dread' hazard	Common
Are the consequences given once and for all?	Irreversible	Reversible

experienced and perceived risk. In performing risk analyses, engineers should always keep these subjective aspects in mind so as to improve communication with different individuals/groups and be able to achieve mutual understanding of complicated safety issues, etc.

1.8 ACCEPTABLE RISK

Some might argue that any risk is unacceptable. This view is questioned by Rowe (1983), who gives the following reasons for the opposite standpoint (i.e. that some risks are indeed acceptable):

- Threshold condition: a risk is perceived to be so small that it can be ignored.
- Status quo condition: a risk is uncontrollable or unavoidable without major disruption in lifestyle.
- Regulatory condition: a credible organization with responsibility for health and safety has, through due process, established an acceptable risk level.
- *De facto* condition: a historic level of risk continues to be acceptable.
- Voluntary balance condition: a risk is deemed by a risk-taker as worth the benefits.

Rowe (1983) further outlines three different models for how an acceptable level of risk is established in society:

1. *Revealed preferences:*
    - By trial and error society has arrived at a near optimum balance between risks and benefits.
    - Reflect a political process where opposing interests compete.

2. *Expressed preferences:*
   - Determine what people find acceptable through a political process.
   - The drawback is that people may have an inconsistent behaviour with respect to risk and don't see the consequences of their choices.
3. *Implied preferences:*
   - Might be seen as a compromise between revealed and expressed preferences.
   - Are a reflection of what people want and what current economic conditions allow.

Our attitude to risk is also reflected by our views on why accidents happen. Art, literature and oral tradition all reflect a number of beliefs that are rooted in culture and religion (see also Kouabenan, 1998):

- Act of God: an extreme interruption with a natural cause (e.g. earthquake, storm, etc.).
- Punishment: people are punished for their sins by accidents.
- Conspiracy: someone wants to hurt you.
- Accident proneness: some people are pursued by bad luck all the time. They will be involved in more than their fair share of accidents.
- Fate: "What is written in the stars, will happen." You cannot escape your fate.
- Mascots: an amulet may protect you against hazards.
- Black cat: you should sharpen your attention.

1.9 CONFLICT OF INTEREST

As already pointed out earlier in this chapter, in any maritime business activity, and almost any other activity for that matter, there will be conflicts of interests between the different parties involved and affected by that activity. For example, consider an oil company that is planning to establish an oil terminal with refinery capacity in a local community somewhere on the coast. For the different parties affected by this establishment there will be both positive and negative effects, as illustrated by Table 1.9. For the community population the project may, for example, result in infrastructure improvements and an increased number of jobs, both of which are obvious benefits. On the other hand the project may result in negative effects such as air pollution and restrictions on land-use and outdoor life. A problem related to activities such as the oil terminal is that the positive and the negative effects, including the income and costs, may be unevenly distributed between the affected parties.

1.10 EXPERTISE AND RATIONALITY

It should be clear from the preceding observations in this chapter that decisions related to safety are difficult for a number of reasons. As engineers we are inclined to perform rational analyses using computational methods. On the other hand it has been commented that as humans we often have an irrational attitude to risks. In addition, people view risks differently, and there may be conflicts of interests in safety-related issues. The dominant view in the field of safety assessment is that formal risk analysis methods should be used

Table 1.9. Conflicts of interests between the involved parties in the case of an oil terminal/refinery located in a coastal area

Actor	Potential positive effects	Potential negative effects
Oil company	Earn money Enter new markets Increases their market share	Production stops – lost income Some liability for accidents Bad reputation because of pollution
Shipowners	New/alternative trades Earn money	Limited liability for accidents Loose contract Be grey/blacklisted
Employees	Employment Income Improved standard of living	Exposed to accidents Few other job alternatives
Population	Infrastructure improvements More jobs	Air pollution Restriction on land-use and outdoor life
Local economy	Tax income Improved service to the public Increased population	Local economy becomes highly dependent on the terminal Increased wage cost Traditional businesses unable to attract competent personnel Pollution may affect primary industries (fishing, fish farming)
Society at large	Contribution to national economy	Must take most of the cost in case of major accidents and oil spill

on many types of activities, and that the public in general is often ill informed and therefore should have little influence on such complicated matters. Perrow (1999) has questioned this view. Some of his observations of contemporary practices in risk analyses are as follows:

- Expected number of fatalities: whether you die from diabetes or murder is irrelevant.
- Who is at risk? Whether 50 unrelated persons from many communities or 50 persons from a small community of 100 inhabitants die in an accident is also irrelevant.
- People are to a large degree sceptical about nuclear power plants but still continue to smoke. It is irrational to dread the nuclear plants that have shown excellent safety performance, whereas smoking is an undisputed factor of risk in relation to lung cancer. However, this argument totally neglects the fact that smoking is a result of intense marketing and advertisement.
- Risk assessment ignores social class distribution of risk, as may be illustrated by the corporate vice-president's dilemma: By investing USD 50 million in a proposed safety measure, the life of one extra worker can be saved. However, by rejecting the proposal the company will avoid USD 20 million on price increases and be able to give USD 30 million in dividends. The last option is chosen as the price is very high given the depressed labour market.

Perrow (1999) also gives an interesting discussion of three basic forms of rationality: 'absolute', enjoyed by economist and engineers, 'bounded or limited' rationality proposed in cognitive science and organizational psychology, and 'cultural or social' rationality mainly practised by the public. The author delivers many interesting arguments for these alternatives to absolute rationality.

REFERENCES

Abeille, M., et al., 1999, *Preliminary Study on the Feasibility of Substituting Land Transport of Chemical Products by Shortsea Shipping*. Centre d'Estudis del Risc Tecnològic, Universitat Politècnica de Catalunya, Barcelona.

Bergantino, A. and Marlow, P., 1998, Factors influencing the choice of flag: Empirical evidence. *Maritime Policy and Management*, Vol. 25(2), 157–174.

Kouabenan, D. R., 1998, Beliefs and the perception of risks and accidents. *Risk Analysis*, Vol. 18(3), 243–252.

Kristensen, H. O., 2002, Cargo transport by sea and road – technical and economic environmental factors. *Marine Technology*, Vol. 39(4), 239–249.

Perrow, C., 1999, *Normal Accidents*. Princeton University Press, Princeton, NJ.

Rowe, W. D., 1983, Acceptable levels of risk for undertakings. Colloquium *Ship Collisions with Bridges and Offshore Structures*, Copenhagen. IABSE Reports Vol. 41, International Association for Bridge and Structural Engineering, Zürich.

Vassalos, D., 1999, Shaping ship safety: the face of the future. *Marine Technology*, Vol. 36(2), 61–76.

2

MARITIME RISK PICTURE

2.1 INTRODUCTION

An accident can be defined as *an undesirable event that results in damage to humans, assets and/or the environment*. In order to get an understanding of the characteristics of maritime accidents we need insight into the maritime risk picture, and this chapter presents some of the key issues and central elements of this risk picture, including some introductory observations on why accidents happen in the maritime domain. Among other things, this chapter will indicate that accidents generally are complex phenomena. It will also be shown that the level of risk does vary significantly between various maritime activities.

2.2 DEFINITIONS

This textbook will examine a wide range of related issues, such as reliability analysis, risk analysis and safety management, and in this context some definitions of key concepts must be presented (Stephenson, 1991):

- Hazards: possible events and conditions that may result in severity, i.e. cause significant harm.
- RAM analysis: reliability, availability and maintainability analysis.
- Reliability: the ability of a system or component to perform certain defined functions.
- Risk: an evaluation of hazards in terms of severity and probability.
- Safety: the degree of freedom from danger and harm. Safety is achieved by doing things right the first time and every time.
- Safety management: keeping an operation safe through systematic and safety-minded organization and management of both human and physical resources.
- Systems safety: the discipline that utilizes systems engineering and management techniques to make systems safe throughout their life-cycle.

2.3 MARITIME ACTIVITY

Maritime activities have had, and still have, an important role in the business, trade and economy of many countries. The key areas of maritime activities include:

- Maritime transport:
  - Coastal shipping
  - Transport of people both inland and overseas
  - International shipping
  - Cruise shipping
- Fishing
- Marine farming
- Continental shelf operations (i.e. oil and gas):
  - Rig operations
  - Supply services
  - Pipeline laying
  - Underwater activities
- Science and survey

These activities have several positive attributes, such as employment, production, creation of values and fortune, spreading economic consequences, positive influences on currency and exchange transactions, etc. There is, however, a price for these benefits in terms of negative effects. Some of the typical hazards found in maritime activities are outlined in Table 2.1.

Maritime accidents may lead to three different kinds of consequences:

- Harm to human beings: injuries and fatalities.
- Environmental pollution.
- Economic losses: damage or loss of vessel and cargo, lost income, etc.

If we limit our studies to maritime accidents we may distinguish between concept accidents, work accidents and maloperations, as outlined in Figure 2.1.

There is no simple answer to why accidents happen in maritime activities. It will be pointed out later in this chapter that accidents are complex phenomena, and usually no simple solutions exist to prevent them.

2.4 CONCEPT OF ACCIDENT TYPES

Ship accidents are usually classified according to the type of energy release involved. The typical accident phenomena/types are shown in Table 2.2.

In order to understand the nature of ship accidents, one must study the failure mechanisms related to systems or functions. A ship includes several systems and functions that are necessary for it to perform its mission, and some of these are presented in Table 2.3. Both reliability, availability and maintainability (RAM) analysis and risk analysis are

Table 2.1. Threats and hazards in maritime activities

Maritime sector	Hazards
Shipping	Dangerous cargo: fire, explosion, poisoning, environmental damage Ocean environment and weather Substandard ships and substandard shipowners Difficult to control safety due to its international character
Fishing	Relatively small vessels with critical features (e.g. hatches) Ocean environment and weather Operation in coastal waters – grounding and steep waves Partly one-person activities (increases vulnerability if something happens) Development of damage and flooding is fast Lack of training
Offshore	Many new kinds of activities, limited experience and knowledge High pace of development work and construction Continuous development of technology and ways of operation Large concentrations of energy resulting in high fire and explosion risk High utilization of the space on platforms
Diving	Increasing water depth (high pressures, difficult to control) Lack of knowledge about physiological factors Ocean environment – splash zone risks New work processes

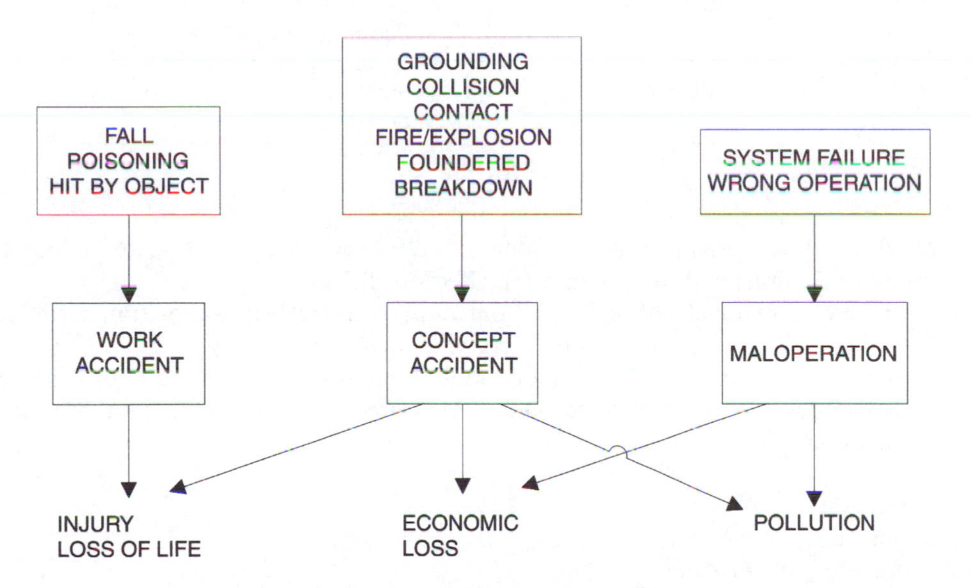

Figure 2.1. Maritime accident types and consequences.

Table 2.2. Accident phenomena

Type	Comments
Collision	Striking between ships
Contact/impact	Striking between ship and other surface objects
Grounding and stranding	Hitting the seabed or shore
Foundering and flooding	Opening and flooding of hull
Hull and machinery failure	Hull or machinery failure is directly responsible for the accident
Fire and explosion	Fire, explosion or dangerous goods release
Missing	
Other miscellaneous	

Table 2.3. Some generic ship systems and functions

Systems	Functions
Accommodation and hotel service	Anchoring
Communications	Carriage of payload
Control	Communications
Electrical	Emergency response and control
Ballast	Habitable environment
Lifting	Manoeuvrability
Machinery and propulsion	Mooring
Management support systems	Navigation
Positioning, thrusters	Pollution prevention
Radar	Power and propulsion
Piping and pumping	Bunkering as storing
Pressure plant, hydraulics	Stability
Safety	Structure

performed on these systems and functions to keep them available (i.e. to reduce the probability of failures) and to minimize the effects of failures.

A common characteristic of accidental outcomes is the release and/or transformation of energy. Figure 2.2 gives examples of energies involved in shipping.

It should be kept in mind that any one failure might lead to different consequences with different degrees of seriousness. Different degrees of seriousness with respect to consequences include:

- An accident
- An incident
- An operating disturbance
- A non-conformance

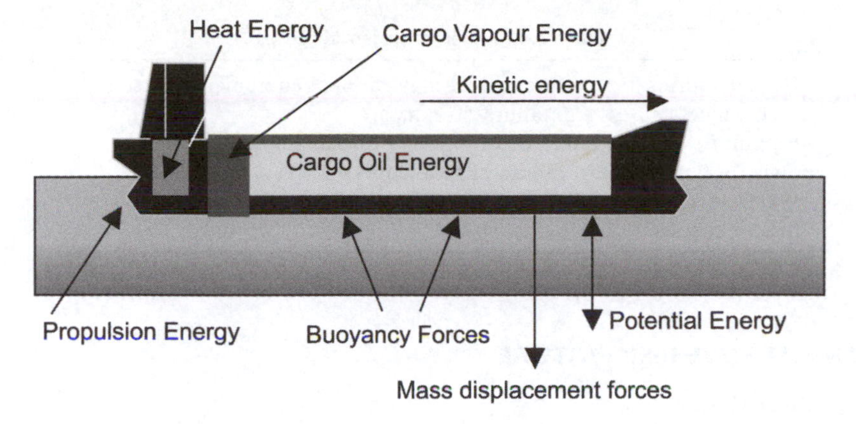

Figure 2.2. Examples of energies involved in shipping.

In the introduction to this chapter, an accident was defined as an undesirable event that results in damage to humans, assets and/or the environment. Incidents are, on the other hand, undesirable events that are detected, brought under control or neutralized before they result in accidental outcomes. If incidents (e.g. small fires in the machinery space of a ship) occur frequently, this indicates an inadequate level of safety. Suddenly one may not be able to bring one of these undesirable events under control, resulting in an accident with harm to personnel, property and/or the environment.

An operating disturbance may take different forms but may be defined as a situation where the operating criteria for a system or component are violated. Typical operating disturbances include:

- Reduced efficiency
- Reduced capacity
- Loss of function
- Operating in emergency mode
- Outside operating performance limits (vibration, wear)
- Temporarily idle

A non-conformance is usually defined as a situation where the operation is outside certain criteria that define what is acceptable.

The causes of accidental outcomes may be highly diverse and are often a combination of several factors. The main groups of accidental causes are listed in Table 2.4. The main objective of performing a risk analysis is to measure the importance of the possible causes for a system and its functions, and to generate and implement safety measures preventing these causes from occurring and/or reducing the consequences if they occur.

The nature of vessel accidents will be discussed at length in a later chapter.

Table 2.4. Generic accidental causes

Human causes (e.g. failure to read navigational equipment correctly)
Mechanical causes (e.g. failure of pumps)
Fire and explosion (e.g. loss of visibility due to smoke)
Structural causes (e.g. failure of bow doors)
Weather-related causes (e.g. high ambient temperature or strong wind)
Miscellaneous

2.5 QUANTITATIVE RISK PICTURE

2.5.1 Vessel Accidents

The different types of ship accidents (identified in Table 2.2) show considerable variation in terms of frequency. The number of total losses for ships greater than 1000 grt (i.e. gross tonnes) was 137 in 2001. The distribution in percent by accident type is shown in Figure 2.3, and from this figure it is clear that the accident type resulting in most losses is foundering, being a result of flooding and loss of hull integrity. Foundering may be the result of a wide range of initiating factors, e.g. extreme weather conditions, failure of hull, engine breakdown, etc.

It is important to consider the underlying data basis of any statistics showing the distribution of different accident types/phenomena. The accident type distribution given in Figure 2.3 was based exclusively on total losses, but in some statistics both losses and serious casualties are combined. Figure 2.4 is an example of the latter type of statistics based on accident and loss data from the period from 1980 to 89. As can be seen from this figure, the inclusion of serious accidents changes the relative importance of different accident types. The most striking difference between Figures 2.3 and 2.4 is that foundering is less dominating in the latter case. This may, however, be expected by the fact that the statistical database used has a large group of accidents not leading to total loss (foundering is, by definition, total loss). It is also interesting to observe that the relative importance of accidents involving grounding/stranding and contact/impact is greater. A dominating category of accidents in Figure 2.4 is 'Hull/Machinery'. This category represents hull and machinery failures that do not result in total loss.

Another complication in comparing accident distributions is the effect of vessel size and type. It is a well-known fact that smaller vessels operating in coastal waters are more prone to certain types of accidents than large vessels operating mainly in open water cross-trades. In Norway there are two ship registers, NOR and NIS and, as shown in Figure 2.5, these two registers may be studied to reveal differences in the accident statistics between smaller coastal vessels and larger vessels operating in international trade. The NOR and NIS registers have the following characteristics:

- NOR: Norway's ordinary register consisting mainly of its native coastal fleet, i.e. primarily smaller vessels.

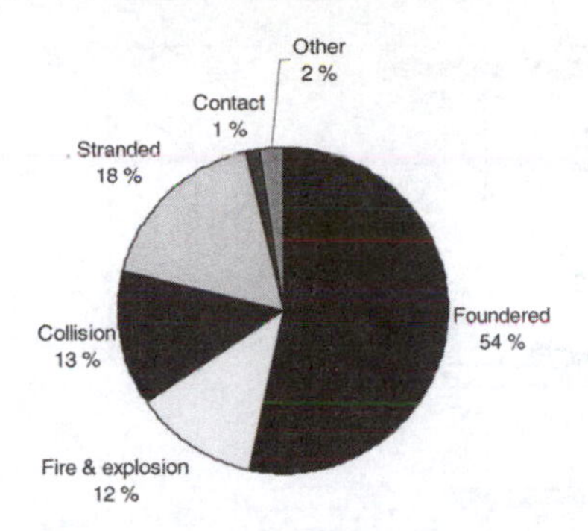

Figure 2.3. Percentage distribution of accident types leading to total loss, world fleet in 2001. (Source: *World Casualty Statistics*, Lloyd's Register of Shipping (1962–93), (1994–98).)

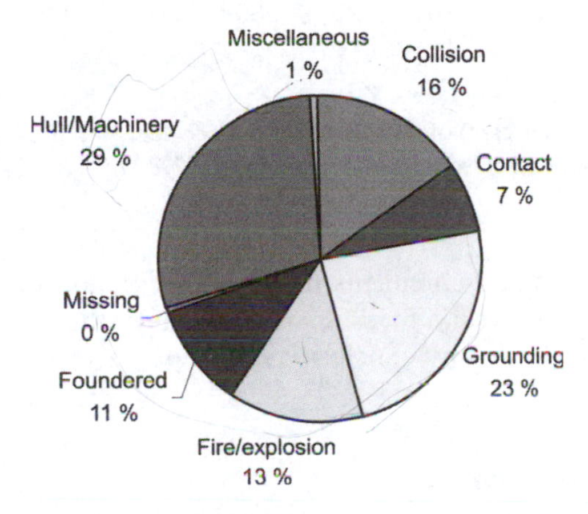

Figure 2.4. Serious ship casualties, world fleet 1980–89. (Source: Lloyd's Register of Shipping (1962–93), (1994–98).)

- NIS: the Norwegian 'open' register consisting primarily of larger vessels operating in international trades.

In Figure 2.5 the total number of accidents studied is somewhat limited due to the fact that only two years were taken as the basis for the accident type distribution. Nevertheless, the following observations can be made:

- The coastal fleet (NOR) is much more prone to grounding. This should not be surprising given that the vessels included in NOR mainly operate close to the coast/shore.
- The international fleet (NIS) experiences, relatively speaking, more collisions. This may, for example, be due to the difficulties related to navigating and operating such large vessels in narrow and busy ports and fairways.
- Relatively, NIS vessels have more fires and explosions than vessels in NOR.

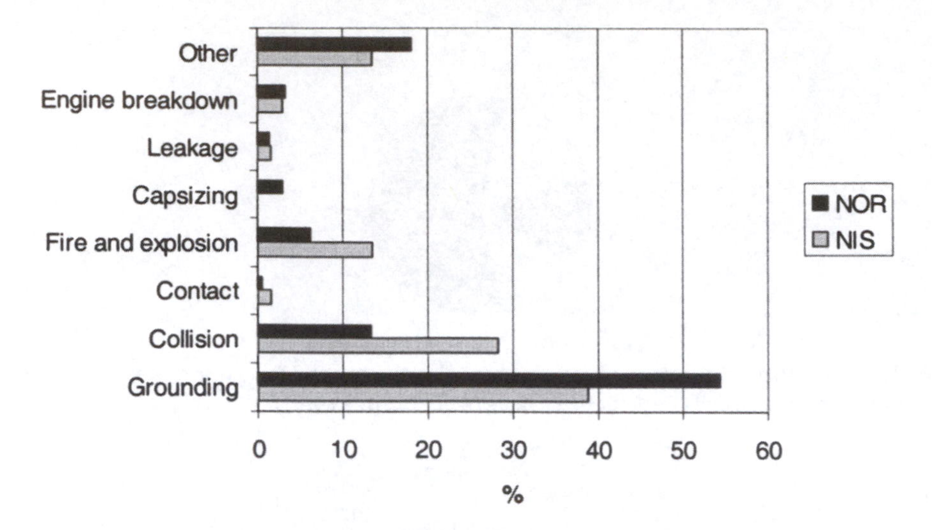

Figure 2.5. Distribution (in per cent) of losses and serious accidents for the Norwegian merchant fleet, 1998–99. (Source: *Official Statistics of Norway: Maritime Statistics 1998–99.*)

From the observations and comments made in this section of the chapter, it should now be clear that it is necessary to evaluate and study the underlying data basis of any statistic before one draws any conclusions both on its results and on accident type distribution.

2.5.2 Trend in Loss Frequency

Over the last decades considerable resources have been spent on reducing the risks involved in shipping. However, the effects of this work are difficult to assess, mainly because isolating the effects of each risk-reducing measure implemented is almost hopeless, given the complexity of cause and effect relations in shipping.

Lancaster (1996) studied the long-term trend of the total loss frequency, and concluded that the annual loss rate had been reduced by a factor of 10 in the twentieth century, from more than 3% in 1900 down to 0.3% in 1990. However, as illustrated by Figure 2.6, the rate of improvement was greatest in the first half of the century, after which the improvement rate levelled out as the potential for further improvements became less and the relative improvements achieved by each safety measure became smaller.

Figure 2.7 gives a more detailed account of how the loss ratio has been reduced for different accident types. The upper line indicates the total loss rate, including all the different accident types. As can be seen, the total loss ratio had setbacks around 1968 and 1980. A further analysis of Figure 2.7 shows that the main contribution to the long-term reduction in loss rate comes from the grounding/stranding category where the loss ratio has shown dramatic improvement.

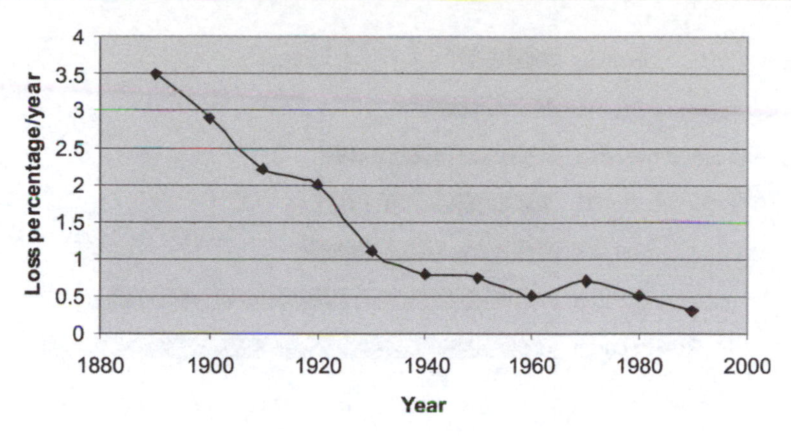

Figure 2.6. Annual percentage of ships lost worldwide.

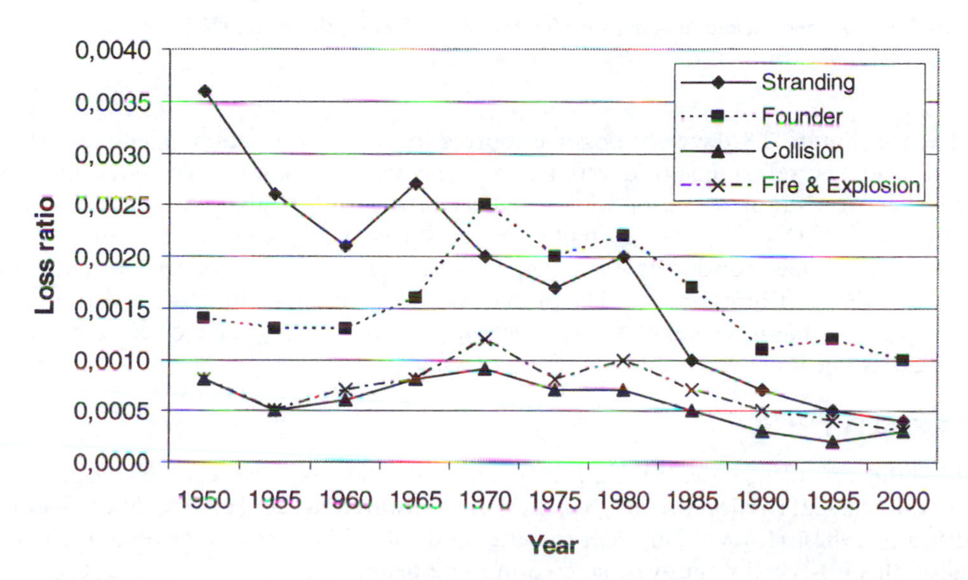

Figure 2.7. World loss ratio time series for ships above 100 GT. The loss ratio is given by annual losses per ships at risk.

Figure 2.8 gives a serious accident rate distribution for different ship types. From this distribution the following observations can be made:

- Ro-Ro (Roll-on Roll-off) cargo vessels are the most accident-prone of all vessel types. This has been explained by operation in congested coastal waters, relatively high-speed operation and weak hull subdivision against flooding.
- Tankers and liquefied gas ships are less vulnerable than general cargo.
- Non Ro-Ro passenger vessels have the best accident rate performance among trading vessels.

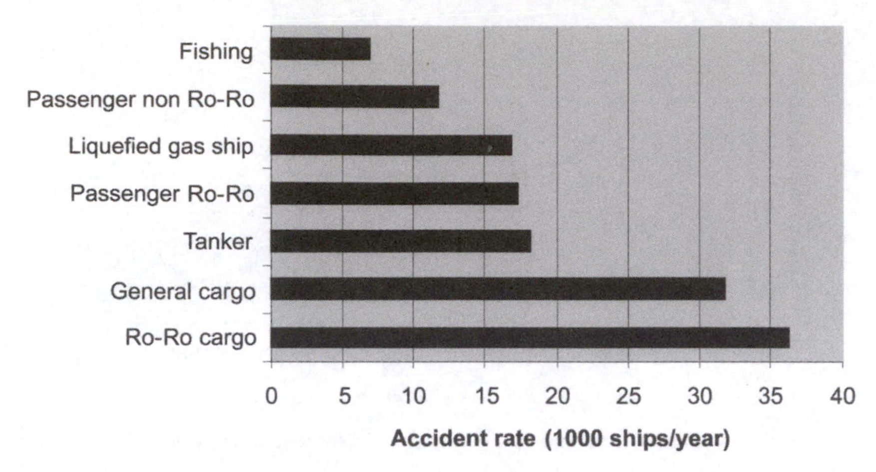

Figure 2.8. Loss and serious accident rate for the world fleet by ship type, 1980–89. (Source: Lloyd's Register of Shipping (1962–93), (1994–98).)

Even in Figure 2.8 there are possible sources for misinterpretation. The low accident rate of fishing vessels should be questioned on the basis of our knowledge about this vessel category, i.e. operation in coastal waters, critical hatches, often subjected to harsh weather conditions, etc. However, an important characteristic of fishing vessels is that the majority of these vessels have considerable inactive periods, whereas trading vessels normally only have a few days off-hire per year. The ratio of accidents to the number of vessels at risk is therefore somewhat misleading. A better measure of exposure might have been the number of accidents per hour of operation (fishing or sailing).

2.6 FATALITY RISK

Only about 5% of all fatalities happen because of accidents. The relatively large public focus on accidents reflects society's considerable awareness of these fatalities (Kårstad and Wulff, 1983). However, in order to better understand how people relate to the risk of fatality, this observation must be studied in more detail.

Accidents may be analysed on the basis of personal characteristics. There is, for instance, a significant correlation between age and the risk of fatality through accidents. For people aged between 5 and 40, accidents constitute the single most important death threat, while they constitute only 5% of the fatalities of people 60 years and above.

There are several alternative ways of measuring accident frequency. One approach is to include all accidents, irrespective of the types of consequences involved. Another approach is to consider only a specific consequence such as fatalities or material and economic losses. Accident frequency varies between different activities, and as shown in Table 2.5, shipping appears to be relatively risky compared to other industrial activities and sectors. Also other maritime activities, such as offshore work (i.e. continental shelf in Table 2.5) and fishing have high fatality rates.

Table 2.5. Fatality frequency in various activities

Industrial activity	Fatalities per 1000 worker-years
Mining	0.9–1.4
Construction	0.3
Industry	0.15
Shipping	1.9–2.1
Continental shelf	2.3
Fishing	1.5

Table 2.6. Transportation risk

Mode of travel	Fatalities per 10^8 passenger-km	Fatalities per 10^8 passenger hours (FAR)
Motorcycle	9.7	300
Pedal cycle	4.3	60
Foot	5.3	20
Car	0.4	15
Van	0.2	6.6
Bus/coach	0.04	0.1
Rail	0.1	4.8
Water	0.6	12
Air	0.03	15

A comparison of different modes of transport is given in Table 2.6 above, and from these statistics it is clear that there are much more dangerous forms of transport than waterborne. There is, for example, less focus on the use of motorcycles than the statistical values should indicate. This can be explained by the fact that the use of a motorcycle is a kind of personal choice, and that the consequence normally is limited to one or two fatalities.

The risk level in the maritime sector is a result of a number of factors relating to the environmental conditions maritime activities are subjected to and the way the work processes are organised. The main factors or hazards were summarised in Table 2.1 earlier in the chapter.

Fatality risks in shipping vary considerably between the different accident types. Figure 2.9 shows the relative distribution of fatality risks on the basis of the number of fatalities, the number of accident cases resulting in fatalities, and the total tonnage involved. As can be seen in this diagram, the accident types leading to the highest number of fatalities are collisions and foundering. This is in accordance with similar observations made earlier in this chapter. Although there are many groundings, these accidents tend to give relatively few fatalities.

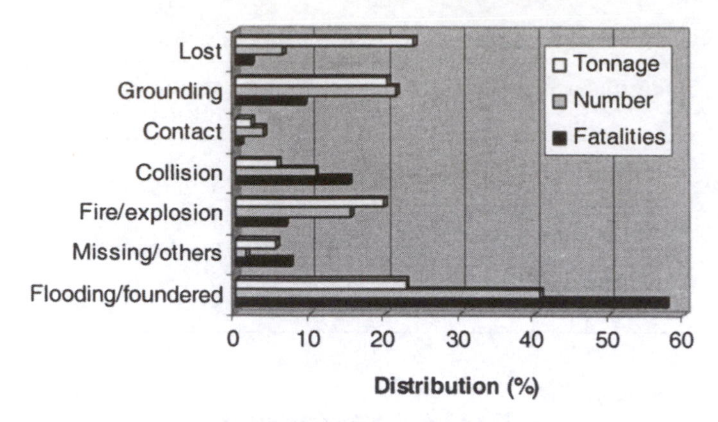

Figure 2.9. Risk parameters for vessel accident types.

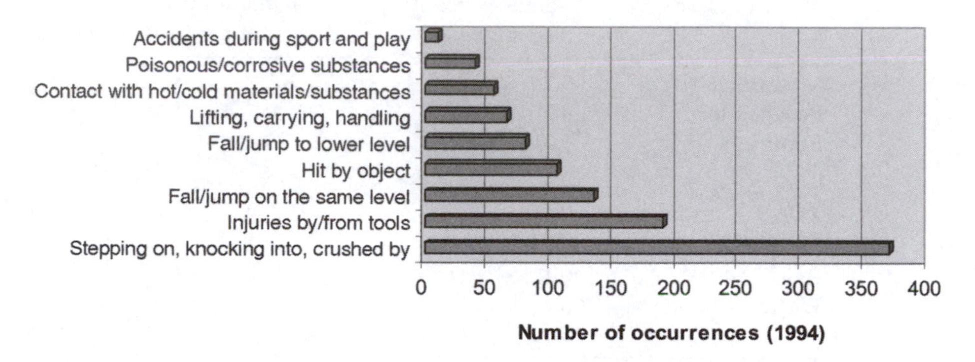

Figure 2.10. Work-related accidents in the Norwegian fleet by category, 1994.

2.6.1 Work Accidents

Fatalities and injuries in shipping are not just related to vessel accidents such as groundings and collisions, they are also related to the work processes on board the vessels. The deck, cargo area, engine room and galley are all seen as dangerous workplaces on a ship. Figure 2.10 gives a breakdown of work-related accident categories/types, and their corresponding number of occurrences, in the Norwegian fleet. It appears that the accident categories of 'stepping on, knocking into, crushed by' and 'injuries by/from tools' are the dominating work-related accidents.

2.7 POLLUTION

In recent times there has been an increasing focus on the environmental aspects of maritime activities. Ships are polluting both the marine environment and the atmosphere, and although there is general agreement about these negative effects of maritime transport, there is considerable uncertainty about the magnitude of the problem. One of the first estimates on this was done in 1985 (see Table 2.7), and this study indicated

Table 2.7. Total input of hydrocarbons into the marine environment in 1985

Source of pollution	Best estimate (million tons annually)	
Natural sources		0.25
Offshore production		0.05
Maritime transportation:		1.50
Tanker operation	0.70	
Tanker accidents	0.40	
Other	0.40	
Atmospheric pollution carried to the sea		0.30
Municipal and industrial wastes and runoff		1.18
Total		3.28

Source: National Research Council, US.

Table 2.8. Estimated world maritime sources of oil entering the marine environment

Source	1990(1)	1981–85(2)	1973–75(2)
Bilge and fuel oil	0.25	0.31	
Tanker operational losses	0.16	0.71	1.08
Accidental spillage:			
Tanker accidents	0.11	0.41	0.20
Non-tanker accidents	0.01	–	0.10
Marine terminal operations	0.03	0.04	0.50
Dry-docking		0.03	0.25
Scrapping of ships	0.01		
Total	0.57	1.50	2.13

Sources: (1) US Coast Guard; (2) National Research Council, US.

that the share of hydrocarbon ocean pollution that could be related to maritime transportation was approximately 45%. The main problem was the spill of oil related to cargo operations and tank cleaning. Massive efforts during the last decade have, however, reduced these sources of oil pollution, as is illustrated above by Table 2.8. It is presently assumed that accidental spills are the major problem. It should, however, be kept in mind that statistics on accidental spills can be somewhat misleading due to the fact that the annual volume of spills varies considerably.

2.8 THE RISK CONCEPT

Risk in the context of engineering is normally presented as the product of the consequences and the probability of occurrence. Quite often, however, the consequences are hard to quantify and may involve some degree of subjectivity. For this reason it is

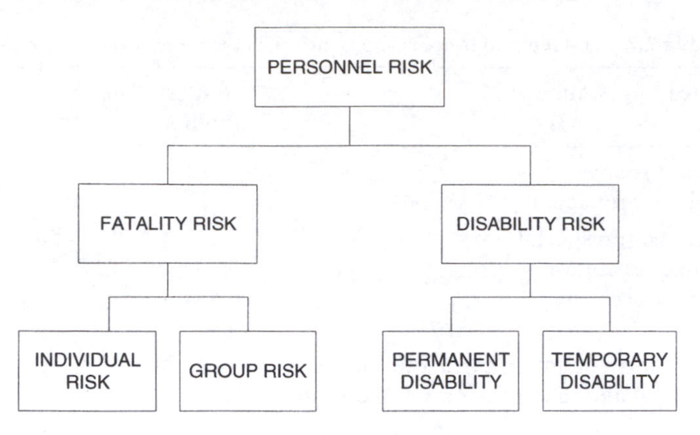

Figure 2.11. Personnel risk.

quite common to present the risk as a probability measure for the various categories of consequences. The categories for personnel risk may be grouped according to Figure 2.11. The abbreviations in Figure 2.11 will be explained in detail below.

Similar approaches may be used for environmental risk and risk associated with damage to assets. There are numerous alternative measures for each consequence, and some examples are presented in Table 2.9.

One should be aware of the fact that the various actors involved in the safety work may apply statistical measures differently. For example, safety managers normally consider the experienced level of safety, while risk analysts are mainly concerned with the estimated/predicted level of risks/safety.

As the total risk picture for a given activity or system can be very complex and involve many different aspects, it is often necessary to break it down into risk scenarios. Equation (2.1) below computes the total risk for a given activity/system as the sum of the risks for each accident type and each phase of the accidental process:

$$R = \sum_i \sum_j p_{ij} \cdot c_{ij} \tag{2.1}$$

where:

R = Total risk
i = The number of scenarios that may lead to a particular consequence (e.g. Table 2.2)
j = Number of phases within each accidental outcome (e.g. initiating event, mitigation, escape, evacuation and rescue)
c_{ij} = Consequence measure for the relevant scenario and phase of the accidental process, e.g. n fatalities, m tons of spill, etc.
p_{ij} = Probability (or frequency) of the relevant consequence c_{ij} for a given scenario and accidental phase

Table 2.9. Risk criteria

Type			Definition
Occupational accidents	LTI-rate	Lost-Time Injury frequency rate	Number of lost-time injuries per 10^6 employee hours
	AIR	Average Individual Risk	Number of fatalities per exposed individual
	IIR	Injury Incident Rate	(Number of reportable injuries in a financial year)/(Average number employed during the year) $\cdot 10^5$
	IFR	Injury Frequency Rate	(Number of injuries in the period)/(Total hours worked during the period) $\cdot 10^5$
	S-rate	Severity rate	Number of working days lost due to lost-time injuries per 10^6 employee-hours. Fatalities and 100% permanent disability account for 7500 days working days lost
	TRI-rate	Total Recorded Injury rate	Total number of recordable injuries (including lost-time injuries, medical treatment injuries and injuries resulting in transfer to another job or restricted work) per 10^6 employee-hours
		Average number of days lost	S-rate/LTI-rate
	FAR	Fatal Accident rate	Number of fatalities per 10^8 working hours
		Fatality rate	Fatalities per 1000 worker-years
	PLL	Potential Loss of Life	Number of fatalities experienced (or predicted) within a given period of time, e.g. the number of lives lost per year in shipping
Work-related	WRD-rate	Work-Related Diseases rate	Number of new cases of possible work-related diseases resulting in absence from work per 10^6 employee hours
		Sick leave percentage	Number of sick-leave days as percent of total number of possible workdays
Pollution		Rate of emissions	Emissions due to accident in kg/m^3 per ton production, e.g. the emissions of fluor in kg per ton produced primary aluminium
Material losses		Loss rate	Number of accidents or losses per produced unit, e.g. collisions per 10^5 nautical miles sailed
		Loss ratio	Number of ships (or tonnage) totally lost per number of ships (or tonnage) at risk, e.g. merchant vessel lost world-wide per total number of merchant vessel
		Relative loss ratio	The loss ratio for an activity divided by the world-wide loss ratio. A loss ratio of one (1) corresponds to the world average, and a higher loss ratio indicates higher losses than the world average

2.8.1 Fatality Risk

Table 2.9 defined a number of fatality risk measures that may be used. The Potential Loss of Life (PLL) measure is a basic measure that may be calculated according to Eq. (2.1). However, this risk criterion has the shortcoming of not incorporating any exposure measure. As outlined in Figure 2.11, it is also necessary to make a clear distinction between individual risk and group risk.

The most commonly used risk measures for individual fatality risk are the Average Individual Risk (AIR) and Fatal Accident Rate (FAR). The AIR measure is calculated by dividing the PLL by the number of people exposed, e.g. the crew size on a merchant ship. In some accident cases only the number of crew on duty are considered. The FAR value is calculated by dividing the PLL by the total man-hours of exposure, and multiplying this measure by a compulsory 10^8 scaling value. The FAR is therefore the expected (or experienced) number of fatalities per 10^8 working hours.

Example

Problem

Accidents involving passenger ferries may result in a large number of fatalities, and hence attract considerable media attention. On the other hand, experts often consider the objective risk of such large-scale accidents as relatively low. From an analysis of the safety level of ferries in the UK since 1950 (Spouge, 1989), it was found that 3 large-scale ferry accidents resulted in an average of 107 fatalities per accident (i.e. 41% of the passengers aboard). Over the period of time studied in the analysis, this gives an average number of about 9 fatalities per year. This average number is, however, not representative for the real distribution of a high number of fatalities on a few number of cases.

In the period of time studied, the UK ferry traffic involved an average of approximately 28 million journeys per year (domestic and international), one return journey per passenger per year, and a typical journey duration was estimated to be 3.5 hours. Given this information, find the Average Individual Risk (AIR) and the Fatal Accident Rate (FAR) for UK ferries.

Solution

Assumptions:
The relatively low number of accidents is representative of the risk picture of UK ferries.

Analysis:
The Average Individual Risk (AIR) rate can be calculated as follows:

$$\mathrm{AIR} = \frac{9[\text{fatalities/year}] \cdot 2[\text{journeys/person}]}{28 \cdot 10^6[\text{journeys/year}]} \approx 6.4 \cdot 10^{-7}[\text{fatalities/person}]$$

The AIR rate for UK ferries may be put into perspective when compared to other UK activities:

- Smoking $5.0 \cdot 10^{-3}$
- All natural causes $1.2 \cdot 10^{-3}$
- Work $2.3 \cdot 10^{-5}$
- Driving $1.0 \cdot 10^{-4}$
- Railway $2.0 \cdot 10^{-6}$
- All accidents $3.0 \cdot 10^{-4}$
- Lightning strike $1.0 \cdot 10^{-7}$

This confirms that the average individual risk for UK ferries is not alarming, compared to other types of transport. However, the AIR value does not include the time of exposure, and in this context the Fatal Accident Rate (FAR) is valuable. The FAR, which is the number of fatalities per 10^8 exposed hours, can be calculated as follows:

$$\text{Exposure time} = 28 \cdot 10^6 \left[\frac{\text{journeys}}{\text{year}}\right] \cdot 3.5[\text{hours}] = 9.8 \cdot 10^6 \left[\frac{\text{hours}}{\text{year}}\right]$$

$$\text{FAR} = \frac{9[\text{fatalities/year}] \cdot 9.8 \cdot 10^6[\text{hours/year}]}{10^8[\text{hours}]} \approx 8.8[\text{fatalities}]$$

The FAR value of 8.8, i.e. 8.8 fatalities per 10^8 exposed hours, can also be compared to other UK FAR values for alternative means of transportation:

- Motorcycle 660
- Aeroplane 240
- Bicycle 96
- Personal car 57
- Railway 5
- Bus 3

The FAR value for UK ferries indicates that ferries are among the safest means of transportation. However, aeroplanes obtain an unfairly high FAR because of the high velocity at which they travel. Therefore, when comparing two alternative ways to travel, the risk per trip is a more reasonable measure than the risk per hour.

In addition to estimating the risk to individuals, attention should also be paid to group risk. Group risk criteria will be explained below.

Group risk criteria can often describe the inherent risk level for an activity or system in a more comprehensive, differentiated and understandable way than most individual risk criteria. The most commonly used technique for presentation of group risk is the *f–N* diagram illustrated by Figures 2.12 and 2.13 below. In *f–N* diagrams, *f* denotes the frequency of accidents causing *N* fatalities. Figure 2.12 shows how *f–N* curves can be used to graphically describe limits of risk acceptance. The curves in Figure 2.12 are established by defining different combinations of consequence (i.e. fatalities) and related frequency that give negligible, acceptable and unacceptable risk, respectively. The hatched area in Figure 2.12 shows the prescribed accepted risk level in the Netherlands. The area above gives higher frequencies and thus increased group risk, and is therefore denoted as unacceptable. In the area below the hatched region the frequencies of occurrence are lower, resulting in lower risk and higher level of safety.

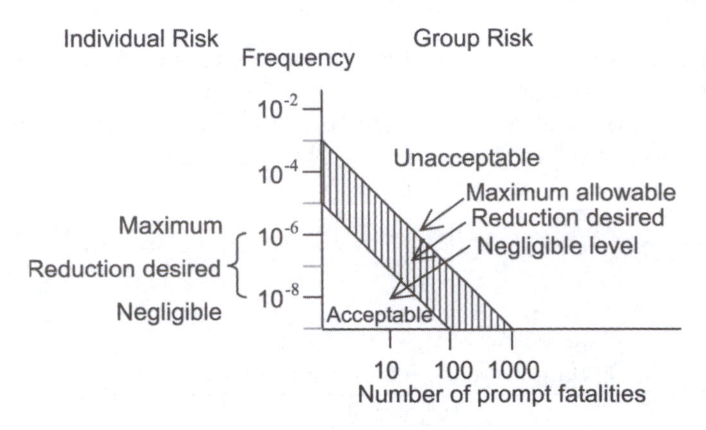

Figure 2.12. Limits of risk acceptance in the Netherlands. (Source: Environmental, 1985.)

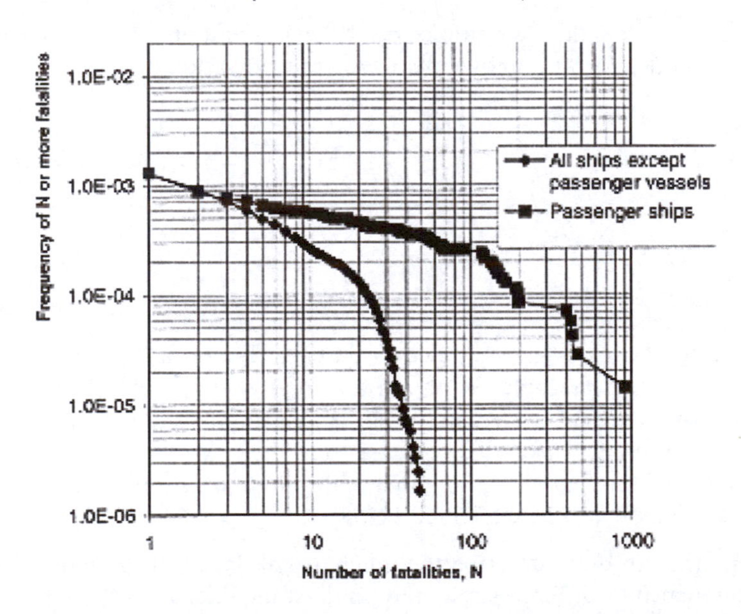

Figure 2.13. Frequency of accidents involving *N* or more fatalities. (Source: DNV, 1998.)

f–N diagrams may describe both the observed risk level for a system or activity and the prescribed (required) risk level. Figure 2.12 is an example of the latter, whereas Figure 2.13 shows observed *f–N* values for passenger ship accidents (i.e. the upper curve) and cargo ships (i.e. the lower curve). It can be observed that for passenger ships 'smaller' accidents involving 1 fatality happen with a frequency of approximately 10^{-3} per ship year, whereas extreme catastrophes with 1000 fatalities happen with a frequency of roughly 10^{-5} per ship year.

2.9 LARGE-SCALE ACCIDENTS

Large-scale maritime accidents, especially those involving fatalities and environmental pollution, get considerable media and public attention, and are often followed by public debate about maritime safety, political discussions regarding the maritime safety regime, and occasionally governmental actions and international regulatory initiatives. The significant attention of such accidents is rooted in the extensive consequences that are perceived publicly as unacceptable. Nevertheless, large-scale accidents normally represent a rather small part of accident occurrences and their contribution to the total risk picture may be relatively low.

There is no generally accepted definition of the term 'large-scale accident', mainly because what is regarded as 'large-scale' may vary between different activities and the fact that we all have a subjective perception of accident consequences. An example may be used to illustrate this: a car accident resulting in five fatalities, all individuals from the same family, will naturally be perceived as a large-scale accident for the remaining family and friends. Society may, however, perceive the same accident as more 'normal', if such a term can be used. A helicopter crash resulting in five fatalities during personnel transport to an offshore installation may, on the other hand, be considered as large-scale by society, hence achieving far more media attention and resulting in public scrutiny of the safety regime for transportation to offshore installations.

Because of the factors described above, it is difficult to give a general objective definition of large-scale accidents, and such criteria must be developed depending on the activity under consideration and public perception. For example, in a Norwegian study large-scale accidents were defined as involving more than five fatalities or economical losses larger than 10 million NOK (approximately 1.5 million USD). Similar quantification can be used for environmental damage and other losses. Table 2.10 gives a summary of large-scale maritime accidents affecting the Norwegian fleet or occurring in Norwegian waters in the period from 1970 to 2000.

The accident and loss of MS *Sleipner* is studied in greater detail below.

Example: The MS *Sleipner* Casualty

What happened?

The fast catamaran ferry MS *Sleipner* (Figure 2.14) had only operated the route between Bergen and Stavanger on the west coast of Norway for about 3 months when it grounded at 19:07 on 26 November 1999. The vessel carried a total of 85 passengers and crew at the

Table 2.10. Large-scale accidents affecting the Norwegian fleet or occurring in Norwegian waters (1970–2000)

Vessel/platform	Key facts
Deep-sea driller	Date: 1 March 1976 Semi-submersible platform Loss of tow resulted in drifting, stranding and loss of buoyancy Consequences: total loss and 6 fatalities
Statfjord A	Date: 1 February 1978 Concrete gravitational platform Fire in leg during installation Consequence: 5 fatalities
Berge Vanga	Date: 29 October 1979 Oil/ore carrier Welding in cargo tanks without sufficient tank cleaning and freeing of gasses resulted in a fire and gas explosion Consequences: foundered in the South Atlantic ocean, 40 fatalities
Alexander L. Kielland	Date: 27 March 1980 Semi-submersible platform Material exhaustion resulted in fracture and loss of main column, followed by ingress of water, heel, and finally loss of stability Consequences: total loss and 123 fatalities
Concern	Date: 4 November 1985 Cement-carrying barge (reconstructed from ship) Cargo shift resulted in capsizing Consequence: 10 fatalities
Soviet submarine	Date: 7 April 1989 Soviet nuclear-powered submarine Caught fire and foundered about 180 kilometres south west of Bjørnøya Consequences: total loss and 41 fatalities
Scandinavian Star	Date: 7 April 1990 Ro-Ro passenger ship Fire started by arsonist in the accommodation area, followed by poor organization of fire fighting, evacuation, and rescue Consequences: 158 fatalities and huge material damages
Sea Cat	Date: 4 November 1991 Fast catamaran ferry Loss of navigational control, struck land Consequences: 2 fatalities, 74 injured, material damages
MS *Sleipner*	Date: 26 November 1999 Fast catamaran ferry Loss of navigational control resulted in grounding. Violation of operational restrictions. Foundered in heavy sea. Poor emergency equipment and organization Consequences: total loss of vessel and 16 fatalities

Type: Austal/Nautica 42 passenger catamaran
Built: 1998/1999
Length: 42,16 meter.
Beam: 13,5 meter
Registered for Passengers: 358
Propulsion power: 2 x MTU, 16v 4000 M79, each of 2320 kW
Speed: 35 knots

Figure 2.14. The fast catamaran ferry MS *Sleipner.*

time of the accident. The weather conditions were rather unpleasant, with estimated gale force winds and about 2 metres significant wave height, when the vessel, 140 metres off course, had a powerful impact with a rock. The vessel's bow was immediately damaged in the impact, and 45 minutes later the vessel slid off the rock and sank to about 150 metres depth. The evacuation equipment and organization failed, resulting in 16 fatalities.

Circumstances and contributing factors

The vessel had a new type of life raft installed that had not been previously used in Norway. By installing advanced emergency equipment the company was allowed to reduce the crew size. Because there had been no hard weather evacuation training, the vessel was not allowed to sail under the existing weather condition. The operational limitations were based on the statistical measure of significant wave height (H_s), which is impossible to measure precisely without special equipment not found onboard.

Immediate causal factors

The immediate causal factors to the accident included the following:

- The bad weather reduced the efficiency of the radar.
- The experienced captain did not detect that the vessel entered two red sectors from lighthouses nearby.
- The vessel hit the rock at a speed of approximately 33 knots (the rock was detected some seconds before the impact, allowing the captain to reduce the speed slightly). Because of the speed involved the passengers hardly felt the impact.
- The crew had poor or little training in emergency situations and evacuation.
- The public address (PA) system failed.
- The emergency rafts could only be released manually by executing 24 operations in the correct order, as the automatic release equipment was not yet installed. As a result only one of the rafts was released.

- The life jackets were of an old design and were difficult to put on and fasten properly, forcing several passengers to jump into the sea without a life jacket fitted.
- The poor weather conditions reduced the rescue efficiency, and the cold water resulted in fast hypothermia.

Basic causal factors

Representatives from the shipping company inspected the vessel three days before the accident and found 34 non-conformities. Still the vessel was considered as seaworthy. In addition, the top-level management of the company was well aware of the violation of the sea state restrictions but nevertheless did not change the practice. They were also aware of the poor training of the crew, and during the accident investigation that followed it was revealed to be unclear who was responsible for the overall safety.

Representatives from the Norwegian Maritime Directorate (NMD, i.e. the Administration) had been aware of the poor life jacket design for 24 years without updating the requirements. NMD's Ship Control Unit had approved the life raft arrangement although it did not meet the functional requirements. In addition, the Administration allowed the vessel to sail, despite the fact that there had been no evacuation and emergency training of the crew.

The Norwegian government had planned to improve the marking of the fairway, and to build a light marker on the rock on which MS *Sleipner* grounded. However, the plan was changed and not completed.

The shore-based Search And Rescue (SAR) base did not monitor the international VHF safety channel (i.e. 16). As a result they had to be alerted by a radio channel, which resulted in longer respond time.

2.10 THE ACCIDENT PHENOMENON

2.10.1 The Accident as a Process

Through its activities/operation a maritime system is exposed to hazardous situations and therefore also to risks of undesirable incidents and accidents. An initiating (or triggering) event, together with contributing factors of operational, environmental and technological aspects, constitutes the so-called casual network leading to an accident. The accidental event itself 'ignites' an escalation process within the system under consideration (e.g. a ship or part of a ship), resulting in physical damage and release of energy, which will expose humans, the activity and the environment to various consequences. To gain insight into the accident phenomena, it is crucial to relate observations and assessments to some sort of model. Figure 2.15 presents the terms necessary to describe the entire accident as a process.

The basic requirement for an accident to happen is that the vessel is in some state of operation and thereby at risk in relation to one or a number of hazards. The causal influence is the element in the model that involves the greatest difficulty with respect to understanding the accident. Despite considerable scientific efforts over the last decades, our knowledge of the causal influences of accidents remains fairly limited. The insufficient

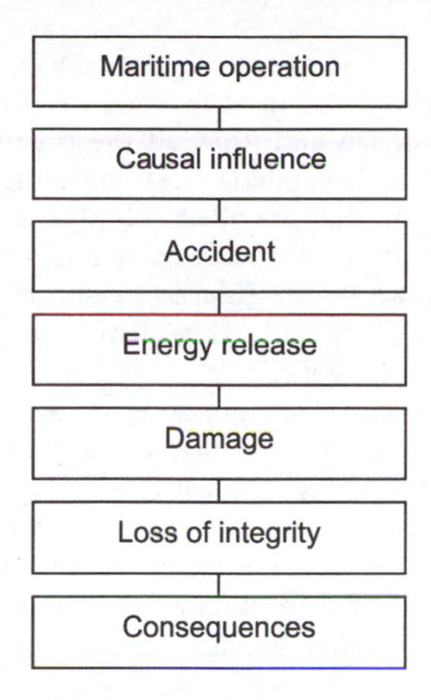

Figure 2.15. The entire accident as a process.

insight is partly related to the problem of combining different scientific disciplines such as engineering, psychology and sociology, and partly related to insufficient analysis models and lack of systematic data.

Although accident causation will be discussed in greater depth in later chapters, some introductory comments are presented here.

2.10.2 Why does it happen?

There exist several theories (with differing insight) as to why maritime accidents occur. Some popular theories include:

- Carelessness
- Deviations from the normal
- 'Act of God'
- New phenomena
- Hazardous activity
- Intoxicated pilot
- Accident-prone
- 'Cowboy' mentality
- Improvising
- Lack of training

The factors presented above may be partly present in some accidents, but this is of little value if the accidental mechanisms are not described in terms of causes that can be influenced, such as system design, equipment failure, planning, operational procedure and organizational management.

In addition to the factors listed above, the concept of human error, normally implying operator error, is an often cited cause and explanation of accidents. By being at the so-called 'sharp end' of the system, the pilot or the operator of a system often seems to be the one to blame. The people at the 'sharp end' are those who directly interact with the hazardous processes in their roles as, for instance, the Master of a passenger vessel. It is at the 'sharp end' that all the practical problems related to the systems are exposed, and it is here that most initiating actions for incidents and accidents occur. The people at the so-called 'blunt end' (e.g. managers, designers, regulators and system architects) are isolated from the actual operation of the processes/systems, but are, however, to a large degree responsible for the conditions met by people at the 'sharp end' because they distribute the resources and create the constraints in which these people work.

Despite its simplicity, the concept of human error in explaining accidents has even reached some level of popularity among conservative engineers. From an engineering point of view, human error may be used to get clear of responsibility for problems not considered as technical. Engineers often seem to have a very positivist or even narrow-minded view of accident causes by focusing merely on problems that can be treated by a technical approach. It is quite simple to write a list of factors explaining the significant human presence in accidents without actually giving any explanation as to why the accident occurred, for example:

- Magnitude of operator-dependent systems
- Humans have restricted capabilities
- Lack of oversight in complex systems
- Inadequate design
- Lack of risk insight

A classical task in system design is to distribute the functionality between operators (i.e. humans) and machines (e.g. instrumentation, computers, etc.). This will be discussed in later chapters and not considered in detail here.

2.10.3 Causal Factors

In an analysis of accidents for Norwegian ships (Karlsen and Kristiansen, 1980), the main causal factors for collisions and groundings were identified and grouped as follows:

- External conditions (i.e. the influence of external forces such as poor weather and waves, reduced visual conditions, etc.).
- Functional failure (i.e. failure or degradation of technical equipment, functions and systems).
- Less than adequate resources (i.e. inadequate ergonomic conditions, planning, organization and training).
- Navigational failure (i.e. failure in manoeuvring and operation, poor understanding of situation, etc.).
- Neglect (i.e. human failure, slips/lapses, and violations or deviation from routines, rules and instructions).
- Other ships (i.e. the influence of failures made by other ships).

Such a listing of causal factors is a rough, but still very useful, simplification of the true characteristics and nature of accidents. Based on a study of 419 grounding accidents for ships greater than 1599 GT, Table 2.11 below indicates the key problem areas of such accidents using the causal factor framework described above.

On the basis of the material presented in Table 2.11, the following general observations can be made:

- External conditions relating to weather and sea are often contributing factors to grounding accidents.
- Functional/technical problems are relatively seldom the main causal factor.
- Problems related to work conditions, human performance and neglect (i.e. so-called 'soft' problems) are the core causal factors.

Table 2.11. Frequency of causal groups related to grounding accidents for ships over 1599 GT

Causal area	Causal group	Frequency	
		Absolute	Per cent
External conditions	Ext. cond. influencing navigational equipment	8	1.9
	Less than adequate buoys and markers	27	6.4
	Reduced visual conditions	53	12.5
39.9%	Influence of channel and squat effects	79	18.9
Functional failure	Functional failure in ship systems	24	5.7
	Functional failure in navigational equipment	8	1.9
	Failure in remote control of ship systems	3	0.7
8.8%	Failure in communication equipment	2	0.5
Less than adequate resources	Bridge design	1	0.2
	Less than adequate charts and manuals	34	8.1
	Failure in bridge organization and manning	35	8.4
	Failure in bridge communication conditions	5	1.2
18.9%	Less than adequate competence or training	4	1.0
Navigational failure	Failure in navigation and manoeuvring	49	11.7
	Failure in observation of fixed markers	35	8.4
	Failure in observation of equipment	10	2.4
22.9%	Failure in understanding traffic situations	2	0.5
Neglect	Failure in watch performance	24	5.7
8.1%	Individual human conditions	10	2.4
Other ships	Functional failures and shortcomings	—	—
1.4%	Navigational failure	6	1.4
Sum		419	100

The grounding of the tanker *Torrey Canyon* in 1967 was a shock to the maritime industry, the political system and the public at large. The severe environmental consequences of the accident marked the beginning of a much stronger focus on the environmental aspects of shipping, a focus that ever since has increased in scope and strength. The accident is summarized in the example below. The tragic fact was that this catastrophe was a wholly human made accident.

Example: The MT *Torrey Canyon* Casualty

What happened?

On the morning of March 18, 1967 the tanker *Torrey Canyon* grounded on the Seven Stones, east of the Isles of Scilly, at full speed of 17 knots (Figure 2.16). During the rescue operation the 120,890 dwt (i.e. dead-weight tons) tanker broke into three parts, and consequently most of the 119,328 tons of cargo was lost, creating an environmental catastrophe. Attempts were made to reduce the oil spill by chemicals, napalm and other explosives. The ship was totally lost, but fortunately the whole crew was put ashore the next day with no injuries suffered.

Circumstances and contributing factors

At the time of the accident the visibility was good and the weather calm, but there were some easterly sea currents present. However, during the rescue operation several storms arose. The fairway was marked with lights and buoys.

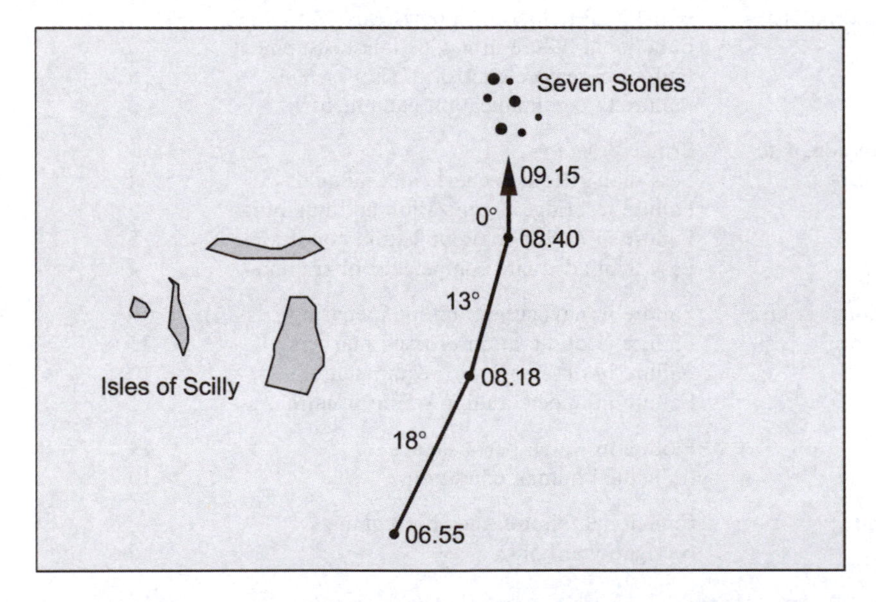

Figure 2.16. The course of MT *Torrey Canyon* before grounding on the Seven Stones.

Immediate causal factors

The navigational officers decided to go east of the Isles of Scilly at full speed. The eastern current was miscalculated, which resulted in incorrect fixing of the ship's position. A fishing vessel caught the tanker by surprise, forcing the crew to go further east than originally planned. This course was, however, too far to the east in the fairway, and the decision to change course was taken too late, as the navigational officers did not recognize and understand the hazardous situation that was about to develop. The grounding resulted in a 610 feet (186 metres) long fissure on the tanker's starboard side that immediately resulted in oil spills.

Basic causal factors

The navigational officers were not familiar with the restricted fairway, but they nevertheless sailed using the autopilot.

The tanker was pressed for time in reaching its next destination because of tidal conditions. Arriving a few minutes too late would result in a minimum of 12 hours delay. The originally planned route was to sail around the Isles of Scilly, but in order to gain time this plan was not held. The Master had been giving incomplete orders regarding this, and the navigational officers further misinterpreted the orders given. In addition, the shipping company had no clear policy on the prioritization between time schedule and safety concerns.

The rescue operation initiated after the grounding was poorly organized and the operation failed several times. During the operation several explosions and fires occurred, and the rescue was further complicated by storms developing in the area of the accident.

Analysis of the accident involving MT *Torrey Canyon* and other maritime accidents reveals a number of interesting accident characteristics that can be summarized by Table 2.12.

As has just been pointed out, there is seldom a single explanation as to why an accident happens. Nevertheless, there has been a more or less continuous search for more general models and theories to be used in explaining accidents. Some of the more popular explanation theories in the shipping domain include:

1. *Flag or registration:* Maritime administrations have a key responsibility in controlling and ensuring the safety standard of shipping.
2. *Age:* Both the technical and manning standards seem to deteriorate with the age of vessels.
3. *Activity level:* The number of maritime accidents in an area is proportional to the traffic volume.

These theories will be discussed briefly in the following sections.

2.10.4 Flag Effect

By studying different flags of registration, some of the effects of different management styles may be revealed. Figure 2.17 presents the loss ratio for ships greater than 100 GT,

Table 2.12. C[illegible]tics of maritime accidents

Characteristic	Description
Routine	Failures or d[illegible]re usually related to routine activities and situations rather tha[illegible]l situations, which rarely are triggering factors to accidents
Several causal factors	There is se[illegible] a single cause to an accident. Usually accidents occur because [illegible]ailures and errors
Process	Causal factors often interact with each other. The accidental process may have been initiated long before the more dramatic events develop
Gradual progress	Accidents usually do not occur instantly. Failures and functional degradation often develop over time, and whether they result in an accident or a hazardous situation may only be a matter of chance
Operator failure	Operator (or human) errors are present in terms of omissions and commissions. An operator is at the 'sharp end' of a system, and is therefore often actively involved in accidents affecting that system
Situation-related	Accidents are situation-related, and it is the combined effect of all situational conditions that is critical in the accidental process. Situational conditions include external conditions, total workload, the competence and experience of the operators, work environment, time of day, etc.
Focus on outcome	It is often more easy to identify failures in the last stages of an accident than in the initiating phase. As a result of this, decision-makers often tend to look for measures that limit the consequences rather than avoiding the accident altogether

plotted for various flags and regions. It can be seen that there is a significant difference in loss ratio between different flags. Mediterranean flags, for instance, had about twice as high a loss ratio as North European flags. In the period from 1984 to 1995 there was a factor of about 14 between the best and worst loss ratios for the individual flags. This is illustrated in Figure 2.18, which is based on a study by Kristiansen and Olofsson (1997).

Owing to the fact that national maritime administrations show different will and ability to enforce international safety regulations, a group of mainly European countries have, upon agreement, initiated unannounced inspections of vessels. This agreement is known as the Paris Memorandum of Understanding on Port State Control (Paris MOU). In the case of serious shortcomings on technical standards, certificates, etc., a ship may be held in port for rectification of these shortcomings. This is commonly known as a detention. Figure 2.19 shows the detention rate for selected countries (i.e. flag states), and confirms the earlier finding that different countries of registration show different performance.

2.10.5 Age Effect

It has been a popular view that the safety standard of a vessel deteriorates with age. One may immediately agree with the argument that corrosion, as well as wear and tear, reduces the function and integrity of hull and machinery. However, one may also counter

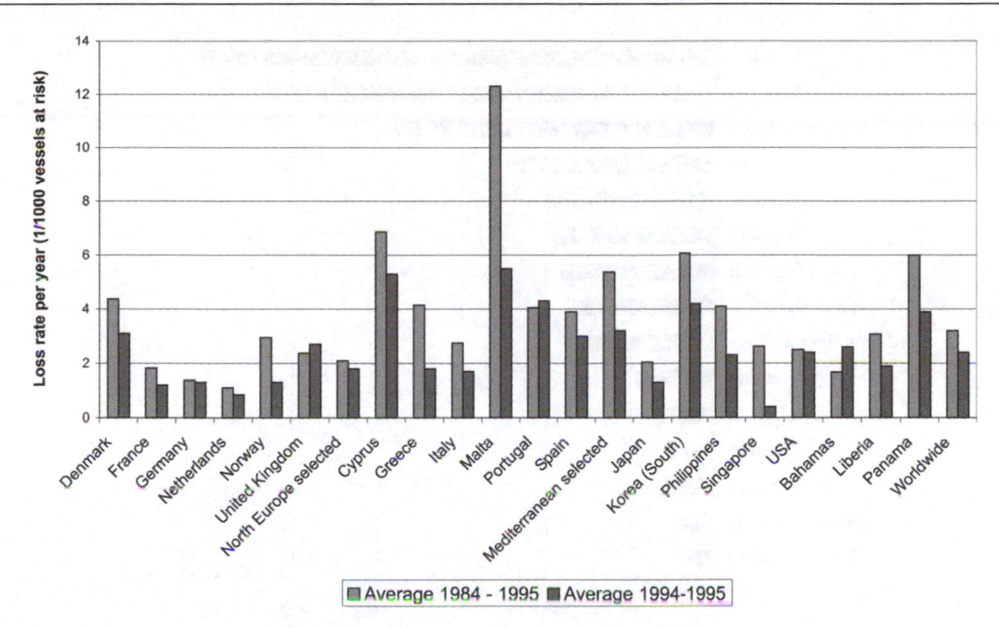

Figure 2.17. Loss rate for different flags.

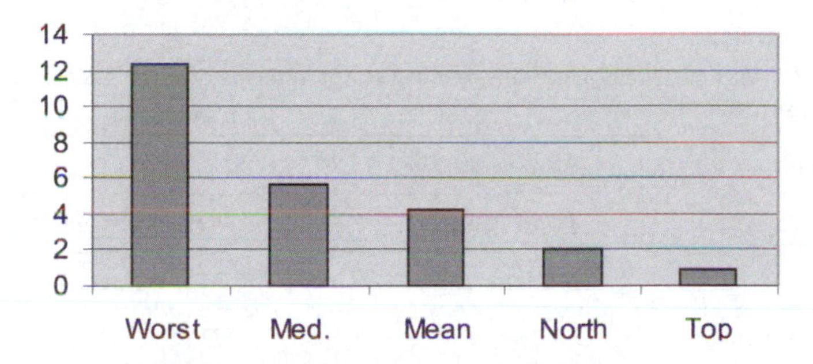

Figure 2.18. Loss rate of European fleets, 1990–93: losses per year per 1000 vessels at risk. (Worst = worst performing states, Med. = average for Mediterranean states, Mean = mean for all European flag states, North = average for North European states, Top = best performing state.)

this hypothesis by arguing that the age effect is minimized through maintenance and repair.

Ponce (1990) has studied the effect of age on total losses for selected fleets, and the results of this study are shown in Table 2.13. Apart from the fact that there is a certain correlation between the median age for vessels lost and the median age for the fleet at risk, it can also be shown (see Figure 2.20) that vessels lost are an average older than the fleet. Roughly estimated, the ships lost are 5 years older than the respective fleet at risk.

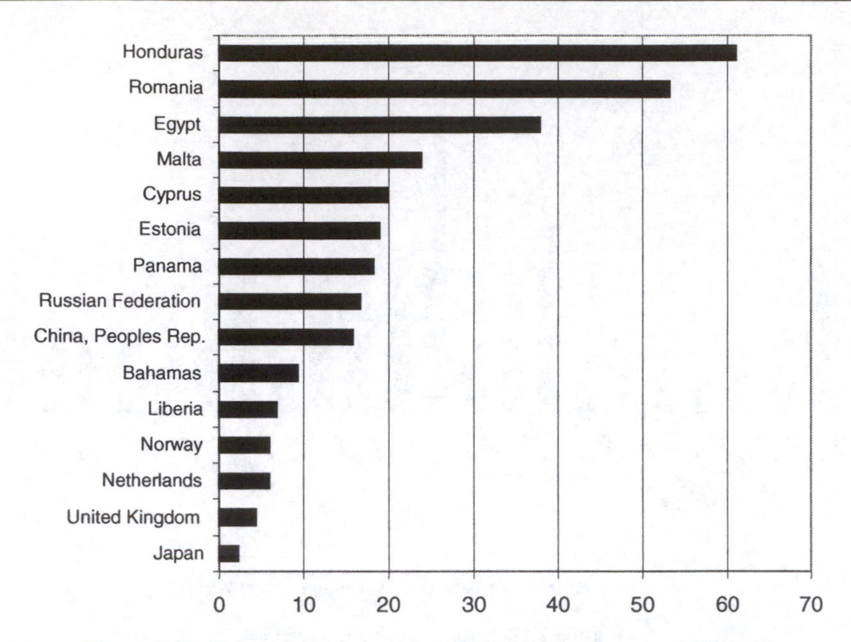

Figure 2.19. Detention rate in percent of inspections, average 1995–97.

Table 2.13. Age distribution for different flag states in terms of total losses and vessel population

Flag state	Age (years)			
	Total losses		Vessel population	
	Median age	Rank	Median age	Rank
United States	25.33	1	21.29	1
Greece	21.05	2	15.77	3
United Kingdom	20.86	3	10.26	6
Panama	20.34	4	13.16	4
Canada	19.5	5	17.22	2
World	18.64	6	11.89	5
Norway	17.83	7	7.55	8
Liberia	14.5	8	8.8	7
Fed. Rep. Germany	11.17	9	6.97	10
Japan	10.93	10	7.48	9

Faragher et al. (1979) studied the effect of age on the casualty rate for structural failure and machinery breakdown casualties. It can be concluded from Figure 2.21 that there is a clear correlation between age and accident rate for both accident forms. Apart from the statistical variation, the correlation was quite high for both models (about 70%).

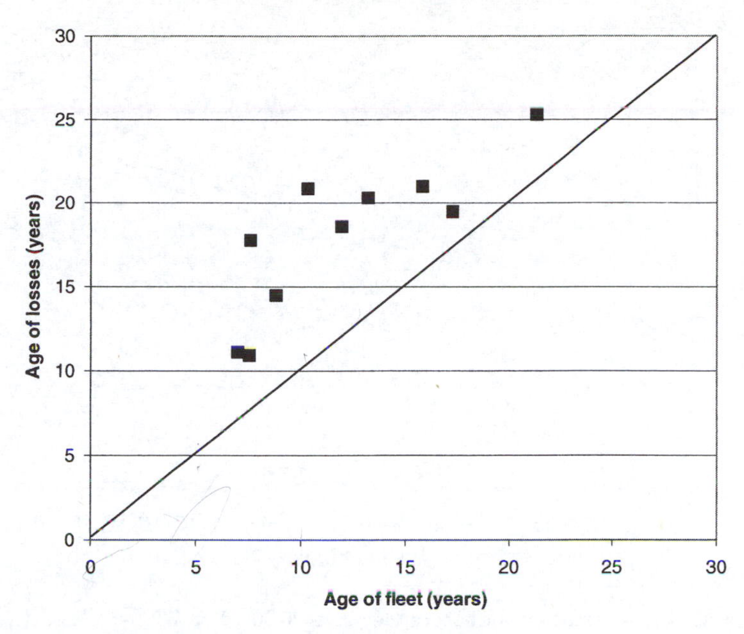

Figure 2.20. Age of lost ships versus age of fleet at risk for selected flag states (scatter diagram).

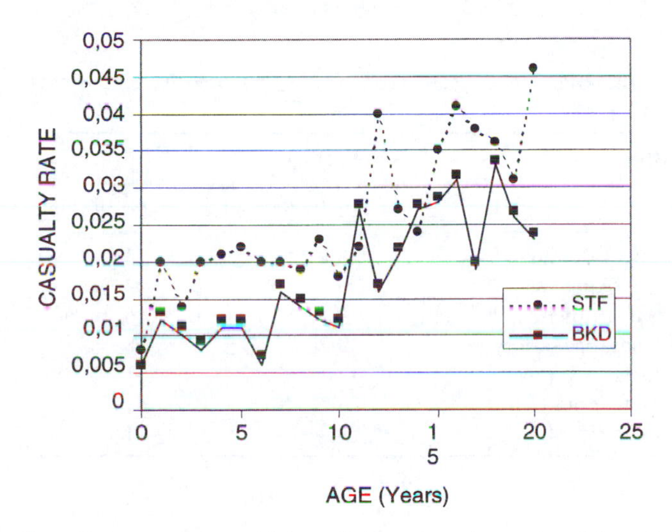

Figure 2.21. The effect of age on structural failure (STF) and machinery breakdown (BKD), world tanker fleet above 1000 grt. Casualty rate in percent of vessel population per year.

Roughly estimated, there is an increase in the casualty rate in the order of 5 to 6 times between new and old vessels.

Thyregod and Nielsen (1993) have studied the age effect for the total yearly casualty rate. The analysis for bulk carriers is shown in Figure 2.22. The data basis for

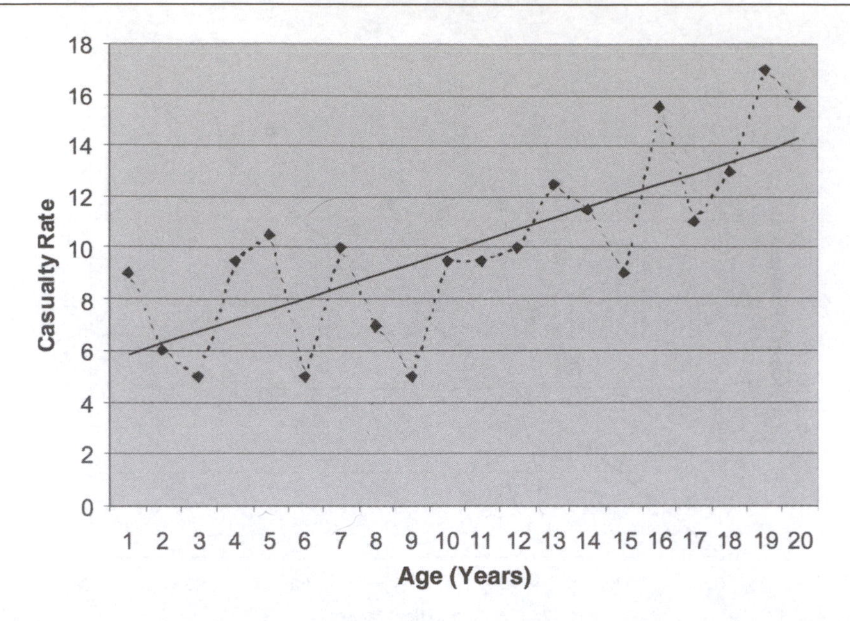

Figure 2.22. Casualty rate versus age for bulk carriers over 10,000 dwt, 1984–92. Casualty rate per 1000 vessels at risk per year.

Table 2.14. Tanker groundings in US ports 1969–76, vessels over 5000 grt

Port	Port calls	
	In 1000	Groundings
Puget Sound	3.8	3
Los Angeles	9.7	3
San Francisco	9.3	16
Chesapeake Bay	9.2	18
Delaware Bay	17.1	51
New York	27	81
Gulf Coast	29.2	81

this analysis was major casualties for bulk carriers greater than 10,000 dwt as reported by the Institute of London Underwriters. Despite the yearly variation in casualty rate, there is a distinct trend with age. In the course of 20 years the casualty rate doubles. However, the increase is not as strong as for structural failure and machinery breakdown (see Figure 2.21). This means that certain accident forms are less dependent on ship age.

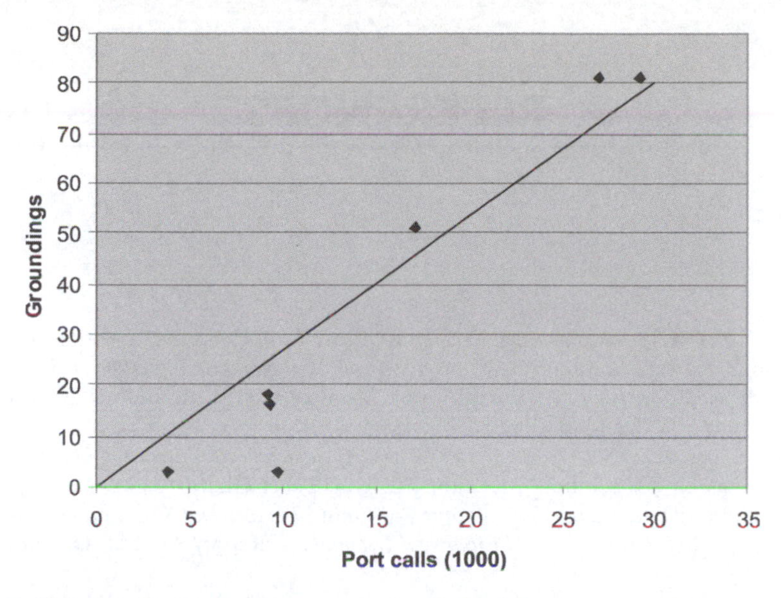

Figure 2.23. Groundings versus port calls for tankers in major US ports.

2.10.6 Effect of Activity Level

Faragher et al. (1979) studied the age effect for all casualty types and found that so-called impact accidents (i.e. groundings, collisions and ramming) were only marginally dependent on age. They concluded that the activity level was a better parameter in explaining accident frequency for impact accidents. By comparing accident rates in American ports they found a strong correlation between groundings and the number of port calls per time-unit. Data for the seven ports studied are shown in Table 2.14. It can be seen from the plot in Figure 2.23 that there is a clear dependence between number of port calls and number of groundings. The most distinct outlier is the port of Los Angeles, where we should have expected a much higher number of groundings.

REFERENCES

DNV, 1998, *FSA of HLA on Passenger Vessels*. Report No. 97-2053, Det Norske Veritas, Oslo.

Environmental . . ., 1985, *Environmental program of the Netherlands 1986–1990*. Ministry of Housing, Physical Planning and Environment, Ministry of Agriculture and Fisheries, Ministry of Transport and Water Management, The Hague.

Faragher, W. E., et al., 1979, *Deep Water Ports Approach/Exit Hazard and Risk Assessment*. Planning Research Corporation. McLean, VA. U.S. Coast Guard Report No. CG-D-6-79.

Karlsen, J. E. and Kristiansen, S., 1980, *Analysis of Causal Factors and Situation Dependent Factors. Project: Cause Relationships of Collisions and Groundings*. Report 80-1144, Det Norske Veritas, Høvik, Norway.

Kårstad, O. and Wulff, E., 1983, *Safety on the Continental Shelf*. (In Norwegian: Sikkerhet pa sokkelen). Universitetsforlaget, Oslo.

Kristiansen, S. and Olofsson, M., 1997, *SAFECO – Operational Safety and Ship Management. WP II.5.1: Criteria for Management*. Marintek Report No. MT23 F97-0175, Trondheim.

Lancaster, J., 1996, *Engineering Catastrophes: Causes and Effects of Major Accidents*. Abington Publishing, Cambridge, UK.

Lloyd's Register of Shipping, 1962–93, *Statistical Tables*, Lloyd's Register of Shipping, London.

Lloyd's Register of Shipping, 1994–98, *World Casualty Statistics*, Lloyd's Register of Shipping, London.

Lloyd's Register of Shipping, 1994–98, *World Fleet Statistics*, Lloyd's Register of Shipping, London.

Norwegian Shipowner's Association, http://www.rederi.no/en/library/

Ponce, P. V., 1990, An analysis of marine total losses worldwide and for selected flags. *Marine Technology*, Vol. 27(2), 114–116.

Spouge, J. R., 1989, The safety of Ro-Ro passenger vessels. *Transactions of the Royal Institute of Naval Architects*, Vol. 131, 1–27.

Stephenson, J., 1991, *System Safety 2000 – A Practical Guide for Planning, Managing, and Conducting System Safety Programs*. Van Nostrand Reinhold, New York.

Thyregod, P. and Nielsen, B. F., 1993, *Trends in Marine Losses and Major Casualties 1984–92*. IMSOR – The Institute of Mathematical Statistics and Operations Research, Technical Report No. 8. The Technical University of Denmark, Lyngby, Denmark.

3

RULES AND REGULATIONS

3.1 INTRODUCTION

This chapter will give an outline of the regulation of safety in seaborne transport. The control of safety is primarily based on the rules (conventions and resolutions) given by the United Nations agency the International Maritime Organization (IMO). These rules have international application but some reference will also be made to national regulations by taking the Norwegian legal regime as an example.

When we use the term *safety*, it will encompass:

- Safety and health of persons
- Safety of vessel
- Environmental aspects

Safety is regulated on the basis of different *legal sources*, the key ones of which are the following:

- International laws and regulations
  - UN Law of the Seas (UNCLOS)
  - European Union (EU) Directives
- National laws and regulations
- Case law (court rulings)
- National territorial zones
- IMO conventions and resolutions
- Classification construction rules
- Port State control MOU guidelines

It should also be kept in mind that there are a number of *actors* that have an impact on safety. The primary ones are:

- Flag and Port State control (Maritime Directorate)
- International Maritime Organization (IMO)

- Classification Societies
- Insurance companies
- Charterer, cargo owner

3.1.1 The Structure of Control

Seen from a national point of view, the regulation of safety is based on a set of international rules that is adopted by the legislative assembly (Parliament) (see Figure 3.1). The concrete rules and regulations are written or translated by the responsible government branch (Foreign and International Trade Department). The role of the Maritime Administration is to ensure that regulations are followed by the shipowners through proper control and certification. This is what is termed Flag State control (FSC). The figure also shows that the Classification Society has a role in the certification process, although this is primarily related to the insurance of the vessel, cargo and third-party interests.

The control of safety in shipping is complex for a number of reasons:

- International, regional and national laws and regulations.
- Control is exercised by a number of agencies.
- Control affects the various life-cycles of the vessel.

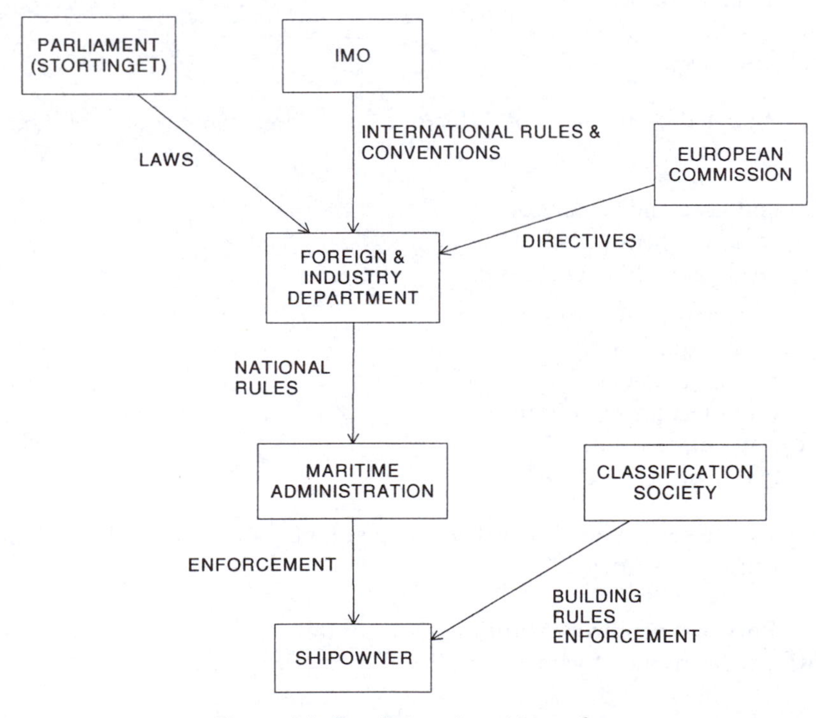

Figure 3.1. Regulation of maritime safety.

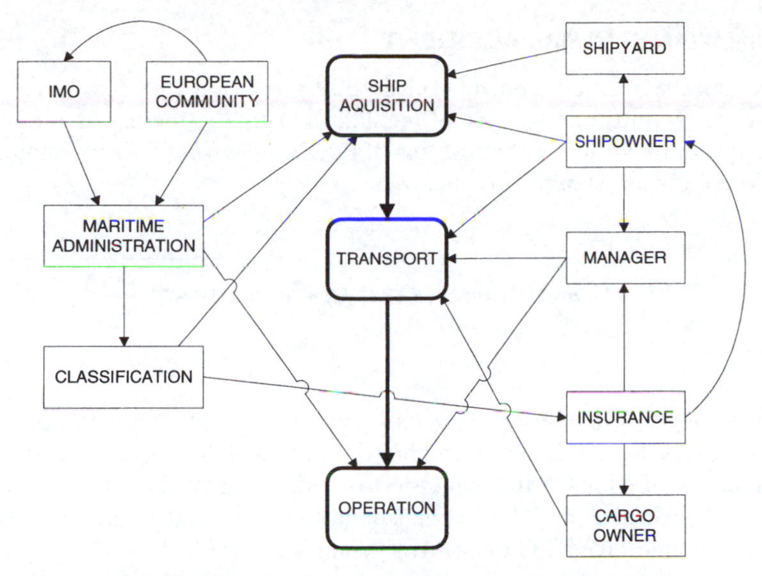

Figure 3.2. Actors and interactions in safety control.

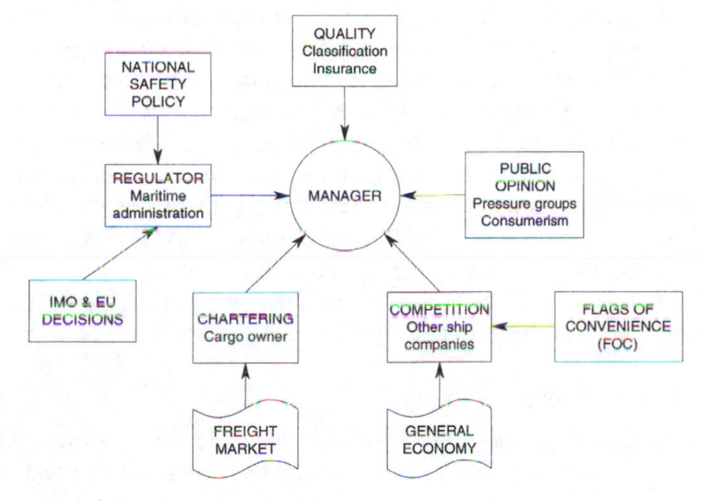

Figure 3.3. Safety subject to regulation and competition.

A simple outline of the number of actors and interactions is shown in Figure 3.2. It should also be kept in mind that shipping as an internationally oriented business is highly competitive and is also influenced by dramatic economic cycles. Seen from the shipowner's perspective, the safety standard is a result of the cross-pressure between control and commercial competition (see Figure 3.3).

3.2 *SCANDINAVIAN STAR* FIRE ACCIDENT

The various aspects of safety control will be outlined in greater detail in the following sections. In order to illustrate the role of regulation in safety, the discussion will illustrate its relevance by commenting on the findings of the *Scandinavian Star* investigation (NOU, 1991). The key facts are summarized below.

SCANDINAVIAN STAR FIRE ACCIDENT

Abstract

The passenger ferry *Scandinavian Star* had just entered service on the route between Oslo and Fredrikshavn. During the night of 7 April 1990 at least two fires were set aboard the vessel on its first trip from Oslo to Frederikshavn. The first fire was put out before any damage was done. The second fire, however, escalated and resulted in a fully developed fire which killed 158 of the 482 persons onboard.

Summary of circumstances

The vessel had recently been taken over by new owners. The transition from one-day cruises in the Caribbean to ferry service in Scandinavia required considerable reconstruction and new facilities. This work was mainly executed by the crew in the weeks before the vessel entered into service. The repairs had not been finished when the vessel left on the first trip from Oslo. A consequence of this was that the crew was unfamiliar with the ship and had not been given emergency training. The emergency plan was adopted from the previous operator. It was, however, based on different operational conditions, i.e. twice as many crew members. In addition the crew, who were mostly Portuguese, to a large extent did not understand Scandinavian or English. The emergency equipment and systems were not up to date: a lack of fire doors, sprinkler system and lifeboats not maintained, deficiencies in alarm system and poor technical arrangement of escape ways.

Event description summary

Two fires were ignited, most likely by a pyromaniac. The first fire was put out by the crew immediately. The second fire, however, escalated quickly and filled the corridors of the different deck levels with poisonous smoke only a few minutes after the ignition. The smoke consisted of carbon monoxide and hydrogen cyanide and killed the exposed individuals within a few minutes. Both the active and passive fire protection failed to a considerable degree. Critical fire doors were not locked and resulted in an air draft that speeded up the spread of the fire. The fire alarm was deficient and forced the badly trained crew to alert the passengers by going through the cabin sections.

Basic causal factor summary

The main objective of investing in *Scandinavian Star* was for the shipping company to save tax on the profit of another ship sale. This required that the *Scandinavian Star* be put into service by 1 April. At this time, however, she was not yet ready and prepared for service. The Master had not carried out the required emergency drill and the critical emergency plans had not been prepared. The fire-fighting and detection systems were poorly maintained and failed. The material in the cabins and corridors had a high heat value and released poisonous gas when ignited.

3.3 INTERNATIONAL MARITIME ORGANIZATION (IMO)

The main principle in the regulation of shipping are harmonized national rules based on international conventions and resolutions given by the IMO. This is an organization under the United Nations system. Its prime function is to establish rules based on participation by the member states. IMO has a complex set of committees that draft and revise regulations which are adopted by the General Assembly. A new regulation has to be ratified by a minimum number of states before it enters into force. IMO has no power to enforce the international safety regulations. This is the task of the member states in their role as so-called Flag States.

3.3.1 SOLAS

The main objective of the SOLAS Convention (Safety of Life at Sea) is to specify minimum standards for the construction, equipment and operation of ships (SOLAS, 2001). The present version of the SOLAS Convention was adopted in 1974 and was later revised and supplemented with so-called Protocols. It entered into force in 1980. SOLAS-74 has 12 articles and 12 chapters with the following specific requirements:

- I. General provisions
- II.
    1. Construction – Subdivision and stability, machinery and electrical installations
    2. Construction – Fire protection, detection and fire extinction
- III. Life-saving appliances and arrangements
- IV. Radiotelegraphy and radiotelephony
- V. Safety of navigation
- VI. Carriage of grain
- VII. Carriage of dangerous goods
- VIII. Nuclear ships
- IX. Management for the safe operation of ships (ISM Code)
- X. Safety measures for high speed craft
- XI. Special measures to enhance safety
- XII. Additional safety measures for bulk carriers.

The convention has been amended a number of times since its adoption in order to be in accordance with the development of new technology and new safety knowledge. The regulation is to a large degree prescriptive by specifying solutions in minute technical detail. Performance criteria are only applied to a limited degree. This has two main drawbacks: technical solutions specified in SOLAS may become obsolete even before it enters into force, and the lack of focus on performance criteria does not stimulate the designer to find or invent better solutions.

The SOLAS-74 Convention has been ratified by most nations. In order to become effective, the convention has to be translated into the official national language and be formally adopted by the government branch. The implementation of SOLAS-74 by the Norwegian Flag State is given in:

- Regulation of 15 June 1987 No. 506 on inspection for issuing of certificates for passenger and cargo vessels and barges, etc.
- Regulation of 15 September 1992 No. 695 on building of passenger and cargo vessels and barges.

SCANDINAVIAN STAR FIRE

The surveys performed by Lloyd's Register and the Nautical Inspector of the Bahamas were evidently unable to detect a number of faults and non-conformities:

- Workshop and stores located on the car deck
- Missing fire door on deck 6
- Missing alarm klaxons
- Sprinkler heads on car deck blocked with rust
- Partly inadequate sound level of alarms
- The emergency signposts was incorrectly located and not in a Scandinavian language

Scandinavian Star was built in 1971 and did therefore comply with SOLAS 1960. This meant that the vessel did not have state-of-the-art fire equipment such as:

- Sprinkler system in all accommodation areas
- Fire alarm system with both heat and smoke detectors
- Automatic closing of fire doors and use of smoke-proof doors
- Fire-resistant (non-combustible) material in interior panels and maximum value on generation of toxic gases
- Separate control of ventilation in each accommodation section
- A uniform and more functional design for signs showing evacuation routes
- A requirement to undertake an evacuation analysis in the design phase

3.3.2 International Convention on Load Lines, 1966

It has long been recognized that limitations on the draught to which a ship may be loaded make a significant contribution to her safety. These limits are given in the form of a freeboard, which, besides external weather-tightness and watertight integrity, constitute the main requirement of the Convention.

The first International Convention on Load Lines (ILLC, 2002), adopted in 1930, was based on the principle of reserve buoyancy. It was also recognized then that the freeboard should ensure adequate stability and avoid excessive stress on the ship's hull as a result of overloading.

The regulations take into account the potential hazards present in different geographical zones and different yearly seasons. The technical annex contains several additional safety requirements concerning doors, freeing ports, hatchways and other items. The Convention includes Annex I with the following four chapters:

I. General
II. Conditions of assignment of freeboard
III. Freeboards
IV. Special requirements for ships assigned timer freeboards

Annex II covers zones, areas and seasonal periods, and Annex III certificates, including the International Load Line Certificate. The ILLC Convention is adopted by Norway through Regulation of 15 September 1992 No. 695 on building of passenger and cargo vessels and barges.

SCANDINAVIAN STAR FIRE

The vessel had to comply with the requirements of the ILLC 1966.

The Commission did not find any factor relating to the freeboard that had any bearing on the disaster.

3.3.3 STCW Convention

The International Convention on Standards of Training, Certification and Watchkeeping for Seafarers (STCW) was the first to establish basic requirements on training, certification and watchkeeping for seafarers at an international level. The technical provisions of the Convention are given in an Annex containing six chapters:

1. General provisions.
2. Master-deck department: This chapter outlines basic principles to be observed in keeping a navigational watch. It also lays down mandatory minimum requirements for the certification of masters, chief mates and officers in charge of navigational watches on ships of 200 grt or more.

3. Engine Department: Outlines basic principles to be observed in keeping an engineering watch. It includes mandatory minimum requirements for certification of officers of ships with main propulsion machinery of 3000 kW.
4. Radio Department.
5. Special requirements for tankers.
6. Proficiency in survival craft.

The 1995 amendments represented a major revision of the 1978 Convention (STCW, 1996). The original Convention had been criticized on many counts. It referred to vague phrases such as 'to the satisfaction of the Administration', which admitted quite different interpretations of minimum manning standards. Others criticized that the Convention was never uniformly applied and did not impose strict obligations on the Flag States regarding its implementation.

SCANDINAVIAN STAR FIRE

It was established by the Commission that the crew of 90 persons was sufficient to meet the safety requirements, even by Scandinavian practice. However, their competence was not adequate:

- Many of the deck officers lacked safety training or had not attended courses for a long time.
- The requirement that 48 crew members should be certified as lifeboat-men (verified competence to handle lifeboats and liferafts) was not met.
- Some of the Portuguese crew members did not speak or understand English or a Scandinavian language.

3.3.4 MARPOL

Both SOLAS and ICCL have an indirect effect on preventing pollution from ships. However, there was a dramatic development of specialized tankers after the Second World War in terms of ship size and complexity of operation. The International Convention for the Prevention of Pollution from Ships (MARPOL) seeks to address the environmental aspects related to design and operation of these ships more directly (MARPOL, 2002).

The Convention prohibits the deliberate discharge of oil or oily mixtures for all seagoing vessels, except tankers less than 150 gross tons and other ships less than 500 gross tons, in areas denoted 'prohibited zones'. In general these zones extend at least 50 n. miles from the coastal areas, although zones of 100 miles and more were established in areas which included the Mediterranean and Adriatic Seas, the Gulf and Red Sea, the coasts of Australia and Madagascar, and some others.

MARPOL introduces a number of measures:

- Segregated ballast tanks (SBT): ballast tanks only used for ballast as cargo oil is prohibited. Reduces cleaning problem.
- Protective location of SBT: SBT arranged in bottom or sides to protect cargo tanks against impact or penetration.
- Draft and trim requirements: to ensure safe operation in ballast condition.
- Tank size limitation to limit potential oil outflow.
- Subdivision and stability in damaged condition.
- Crude oil washing (COW).
- Inert gas system (IGS) for empty cargo tanks.
- Slop tanks for containing slop, sludge and washings.

The implementation of MARPOL is based on a complex scheme where ship size and whether it is an existing or a new building determine which requirements apply. An interesting outline of the tankship technology has developed, and the present environmental challenges are given in NAS (1991).

SCANDINAVIAN STAR FIRE

The vessel had to comply with the requirements of MARPOL 73.

The Commission did not find any factor relating to pollution prevention that had any bearing on the disaster.

3.3.5 The ISM Code

The introduction of the International Management Code for Safe Operation and Pollution Prevention (ISM, 2002) represented a dramatic departure in regulatory thinking by the IMO. It acknowledges that detailed prescriptive rules for design and manning have serious limitations. Inspired by principles from quality management and internal control, the ISM Code will stimulate safety consciousness and a systematic approach in every part of the organization both ashore and onboard.

The ISM Code itself is a fairly short document of about 9 pages. The main intention with ISM is to induce the shipping companies to create a safety management system that works. The Code does not prescribe in detail how the company should undertake this, but just states some basic principles and controls that should be applied. The philosophy behind ISM is commitment from the top management, verification of positive attitudes and competence, clear placement of responsibility and quality control of work processes.

The Code states the following objectives for the adoption of a management system:

1. To provide for safe practices in ship operation and a safe working environment;
2. To establish safeguards against all identified risks; and

3. To continuously improve the safety management skills of personnel ashore and aboard, including preparing for emergencies related both to safety and environmental protection.

The Code has 13 chapters, which are listed in Table 3.1. When addressing the effect of ISM on safety, there are two key aspects: the material content of the regulation, and what is an acceptable *compliance* with the Code. In order to implement ISM correctly, certain elements are required:

- Documentation of how the ISM Code will be implemented.
- External verification and certification.
- Reporting (logging) of the safety management processes.
- Internal (company) verification.

Apart from this, the *Guidelines on Implementation of ISM* (ISM, 2002) is fairly vague on how to verify that a *safety management system* (*SMS*) conforms with the Code. It

Table 3.1. Organization of the ISM Code elements

Management function	Chapter	ISM element
Objective, policy	1.2.2	Provide safe practices, establish safeguards and continuously improve skills
	2	Safety and environmental protection policy
Requirements	1.2.3	Compliance with rules and regulations Other IMO Conventions: SOLAS, STCW, MARPOL, COLREG, Load lines, etc.
	1.3	Functional requirements: policy, instructions, authority, communication, accident reporting, emergency preparedness, audits
Controls	3	Company responsibilities and authority
	4	Designated persons
	6	Resources and personnel
	7	Development of plans for shipboard operations
	8	Emergency preparedness
	10	Maintenance of the ship and equipment
Safety management system	11	Documentation
Implementation of controls	5	Master's responsibility and authority
Monitoring of the system	9	Reports and analysis of non-conformities, accidents and hazardous occurrences
	12	Company verification, review and evaluation
The periodic system review	13	Certification, verification and control

admits that certain criteria for assessment are necessary, but also warns against the emergence of prescriptive requirements and solutions prepared by external consultants. The obvious philosophy behind this attitude is that the SMS should be an integral part of the management thinking of the company. In that sense the SMS should reflect the objectives of the Code but otherwise be implemented in such a way that it is viewed as an element of the culture, organization and decision-making processes of the company.

The ISM Code specifies certain requirements for the safety management system (SMS) of the operating company. In order for the SMS to work, certain distinct functions have to be in place. The core of the SMS is made up of certain *controls* which are defined in terms of (see ISM, 2002):

- Responsibility and authority.
- Provision of resources and support.
- Procedures for checking of competence and operational readiness, training, and shipboard operations.
- Establishing minimum standards for the maintenance system.

Another key feature of the ISM concept is the definition of a *monitoring function*, which is based on audits and reporting of events. The audit will ensure that errors and shortcoming in the SMS are corrected and that the system is updated in view of new requirements and experience gained. The auditing and event reporting will also address operational errors and failures directly and thereby lead to corrective action in terms of modified systems and improved procedures.

Chapter 13 states that the company should have a *certificate of approval* which documents that the SMS is in accordance with the intentions and specific requirements of the ISM Code. It should be kept in mind that ISM has a relation to existing or traditional regulatory approaches for design, equipment, training and emergency preparedness. The Code should be understood in the context of existing safety regulations that have already been mentioned: SOLAS, ILLC, MARPOL, COLREG and STCW. ISM does not address any of the specific requirements in these conventions, but just assumes that the management system should ensure that they are met.

The ISM Code will be discussed further in Chapter 15 on safety management.

SCANDINAVIAN STAR FIRE

The ISM Code first took effect for passenger vessels in the summer of 1998. Therefore, it is only possible here to discuss the relevance of the ISM Code in light of the management shortcomings that were associated with the disaster:

- Lack of safety policy, cross-pressure from management (chapter 2)
- No overall management plan for verifying safety functions (chapter 12)

- No designated person to coordinate the safety work (chapter 4)
- Hazard identification and risk assessment: establish safeguards (chapter 1.2.2.2) and identify critical systems (chapter 10.3)
- A number of faults were not detected or corrected; not complying with rules and regulations (chapter 1.2.3 and chapter 10)
- Incomplete emergency plans and no training or drills (chapter 8)
- Lack of leadership by the Master in the emergency situation (chapter 8)
- Lack of competence in fire-fighting and supervision of evacuation (chapter 8)

3.4 FLAG STATE CONTROL

As already pointed out, the set of internationally accepted safety rules and regulations are not enforced by the IMO but by the so-called Flag States. The national maritime administration is acting as Flag State on behalf of the country in question. Based on plans, technical documentation and inspections, a ship is subject to registration and awarded the necessary safety-related certificates.

3.4.1 The Seaworthiness Act

Each country has to give a legal basis for exercising this role as Flag State. In Norway the competence of the Maritime Administration is laid down in the Seaworthiness Act (Falkanger et al., 1998). The law regulates shipping activity in relation to the public sphere and also defines the role of the national Maritime Administration (in Norway, the Maritime Directorate). The key functions specified by the law are:

- Safety control activity in general
- The competence of the Maritime Directorate
- Investigation of accidents (Sea Court)
- Inspection and detention (withholding a vessel)
- Certificates
- Safety and occupational health-related activities onboard
- Equipment standard
- Cargo condition and safety
- Manning and working hours
- Control of passenger vessels
- Responsibility of Master and Owner

Section 2 of the Seaworthiness Act defines *seaworthiness* as follows:

> A ship is considered unseaworthy when, because of defects in hull, equipment, machinery or crewing or due to overloading or deficient loading or other grounds, it is in such a condition, that in consideration of the vessel's trade, the risk to human life associated with going to sea exceeds what is customary.

The law basically applies to vessels greater than 50 gross register tons, but the Administration (Flag State) may decide that other vessels also have to be built in accordance with the rules under the law.

The jurisdiction of this law is in principle limited to Norwegian vessels. The maritime administration acts in this manner as *Flag State*. However, international law has developed during the last decades and today accepts that a nation may exercise some control and, if necessary, detain a foreign vessel viewed as a risk to human life (passenger transport) and coastal environment (oil pollution). The maritime administration in that sense acts as a *Port State*. We will return to this role later.

Shipping activity in Norway or more precisely Norwegian national register vessels (NOR) are subject to both private and public law. The international register in Norway (NIS) is regulated through a separate act. Some of the key laws are (Sjøfartsdirektoratet, 1988):

1. The Maritime Code (Sjøloven av 24. juni 1994 nr. 39).
2. The Seaworthiness Act (Sjødyktighetsloven av 9. juni 1903 nr. 7).
3. The Seaman's Act (Sjømannsloven av 30. Mai 1975 nr. 18).
4. Norwegian International Register Act (NIS-loven av 12. juni 1987 nr. 48).

3.4.2 Delegation of Flag State Control

Some Flag States accept foreign vessels and have become what is commonly termed international or offshore registers. The standard of some of these registers has been questioned and they have been branded as *Flags of Convenience* (FOC). They are suspected to offer registration to foreign owners mainly for economic reasons and are viewed as having a lenient enforcement of safety regulations. Another characteristic is the lack of or minimal maritime administration. A common practice is to delegate the control to an independent certifying authority, primarily classification societies and even consultants.

> **SCANDINAVIAN STAR FIRE**
>
> The ship was surveyed by Lloyd's Register on behalf of the Bahamas in the first days of January a few months before the accident. The inspector spent half a day onboard on this occasion. It was later found that the vessel had a number of faults or defects that were not detected during the survey. The concrete items were discussed under the SOLAS section in this chapter. Based on the survey of LR, a new *SOLAS Passenger Ship Safety Certificate* was issued. The Nautical Inspector of the Bahamas had no remarks to the survey made by LR. The Flag State did not survey the vessel after the modification of the interior and start-up of the new service.

3.4.3 Effectiveness of Flag State Control

Flag State control (FSC) has for years been a key principle in the safety control of shipping. Based on internationally accepted rules, the safety is to be ensured by the maritime authority of the nation of registration of the vessel. It has, however, become evident that different Flag States have varying competence and motivation to undertake their role. This was clearly demonstrated in a small survey of the *SAFECO I* project (Kristiansen and Olofsson, 1997). Table 3.2 shows the loss rate for some selected Flag States. It is clear that the annual loss rate may vary by a factor of more than 10. This great variation can even be observed among European Flag States, as shown in Figure 3.4.

3.4.4 The Flag State Audit Project

The Seafarers International Research Centre (SIRC) at Cardiff University has recently undertaken an assessment of the performance of the main Flag States

Table 3.2. Total loss rate by flag for vessels greater than 100 grt

Flag	Fleet size 1993	Loss rate per 1000 shipyears 1994–95
Denmark	599	3.1
France	769	1.2
Germany	1234	1.3
Netherlands	1006	0.84
Norway	1691	1.3
United Kingdom	1532	2.7
North Europe selected	6831	1.8[a]
Cyprus	1591	5.3
Greece	1929	1.8
Italy	1548	1.7
Malta	1037	5.5
Portugal	307	—
Spain	2111	3.0
Mediterranean selected	8523	3.2[a]
Japan	9950	1.3
Korea (South)	2085	4.2
Philippines	1469	2.3
Singapore	1129	0.4
USA	5646	2.4
Bahamas	1121	2.6
Liberia	1611	1.9
Panama	5564	3.9
Worldwide	80655	2.4

[a]Weighted estimation on the basis of the selected countries.
Source: *World Fleet Statistics* and *Casualty Return*, Lloyd's Register, London.

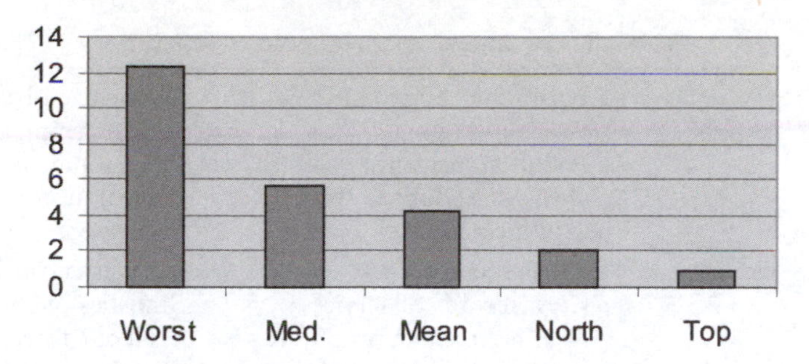

Figure 3.4. Loss rate of European fleet segments, 1994–95: loss rate per 1000 ship-years. (Med. – Mediterranean countries, North – Northern and Central Europe.)

(Alderton and Winchester, 2001). Some shipowners prefer to register their fleet under a flag other than the national one. This has been a practice for years but has gained renewed importance during the present trend toward globalization and deregulation of industry and trade. Some of these flags lack both motivation and competence to enforce the international safety standards set by IMO. These flags have been termed Flags of Convenience (FOC). However, today it seems too simple to distinguish between national flags and FOCs. The International Transport Workers' Federation therefore commissioned a study of the performance of the various flags operating today.

The first step in the study was to define a set of criteria for ranking flags. It was decided to create an index (FLASCI) based on a weighted ranking of following factors:

1. The nature of the maritime administration
2. Administrative capacity
3. Maritime law
4. Seafarers' safety and welfare
5. Trade union law
6. Corruption
7. Corporate practice

The relative weighting and detailed factors assessed are summarized in Table 3.3. Data were retrieved from a literature search, and review of Internet sources and other available information on the Flag States such as Port State control statistics. The FLASCI scores are summarized in Table 3.4.

The Flag States got scores of between 19 and 84 and inspection of the findings suggested that the Flag States might be grouped into five categories, as shown in Table 3.4. The study clearly shows that flags show greater variation in performance than has generally been accepted. Some of the main findings were:

- Some of the so-called second registers perform as well as the best national registers: Norway (NIS), Denmark (DIS), Germany (GIS) and France (Kerguelen Islands).

Table 3.3. Flag State Conformance Index (FLASCI)

Flag State fleet 15%	Port State control rates Casualty rates Pollution incidence Own-citizen labour force participation	Own-citizen beneficial ownership Abandonment of crews Appearance in crew complaints DB
FS administrative capacity 30%	Death records Crew records of service Health screening procedures and records Accessibility of consular services Enforcement of IMO and ILO Conventions	Casualty investigation capacity Statistics of ships, owners and labour force Certification of seafarers Involvement intraining and education
FS maritime law 20%	Ratification of IMO and ILO Conventions Provisions of maritime legal code Publication of relevant law reports	Specialist law practitioners Location of registry 'Ownership' of registry
Miscellaneous maritime 5%	Maritime welfare support and maritime charities Maritime interest groups	Government ministries with maritime remit Stock exchange maritime listings State-owned shipping
Trade union law 10%	Legal rights for migrant labour Independent trade unions Mediation/arbitration procedures	Provision for trade union recognition Enforcement of trade union recognition procedures
Corruption 10%	Probity of public officials Misapplication of public funds	Integrity of political institutions and legal process Corporate integrity
Corporate practice 10%	Regulation of financial institutions Regulation of non-resident companies	Regulation of accounting standards Legal definition of corporate public responsibility

Source: Alderton and Winchester (2001).

- A few of the established FOCs are performing relatively well: Bermuda (63). Other FOCs such as Bahamas (43) and Liberia (43) are ranked lower but are still better than the worst performing.
- There seems to be a clear correlation between low performance and short operation as flag (new entrants). Port State control of these flags shows a quite high detention rate as shown in Table 3.5.

Table 3.4. Ranking of selected Flag States

Category	Selected flags (score)	Score range
Traditional maritime nations	NOR (84), UK (80), DIS (77), NIS (77),	84–72
Centrally operated second registers	Netherlands (76), GIS (75), Kerguelen Islands (72)	
Semi-autonomous second registers	Hong Kong (64), Bermuda (63), Latvia (60), Cayman Islands (62), Estonia (58)	64–58
Established open registers (seeking EU membership)	Cyprus (50), Malta (49), Russia (48), Bahamas (43), Liberia (43), Panama (41)	50–41
National registers		
New open registers	Marshall Islands (36), Ukraine (36), Honduras (35), Lebanon (35)	36–35
New entrants to the open register markets	St. Vincent and Grenadines (30), Bolivia (30), Belize (27), Equatorial Guinea (24), Cambodia (19)	30–19

Source: Alderton and Winchester (2001).

Table 3.5. Detention rate for 'new entrant' flags

	Belize	Bolivia	Cambodia	Equatorial Guinea
Asia–Pacific MOU (average 7%)	24.7%	No data	30%	11.1%
Paris MOU (average 9%)	31.4% Blacklisted	70%	24.8% Blacklisted	14.3%
USCG (average 5%)	50.6% Targeted	No data	Too few inspections	28.6% Targeted

Source: Alderton and Winchester (2001).

The last point can be explained by the apparent dynamics in the 'market' of Flag States. FOCs will, after some time when they are more established, be under pressure to improve their performances. As they eventually do this, it will open a market for new flags that will offer a more lenient safety regime. The SIRC study also showed that the fleets of the new entrants have a much higher growth rate than the average rate for the world fleet.

The SIRC study also analysed the working conditions on board and it was confirmed to be a less attractive climate on new entrant flag vessels. This is discussed in Chapter 12 on occupational safety.

3.5 PORT STATE CONTROL

3.5.1 UNCLOS

The basis for international shipping is the principle of freedom of the seas. The international legal basis is defined in the *United Nations Convention on the Law of the Sea* or *UNCLOS* (AMLG, 2004). The principle has the following key elements:

- Ships may sail without restriction in all waters on innocent passage (Article 17).
- The country of registration (Flag State) has the sole jurisdiction over the ship (Article 91).
- Other countries have limited jurisdiction even in own territorial sea.

The coastal state has at the outset the following rights:

- The outer limit of the territorial sea is 12 nm from the coast (baseline) within which it has full jurisdiction.
- The exclusive economic zone stretches out to 200 nm:
  - Very limited control jurisdiction.
  - Certain rights to take measures to preserve the marine environment.
  - However, the control should be exercised in accordance with international practice or non-discrimination against foreign vessels (Article 227).

The above means that the coastal states have to exercise their rights with respect to pollution hazards with delicacy. This becomes even more complicated when a state has both a substantial international trading fleet and a threatened coast. A good example is one of the initiatives of Spain and France in the aftermath of the *Prestige* accident. In an EU communication the following is stated:

> ...INVITES Member States to adopt measures, in compliance with international law of the sea, which would permit coastal States to control and possibly to limit, in a non-discriminatory way, the traffic of vessels carrying dangerous and polluting goods, within 200 miles of their coastline...

This position has been strongly opposed by INTERTANKO, which stresses that any measure in this area must adhere to international law and more specifically UNCLOS.

3.5.2 MOU PSC

The basic principle is that under the international safety conventions a certificate issued by Flag State A is equivalent to a certificate issued by state B. However, a Port State may challenge a certificate if there are indications that the condition of the foreign vessel is not in accordance with the particulars of the certificate.

The legal basis for Port State control (PSC) in Europe is found in the so-called Paris MOU (MOU, 2004), the 'Memorandum of Understanding on Port State Control' signed in 1982 by 19 European states and Canada.

The introduction of PSC was initially heavily opposed by shipping interests who feared that it would have a negative impact on the principle of equal market access and free competition. But in the end all involved parties acknowledged the shortcomings of Flag State control and the necessity of giving Port States authority to control shipping in their own waters.

The MOU has been given legal basis in national and international law, for instance in Norway by Regulation of 1 July 1996, No. 774, regarding control of foreign vessels, and similarly in Europe by Council Directive 95/21/EC of 19 June 1995 (Directives of the European Commission have status as law).

The objective for each Port State is to control 25 per cent of the foreign flag ships calling at their ports on an annual basis. An inspection may result in:

- *Deficiency:* a non-conformity, technical failure or lack of function. A deadline for correction will be given.
- *Detention:* a serious deficiency or multitude of deficiencies that must be corrected before the vessel is allowed to leave the port.
- *Banning:* ships having a multitude of detentions or lacking an ISM certificate may be banned from European waters.

Since the Paris MOU was established, a number of similar MOUs have been set up in other parts of the world. More information is available on the EQUASIS homepage (EQUASIS, 2004).

The findings and actions of Paris MOU are published in yearbooks (MOU, 2004). A summary of the number of inspections and relative frequency of deficiencies and detentions is shown in Figure 3.5.

Table 3.6 gives a summary of the relative deficiency rate for specific inspection areas. The following areas have a relative high frequency:

- Life-saving appliances
- Safety in general
- Safety of navigation

The Flag States show quite different performance in terms of deficiencies and detentions. The worst performing states are shown in Figure 3.6. Both 'classical' FOCs and new entrants have a quite high detention rate: Honduras, Belize and St. Vincent & Grenadines.

SCANDINAVIAN STAR FIRE

It became clear as a result of the accident investigation that neither the Danish nor the Norwegian Maritime Administration had been active in any way in connection with the start-up of the line between Oslo and Fredrikshavn. In fact it has been speculated whether the administrations were aware of the existence of the vessel at all. It has also been put forward as a theory that the administrations were reluctant to exercise their Port State control authority for fear of reprisals towards own-flag ships abroad.

The vessel was subject to Port State control in the USA in January but the Commission report does not refer any findings.

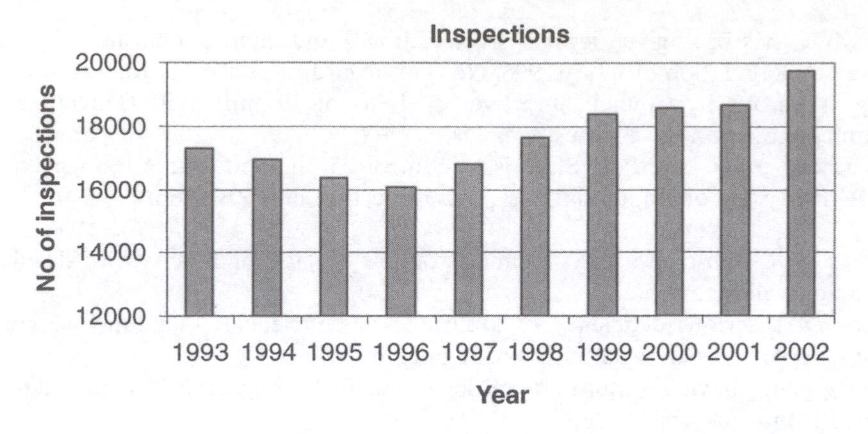

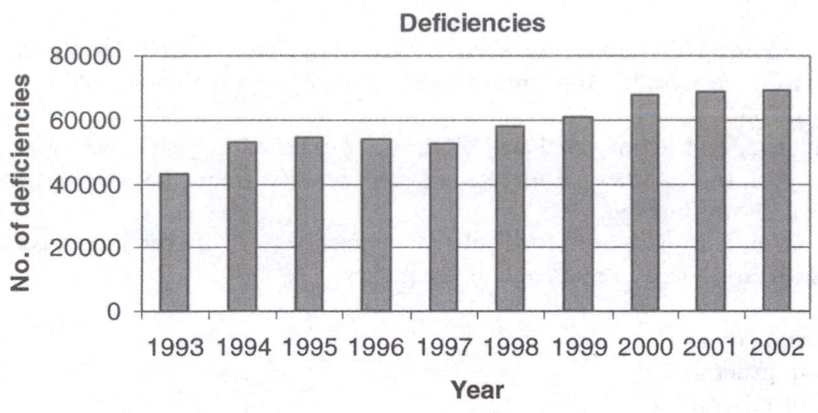

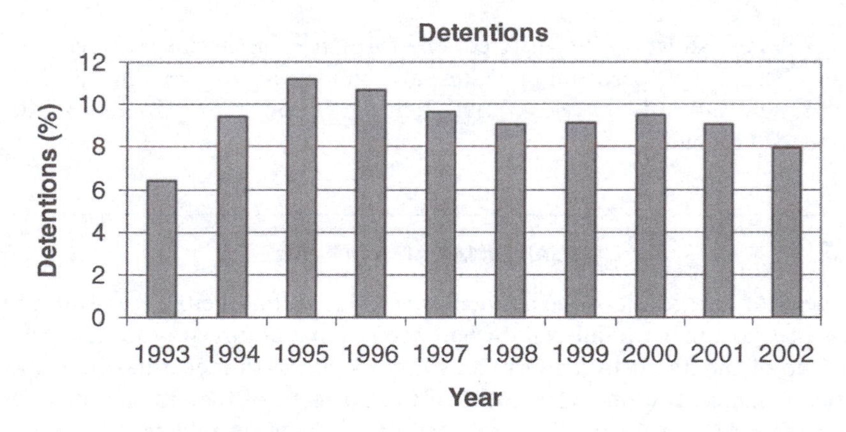

Figure 3.5. Port State control findings by Paris MOU. (Source: MOU, 2004.)

Table 3.6. Deficiency rate in % for inspection areas

	No. of deficiencies			Def. in % of total number			Ratio of def. to inspections × 100			Ratio of def. to indiv. ships × 100		
	2000	2001	2002	2000	2001	2002	2000	2001	2002	2000	2001	2002
Ship's certificates and documents	3465	3581	3369	5.1	5.2	4.88	18.8	19.2	17.04	30.8	30.7	28.50
Training certification and watchkeeping for seafarers	1179	1302	5522	1.7	1.9	7.99	6.4	7.0	27.94	10.5	11.2	46.71
Crew and Accommodation (ILO 147)	1963	2113	1853	2.9	3.1	2.68	10.7	11.3	9.37	17.5	18.1	15.67
Food and catering (ILO 147)	1031	876	664	1.5	1.3	0.96	5.6	4.7	3.36	9.2	7.5	5.62
Working space (ILO 147)	678	703	602	1.0	1.0	0.87	3.7	3.8	3.05	6.0	6.0	5.09
Life-saving appliances	10942	10516	9009	16.2	15.3	13.04	59.5	56.3	45.58	97.3	90.2	76.20
Fire safety measures	8789	8547	8158	13.0	12.4	11.81	47.8	45.8	41.27	78.1	73.3	69.00
Accident prevention (ILO 147)	1506	1586	1429	2.2	2.3	2.07	8.2	8.5	7.23	13.4	13.6	12.09
Safety in general	9243	8951	9306	13.7	13.0	13.47	50.2	47.9	47.08	82.2	76.8	78.71
Alarm, signals	330	326	301	0.5	0.5	0.44	1.8	1.7	1.52	2.9	2.8	2.55
Carrtage of cargo and dangerous goods	836	1323	1028	1.2	1.9	1.49	4.5	7.1	5.20	7.4	11.3	8.69
Load lines	3816	3906	3507	5.6	5.7	5.08	20.7	20.9	17.74	33.9	33.5	29.66
Mooring arrangements (ILO 147)	878	1109	1060	1.3	1.6	1.53	4.8	5.9	5.36	7.8	9.5	8.97
Propulsion and aux. machinery	3671	3713	3606	5.4	5.4	5.22	20.0	19.9	18.24	32.6	31.8	30.50
Safety of navigation	8055	8315	6769	11.9	12.1	9.80	43.8	44.5	34.25	71.6	71.3	57.25
Radio communication	2638	2703	2421	3.9	3.9	3.50	14.3	14.5	12.25	23.5	23.2	20.48
MARPOL, annex I	4875	5116	4421	7.2	7.4	6.40	26.5	27.4	22.37	43.3	43.9	37.39
Oil tankers, chemical tankers and gas carriers	212	151	202	0.3	0.2	0.29	1.2	0.8	1.02	1.9	1.3	1.71
MARPOL, annex II	71	43	64	0.1	0.1	0.09	0.4	0.2	0.32	0.6	0.4	0.54
SOLAS-related operational deficiencies	1132	1262	1353	1.7	1.8	1.96	6.2	6.8	6.85	10.1	10.8	11.44

Table 3.6. Continued

	No. of deficiencies			Def. in % of total number			Ratio of def. to inspections × 100			Ratio of def. to indiv. Ships × 100		
	2000	2001	2002	2000	2001	2002	2000	2001	2002	2000	2001	2002
MARPOL-related operational deficiencies	618	456	341	0.9	0.7	0.49	3.4	2.45	1.73	5.5	3.9	2.88
MARPOL, annex III	31	13	21	0.0	0.0	0.03	0.2	0.1	0.11	0.3	0.1	0.18
MARPOL, annex V	742	758	701	1.1	1.1	1.01	4.0	4.1	3.55	6.6	6.5	5.93
ISM	929	1239	3210	1.4	1.8	4.65	5.0	6.6	16.24	8.3	10.6	27.15
Bulk carriers, additional safety measures	9	50	51	0.0	0.1	0.07	0.0	0.3	0.26	0.1	0.4	0.43
Other def. clearly hazardous to safety	44	33	4	0.1	0.1	0.07	0.2	0.2	0.24	0.4	0.3	0.41
Other def. not clearly hazardous	52	65	63	0.1	0.1	0.09	0.3	0.3	0.32	0.5	0.6	0.53
Total	67735	68756	69079									

Source: MOU (2004).

(continued)

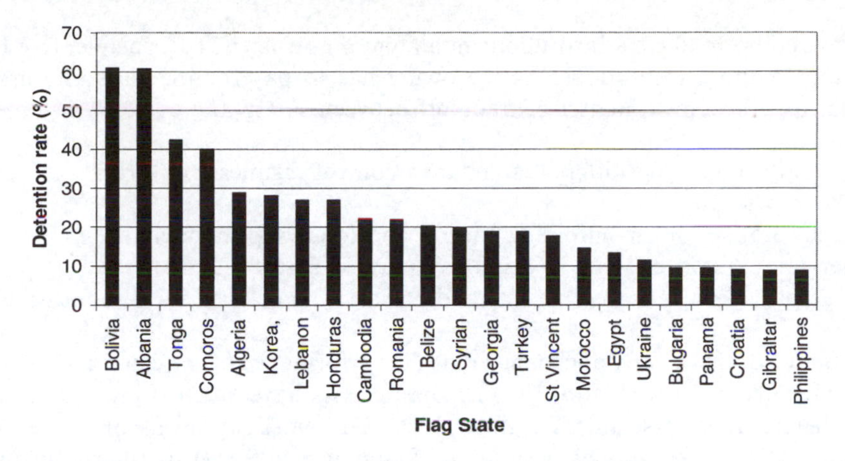

Figure 3.6. Detention rate in % for Flag States above the average rate. (Source: MOU, 2004.)

3.6 CLASSIFICATION SOCIETIES

Classification Societies are independent bodies which set standards for design, maintenance and repair of ships. The Classification building rules cover:

- Hull strength and design
- Materials
- Main and auxiliary machinery
- Electrical installations
- Control systems
- Safety equipment

A vessel is given a *class certificate* on the basis of drawings, engineering documentation, inspections during building and tests. A classed vessel will be surveyed on a regular basis and given recommendations for necessary maintenance and repair in order to keep its class.

The class is the basis for negotiating insurance of the vessel. The class in this sense is a kind of quality check for the insurance company. The Classification Society has otherwise no official role relative to international and national regulation. This is, however, not quite correct, as national regulation (Regulation of 15 September 1992, No. 695, on building of passenger and cargo vessels and barges) lists the following accepted Class Institutions:

1. Det Norske Veritas (DNV)
2. Lloyd's Register of Shipping (LRS)
3. Bureau Veritas (BV)
4. Germanischer Lloyd (GL)
5. American Bureau of Shipping (ABS)

There are about 40 class institutions in all and an owner is, in principle, free to select a class among those institutions. As the owner has to pay for the class and associated services, it may become a matter of trade-off between safety and cost:

- Class institutions offer different standards, control regimes and tariffs.
- They compete on price.
- Some are not serious in enforcing control and follow-up maintenance.
- Owners may 'jump between institutions' to avoid costly maintenance.
- Change of class means that an outstanding survey is delayed for 3 months.

Port State control clearly documents that the performance of the classification societies differs quite substantially (Figure 3.7). The best classes have a detention rate in the order of 1% whereas the worst are as high as 35%. The most notorious ones are: Register of Albania (RS), International Register of Shipping – USA (IS), International Naval Surveys Bureau – Greece (INSB) and Bulgarski Koraben Registar (BKR). Against this background, certain Flag States are contemplating banning vessels classed in specific classification societies.

The serious class institutions are organized in IACS (International Association of Classification Societies). The members cooperate in order to attain a harmonized standard for the serious institutions. Finally, it should be pointed out that the serious institutions maintain a high professional standard and contribute in many ways to the advancement of the safety standard.

As already pointed out, the Classification Society may also undertake control tasks on behalf of a Flag State administration. Presently they also undertake auditing tasks and assignment of ISM Certificates.

SCANDINAVIAN STAR FIRE

As already discussed in relation to Flag State control and SOLAS, there were some shortcomings of the control that also were relevant for the class survey:

- A number of technical faults and missing components.
- No survey was undertaken immediately before the start-up of the operation in Scandinavia.

3.6.1 The Maritime Code (Sjøloven)

The Maritime Code (MC) covers the legal aspects of shipping and ship operation as a commercial activity. The law can be summarized by the following keywords:

- Registration of vessel
- Partnerships
- The Master of the vessel

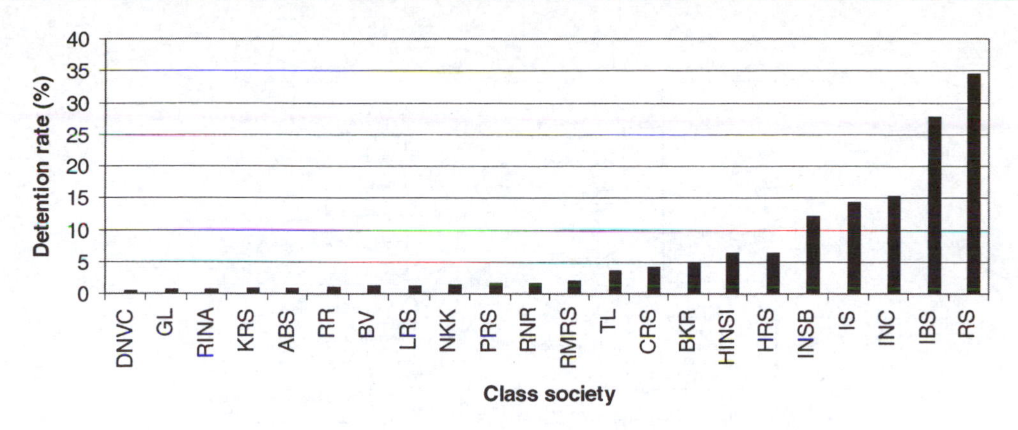

Figure 3.7. Detention rate of classification societies (MOU, 2004).

- Liability, and limitation of liability
- Contracts of affreightment, bill of lading
- Carriage of passengers
- Oil pollution and environmental liability
- Liability for collisions
- Salvage
- Marine insurance
- Maritime inquiries

One of the controversial aspects of the *Scandinavian Star* accident was the unclear ownership and the lack of involvement from Scandinavian maritime administrations.

SCANDINAVIAN STAR FIRE

The ship was owned and operated by a number of companies within the VR DaNo group which in itself was not a legal entity. The ownership was related to K/S *Scandinavian Star*, which chartered the vessel to Project Shipping Ltd on a bare-boat basis. The manned vessel was further transferred on a time-charter to VR DaNo ApS. As far as the investigation could establish, all these companies were related to a group of persons representing a sphere of interests. The identity of the real owner was unclear at the time of the accident and has yet not been fully established. It has been indicated that the Danish owner only fronted for other parties, among which some well-known Scandinavian companies have been mentioned. A more detailed description of the company and owner structure is given in Figure 3.8.

The vessel had registration on the Bahamas under their *Merchant Shipping Act of 1976*. That means that neither the Norwegian nor the Danish Flag State administrations were involved in the registration, and therefore had no direct jurisdiction over the company or the vessel.

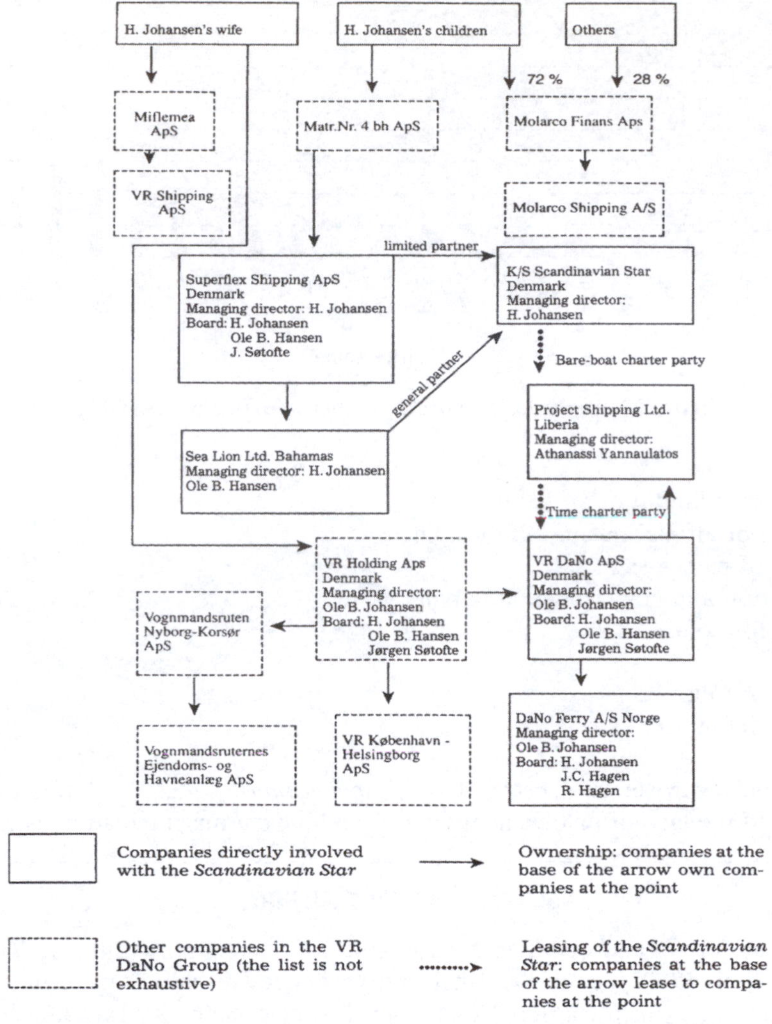

Figure 3.8. Companies in the VR DaNo group.

The Master of a vessel has a special position in comparison to the other crew members. The MC Chapter states that the Master:

- Has the highest authority on board
- Is responsible for seaworthiness
- Is responsible for seaworthiness in relation to the cargo
- Has the power to enter a contract with a salvage tug

This philosophy has a historic background based on the fact that in the age of the sailing ships, the shipowner had no daily control over his vessel and had to rely on the trust and competence of his representative on board, namely the Master.

In today's world of modern communication, the Master has the opportunity to report and confer daily with the manager. Likewise the manager has the complete freedom to instruct and control his vessel in detail. This means that the role of the Master has changed significantly. This fact makes his unique authority and responsibility somewhat outdated.

SCANDINAVIAN STAR FIRE

The Master entered the ship on 23 March, roughly two weeks before the tragedy. During this short period the vessel was subject to continued modification work, manning and preparation of the operation. The clear message from the owner was to get the vessel ready. The Master spent considerable time on checking safety systems on board, and it must have become clear to him that the vessel was not operationally ready with respect to vessel, manning or routines.

There was no indication that the Master took this problem up with the owner or tried to delay the start of the operation. This would have required considerable personal and moral strength. Given the determination of the shipowner and the fact that the Master had got this commission after a considerable period of unemployment, the legal status of the Master becomes less meaningful.

3.6.2 Liability

Liability in case of sea transport is a large and complex topic. Here we shall only comment on liability in relation to passengers and environment.

The Norwegian Maritime Code §418 covers the matter of liability in case of personal injury and death to passengers. Liability will be imposed under circumstances where the injury is caused by fault or neglect of the carrier. An important requirement, however, is that the claimant must prove that:

- The harmful event took place during the voyage, and
- Was the result of fault or neglect by the carrier.

In other words, liability stemming from personal injury is *objective*. Presently §422 limits the compensation to NOK 1,622,500 per passenger.

During the last 20–30 years the world has witnessed a number of serious ship accidents with massive spills like *Torrey Canyon*, *Amoco Cadiz* and *Exxon Valdez*. This soon raised the matter of liability related to environmental harm. The so-called CLC Convention (the intervention and liability convention) was approved in 1969. It represented a radical change in stating that:

- Owners of ships transporting oil as bulk cargo are made strictly liable for oil pollution, with virtually no exceptions, and
- The amounts could only be limited to sums much larger than the general rules.

This means that, contrary to personal injury, in the case of an environmental accident a claim can be put forward without proving negligence.

The United States has introduced its own rules through the Oil Pollution Act (EPA, 2004), which gives the plaintiff almost unlimited right to make the shipowner liable.

Activities on board a vessel may also be subject to prosecution under the Norwegian Criminal Code of 1902. §12 contains rules with respect to personal acts and §48 covers the provisions for companies. The main principle is that the same laws that apply ashore also apply on Norwegian vessels.

SCANDINAVIAN STAR FIRE

The owner of *Scandinavian Star* was made liable and sentenced for fault and neglect mainly for securing adequate operational readiness. However, the fine was symbolic. The passengers and relatives of the deceased were compensated as a result of a joint agreement between the group and SKULD which covered the P & I insurance (protection and indemnity or third-party liability).

It has also been speculated that the shipowner accepted the fine immediately, rather than entering into legal battles over the ownership. If fraudulent circumstances had surfaced as a result of a legal process, substantially larger fines might have been the consequence.

REFERENCES

Alderton, T. and Winchester, N., 2001, The Flag State Audit. *Proceedings of SIRC's Second Symposium*, Cardiff University, 29 June. Seafarers International Research Centre, Cardiff, UK. http://www.sirc.cf.ac.uk/pubs.html

AMLG, 2004, *Admirality and Maritime Law Guide – International Conventions*. http://www.admiraltylawguide.com/conven/unclospart2.html

EPA, 2004, Oil Pollution Act, US Environmental Protection Agency. http://www.epa.gov/region5/defs/html/opa.htm

EQUASIS, 2004, EQUASIS – Public web-site promoting quality shipping. http://www.equasis.org/

Falkanger, T., Bull, H. J. and Brautaset, L., 1998, *Introduction to Maritime Law: The Scandinavian Perspective*. Tano Aschehoug, Oslo.

ILLC, 2002, *International Conference on Load Lines 1966, 2002 Edition*. International Maritime Organization, London.

ISM, 2002, *International Safety Management Code (ISM Code) and Guidelines on Implementation of the ISM Code, 2002 Edition*. International Maritime Organization, London.

Kristiansen, S. and Olofsson, M., 1997, *SAFECO – Safety of Shipping in Coastal Waters. Operational Safety and Ship Management – WPII.5.1: Criteria for Management Assessment*. Report No. MT23 F97-0175, Marintek As, Trondheim, Norway.

MARPOL, 2002, *MARPOL 73/78 Consolidated Edition 2002*. International Maritime Organization, London.

MOU, 2004, The Paris Memorandum of Understanding on Port State Control, *Yearbook 2002*. http://www.parismou.org/

NAS, 1991, *Tanker Spills: Prevention by Design*. Committee on Tank Vessel Design, Marine Board, National Research Council, National Academy of Sciences. The National Academies Press, Washington, DC.

NOU, 1991, *The Scandinavian Star Disaster of 7 April 1990*. Norwegian Official Reports NOR 1991: 1 E, Government Administration Services, Government Printing Service, Oslo.

Sjøfartsdirektoratet, 1998, *Rules of the Norwegian Ship Control* [In Norwegian: *Den Norske Skipskontrolls Regler*]. Erlanders Forlag, Oslo.

SOLAS, 2001, *SOLAS Consolidated Edition 2001*. International Maritime Organization, London.

STCW, 1996, *Convention on Standards of Training, Certification and Watchkeeping for Seafarers, 1996 Edition*. International Maritime Organization, London.

PART II
STATISTICAL METHODS

4

STATISTICAL RISK MONITORING

A paradox of life: The risk taker may be undeservedly lucky, whereas the prudent person may be struck by a catastrophe.

(Attributed to J. Reason)

4.1 INTRODUCTION

In order to manage risk it is of key importance to be able to monitor the safety level of the operation. Key risk parameters are accident frequency and expected consequences in terms of human suffering, environmental damage or economic loss. It is clear that accidents are the result of complex interactions within the system, in relation to the operators and to the environment. This means that both the occurrence and outcome of accidents are to some degree stochastic in nature. It is therefore important that the risk manager has a good understanding of how statistics can be used in the monitoring of accident phenomena. The following presentation will highlight some key topics from statistical theory with strong emphasis on the practical application in risk management.

4.2 STATISTICAL MEASURES

Let us consider a random variable with a known probability density function. The variable may be characterized with certain statistical measures.

4.2.1 Mean, Weighted Mean

The mean of a random variable is also termed the average or expected value. It may be viewed as the centre of gravity of the associated distribution. The most straightforward way to compute the mean accident frequency rate (AFR) is to apply the *sample mean* for N observations with value X_i:

$$\overline{X} = \frac{1}{N}\sum_{i=1}^{N} X_i$$

Observe that we use the symbol μ for the mean of the true population (*population mean*). Recall that a sample is drawn from the true population and may therefore be seen as a subset.

The mean may also be based on grouped observations of the random variable and the *weighted mean* may then be more relevant:

$$\overline{X} = \frac{1}{M}\sum_{i=1}^{M} p_i \cdot X_i$$

where p_i denotes the probability of observing a member of group i with mean X_i and M is the number of groups. Given that a group has N_i observations and the total number of observations is N, we have:

$$p_i = \frac{N_i}{N}$$

Example

The number of accidents among crew groups in a shipping company has been investigated. The results are shown in Table 4.E1 in terms of the accident frequency rate (AFR).

The mean AFR computed as the simple mean is:

$$\begin{aligned}\overline{\text{AFR}} &= \tfrac{1}{6}(5 + 10 + 25 + 15 + 20 + 17)\\ &= 15\end{aligned}$$

The average accident frequency rate for the total seagoing workforce in the company is, in other words, 15 accidents per 200,000 work-hours.

However, this way of computing the mean does not reflect the fact that some of the largest crew groups have an AFR higher than the estimated mean. It may therefore seem

Table 4.EI. Number of work accidents in a company: accident frequency rate (AFR) in terms of number per 200,000 work-hours

Crew category	Master	Mates	Deck ratings	Engineer officers	Engine ratings	Catering, hotel
Accident Frequency Rate (AFR)	5	10	25	15	20	17
Fraction of workforce (%)	5	25	25	20	20	5

more relevant to estimate the *weighted mean* where the relative magnitude of the groups is taken into consideration:

$$\begin{aligned}\overline{\text{AFR}} &= \tfrac{1}{6}(0.05 \cdot 5 + 0.25 \cdot 10 + 0.25 \cdot 25 + 0.20 \cdot 15 + 0.20 \cdot 20 + 0.05 \cdot 17) \\ &= 17\end{aligned}$$

This gives a somewhat higher value for the AFR than the previous estimate.

4.2.2 Median

The median is found by arranging the observations in ascending order and selecting the middle data point. Let us assume the following sample space of five observations:

$$S = \{1, 3, 5, 8, 10\}$$

The median is evidently 5, whereas the mean is 5.4. Another way is to define the median as the value corresponding to 50% probability of exceedance. In general, the mean is preferred to the median as it expresses the 'gravity' point of the sample. However, a useful property of the median is its ability to ignore *outliers*. Assume a data set where one extra observation is added that has a value significantly higher or lower than the initial observations. This extreme value will not change the median value as much as the mean.

4.2.3 Dispersion, Variance, Standard Deviation

An immediate question is how well the mean value reflects the observation data. In other words, how much can an observation be expected to deviate from the mean? This parameter is called the variance and is computed as the mean of the sum of squares of deviations. More often we prefer to use the standard deviation that is given by the square of the variance. The *population standard deviation* is:

$$\sigma = \sqrt{\frac{1}{N}\sum_{i=1}^{N}(X_i - \mu)^2}$$

In a practical situation we have only a limited set of observations of the true population or a sample. As an estimate of the standard deviation we apply the *sample standard deviation* given by following a slightly different expression:

$$s = \sqrt{\frac{1}{N-1}\sum_{i=1}^{N}(X_i - \overline{X})^2}$$

Recall our example of the observation of the accident frequency rate AFR where the simple mean was estimated to be 15. The sample standard deviation can be computed as follows:

$$s = \sqrt{\frac{1}{6-1}[(5-15)^2 + (10-15)^2 + (25-15)^2 + (15-15)^2 + (20-15)^2 + (17-15)^2]}$$

$$= \sqrt{\frac{1}{6-1}[100 + 25 + 100 + 0 + 25 + 4]}$$

$$= \sqrt{254/(6-1)} = \sqrt{50.8} = 7.1$$

The standard deviation for our observations of AFR is 7.1. As a digression it should be pointed out that there is no obvious reason behind the definition of the variance or standard deviation other than that by squaring the deviations one avoids deviations with the opposite sign cancelling each other out in the expression.

4.3 DISCRETE PROBABILITY DISTRIBUTIONS

4.3.1 Definitions

It is useful to make a distinction between discrete and continuous probability models. A *discrete* model has a set of discrete outcomes, as for instance the number of dots on each face of a dice:

$$\Omega = \{X_1, X_2, \ldots, X_6\} = \{1, 2, 3, 4, 5, 6\}$$

A more precise definition is:

$$0 \leq p(X) \leq 1.0$$
$$p(X_1) + p(X_2) + p(X_3) + \cdots = 1.0$$

For a 'fair' dice the probability of each outcome will be the same:

$$p(1) = p(2) = p(3) = p(4) = p(5) = p(6) = \tfrac{1}{6}$$

The probability density function (PDF) for a discrete function takes the graphical form of a histogram as indicated in Figure 4.1.

The cumulative distribution function (CDF) expresses the probability that the outcome X is equal to or less than a given value x:

$$F(x) = P(X \leq x)$$

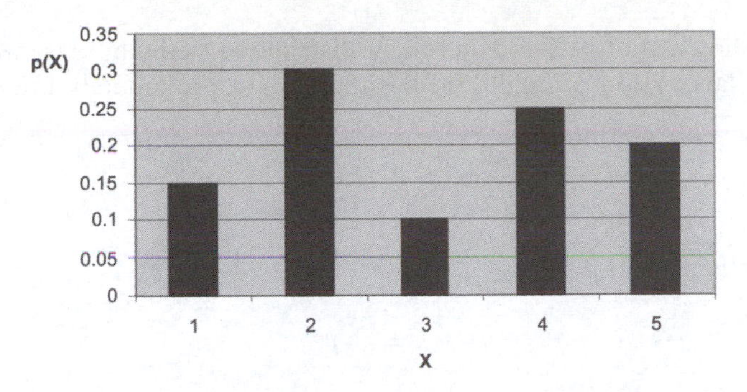

Figure 4.1. Probability density function of discrete distribution.

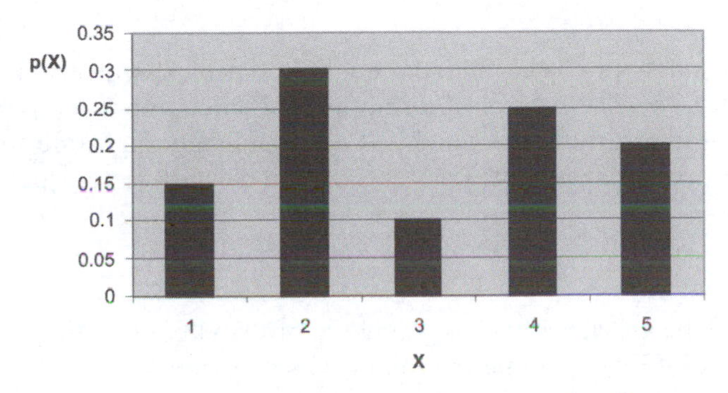

Figure 4.2. Cumulative distribution for discrete distribution.

The format of CDF for a discrete distribution is shown in Figure 4.2. Observe that the CDF is a monotonous function between 0 and 1.0.

4.3.2 The Binomial Distribution

Let us assume that we are performing a series of n independent experiments where the outcome is either a *success* or a *failure*. The probability of success for each experiment is p. The number of successes in n experiments is given by a binomial distribution with the parameters (n, p):

$$p(x) = P(X = x) = \left(\frac{n!}{x!(n-x)!}\right) p^x (1-p)^{n-x}; \qquad x = 0, 1, \ldots, n$$

The expected value and variance of X are given by:

$$E(X) = \mu = n \cdot p \qquad \operatorname{var} X = \sigma^2 = n \cdot p \cdot (1-p)$$

It can also be shown that if X is binomially distributed with the parameters (n, p), and further that n is large and p is small, one has that X is approximately Poisson distributed:

$$p(x) = \left(\frac{n!}{x!(n-x)!}\right) p^x (1-p)^{n-x} \approx \frac{(np)^x}{x!} e^{-np}$$

The expected value is:

$$\lambda = n \cdot p$$

Example

Problem

A component is mass- produced and the quality control has determined that 10% of the output is defective. The control of a shipment of 45 components found that 11 were defective. The question arises whether this is an appallingly high rate for the shipment. Does 11 defective components out of 45 or 24.4% indicate a lower quality than initially established?

Solution

This problem may be answered by computing the probability of getting at least 11 defects in a total sample of 45 by applying the following sum expression:

$$P(X \geq x) = 1 - \sum_0^x \binom{n}{x} p^x (1-p)^{n-x}$$

The first expression is:

$$P(X \geq 11) = \sum_{11}^{45} \left(\frac{45!}{11!(45-11)!}\right) \cdot 0.10^{11} \cdot (1-0.10)^{45-11}$$

Keep in mind that the success probability (p) is identical with the defect probability in this case. The probability of at least 11 defects can be looked up in a table for *cumulative terms* of the binomial probability distribution with the values $(n, x, p) = (45, 11, 0.10)$ and gives 0.004. The probability of having at least 11 defects is 0.4% or, in other words, a fairly remote event. One can conclude that the shipment does not meet the quality standard.

4.3.3 The Poisson Distribution

The Poisson distribution is widely applied in reliability and risk analysis. It is especially useful for describing the number of failures in a given period of time t. Like the binomial distribution it is a discrete distribution by only taking on integer values:

$$P\{C(t)\} = \frac{1}{n!}(\theta \cdot t)^n e^{-\theta t} \tag{4.1}$$

where $C(t)$ = number of failures in the period t, and $n = 1, 2, 3, \ldots$.

By assuming a standardized period t and introducing the parameter λ:

$$\lambda = \theta t$$

Eq. (4.1) can be given in a simplified form:

$$P\{X = x\} = \frac{\lambda^x}{x!} e^{-\lambda}$$

As can be seen from the expression, the Poisson distribution has only one parameter, namely λ. It can further be shown that this parameter expresses both the mean and the variance:

$$\mu = \lambda$$
$$\sigma^2 = \lambda$$

As can be seen in Figure 4.3, the density distribution is asymmetric for low values of λ (0.5) and becomes more and more symmetric as λ increases in value as indicated for

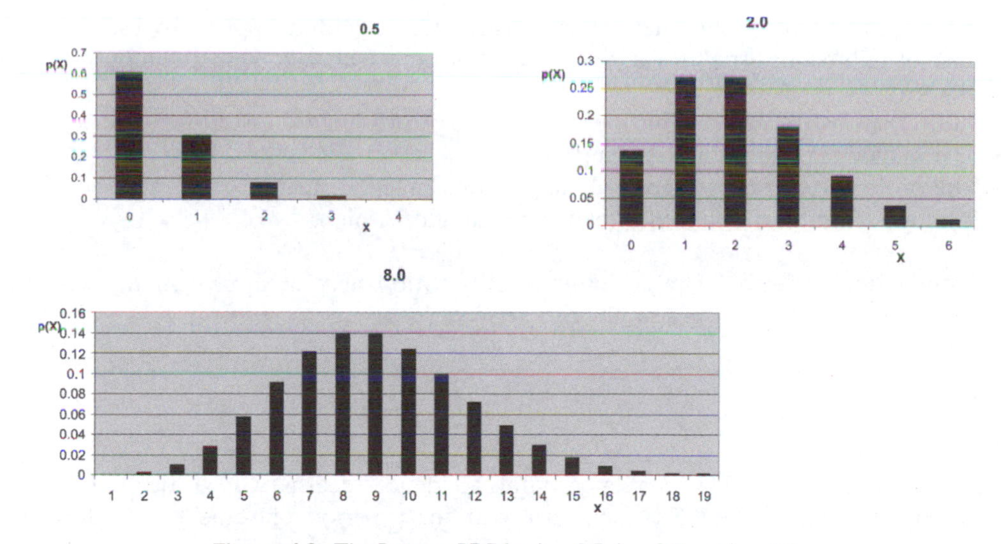

Figure 4.3. The Poisson PDF for $\lambda = 0.5$, $\lambda = 2.0$ and $\lambda = 8.0$.

the λ values 2.0 and 8.0. Apart from changing from asymmetric to symmetric form, the dispersion increases considerably. The standard deviation for $\lambda = 0.5$ is $0.5^{1/2} = 0.707$ whereas for $\lambda = 8.0$ it is $8.0^{1/2} = 2.828$. This means that for the higher values of λ the degree of variation (or uncertainty) becomes pronounced.

The fact that the Poisson distribution becomes symmetrical for higher values of λ makes it suitable for approximation by other and more computable distributions such as for instance the normal distribution. This topic will be discussed later.

The cumulative distribution function (CDF) can be computed by the following expansion:

$$F(x) = P(X \leq x) = e^{-\lambda} + \lambda e^{-\lambda} + \frac{\lambda^2}{2!} e^{-\lambda} + \cdots + \frac{\lambda^x}{x!} e^{-\lambda}$$

or

$$F(x) = 1 - \sum_{x}^{\infty} \left(\frac{\lambda^x e^{-\lambda}}{x!} \right)$$

Both expressions can also be looked up in a Poisson table. It has been found that the distribution applies for phenomena with following characteristics:

- The events are independent of each other in non-overlapping time intervals.
- The probability of an event is proportional to the length of the period.
- The probability of having more than one event in a small time frame is small compared to the probability of having one event.

As already pointed out, the Poisson distribution is often applied to estimate the number of failures or errors for a given period of time. The following assumptions must be satisfied:

- The individual failures are independent events.
- The probability of occurrence of a single event must be small.
- The opportunity of occurrence (exposure) should be high.

It should also be noted that the binomial distribution may be approximated with the Poisson if n is large and p is small. The parameter is given by $\lambda = np$.

Example

Problem

The average number of work-related fatal accidents (deaths) for a fleet of vessels has been estimated to be 0.5 persons per year for a period of some years. However, for the last year the same fleet has reported 3 fatal accidents. The management is therefore concerned about whether the risk level has increased during the last reported period.

Solution

The number of fatalities per year is assumed to be Poisson distributed with mean $\lambda = 0.5$. The probability of having exactly 3 deaths per year is:

$$P\{X = 3\} = \frac{1}{3!}(0.5)^3\, e^{-0.5} = 0.0126$$

Assuming that the risk level is unchanged, the probability of having 3 fatalities is, in other words, 1.3%, which indicates that this is a fairly remote event. However, the correct way of assessing the situation is to estimate the probability of having *at least* 3 fatalities per year:

$$\begin{aligned} P(X \geq 3) &= \sum_{x=3}^{\infty} P(X \geq x) = \sum_{x=3}^{\infty} \left(\frac{e^{-0.5} \cdot 0.5^x}{x!} \right) \\ &= 1 - \sum_{x=0}^{2} \left(\frac{e^{-0.5} \cdot 0.5^x}{x!} \right) \\ &= 1 - 0.6065 - 0.3033 - 0.0758 \\ &= 0.0144 \end{aligned}$$

We see from this result that the probability of having 3 fatal accidents per year is still less than 1.5%. This may therefore be taken by the management as an indication that the risk level of the fleet has in fact *increased.*

A more formal conclusion may be stated as follows:

The null-hypothesis H_0: $\lambda = 0.5$
Significance level: $\alpha = 5\%$ (the accepted risk for rejecting a true H_0)

$$P(X \geq 3) = 0.014, \text{ which is less than } \alpha = 0.05$$

The observation is outside the confidence interval and the conclusion is that

$$H_0 \text{ must be rejected; or in other words, } \lambda \neq 0.5$$

The PDF and CDF for a Poisson distribution with $\lambda = 0.5$ are shown graphically and in the table below.

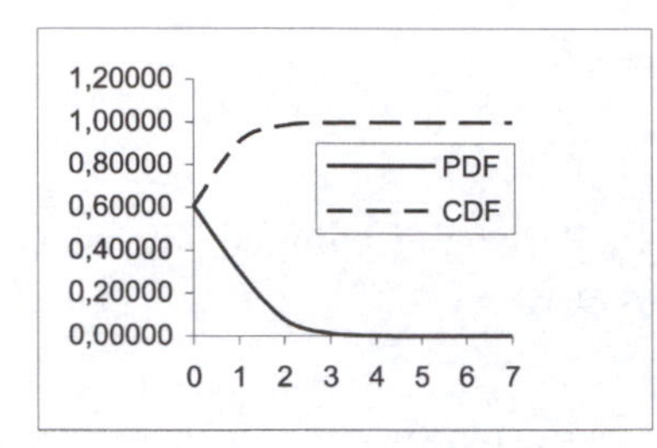

Accidents/year	PDF	CDF
0	0.60653	0.60653
1	0.30327	0.90980
2	0.07582	0.98561
3	0.01264	0.99825
4	0.00158	0.99983
5	0.00016	0.99999
6	0.00001	1.00000
7	0.00000	1.00000

4.3.4 The Uncertainty of the Estimated λ

In certain situations we are concerned with the uncertainty of the estimated parameter of the Poisson distribution, λ. As we recall, the parameter expresses both the mean and the variance.

Let us illustrate this by the following example. A port has kept a close look on the accident record for some years and has established the following time series for the number of accidents and number of calls (ship visits):

	1989	1990	1991	1992	1993	1994
Accidents/year	8	6	7	7	9	11
Calls/year	23,529	27,270	25,925	24,140	25,000	21,430
Accidents/10,000 visits	3.4	2.2	2.7	2.9	3.2	5.6

The port management has been concerned about the seemingly high accident frequency reported for 1994. Again, the question that might be raised is whether this indicates a loss of control over safety for the port.

The previous safety level may be expressed by means of the average accident rate for the 5-year period (1989–93):

$$\text{Mean number of accidents/year:} \quad N_a = 8+6+7+7+9 = 37$$

$$\text{Mean number of calls/year:} \quad N_p = 23{,}529+27{,}270+25{,}925+24{,}140+25{,}000 = 125{,}864$$

$$\text{The mean loss rate is} \quad \lambda = N_a/N_p = 37/125{,}864 = 2.94 \text{ accidents}/10{,}000 \text{ calls.}$$

The 5-year average loss rate for the most recent period (1990–94), which includes the high value for the last year, is:

$$\text{Mean number of accidents/year:} \quad N_a = 6+7+7+9+11 = 40$$

$$\text{Mean number of calls/year:} \quad N_p = 27{,}270+25{,}925+24{,}140+25{,}000+21{,}430 = 123{,}765$$

$$\text{The mean loss rate is} \quad \lambda = 40/123{,}765 = 3.23 \text{ accidents}/10{,}000 \text{ calls.}$$

In other words, it can be concluded that the 5-year average loss rate has increased from 1993 to 1994. The remaining question, however, is whether this increase is significant or not. The following statistical knowledge can be applied.

For the sum of k observations of a variable X that are Poisson distributed, we have that:

$$\sum_k X \text{ is also Poisson distributed with the parameter } k\theta$$

This fact can be applied as follows: The number of accidents in the *first period* was 37. By looking up a Poisson distribution table we find the confidence limits for the *parameter* $\lambda = 37$:

Assumed significance level: $\alpha = 0.05$

Confidence interval: $26.0 < \Sigma X < 51$

The corresponding confidence interval for the average accident rate is computed by division with the accumulated traffic:

$$\frac{26.0}{12.5864} < \theta < \frac{51}{12.5864} \Rightarrow 2.07 < \theta < 4.05 \text{ (accidents/10,000 calls)}$$

We recall that the average loss rate for the recent period was $\theta = 3.23$. This value lies within the confidence limits of the former average value. We can therefore conclude that the increased accident rate in 1994 is not sufficient to say that the average accident rate is increased significantly. By applying 5-year average values, one has in fact taken a conservative position with respect to risk management. This can be demonstrated by applying the earlier simple Poisson model:

Assume following mean accident rate: $\lambda = 2.94 \approx 3$

The probability of having at least 6 accidents in one year is looked up in a table:

$$P(X \leq 5) = 0.916; \qquad P(X \geq 6) = 1 - 0.916 = 0.084$$

It can be seen that the probability of having at least 6 accidents is 8.4%, which is within the confidence interval. This approach also shows that we do not have an indication of an increased risk level in the port.

4.4 CONTINUOUS DISTRIBUTIONS

Another important group of statistical distributions are the so-called continuous distributions, where the outcome may take any real number in a given range.

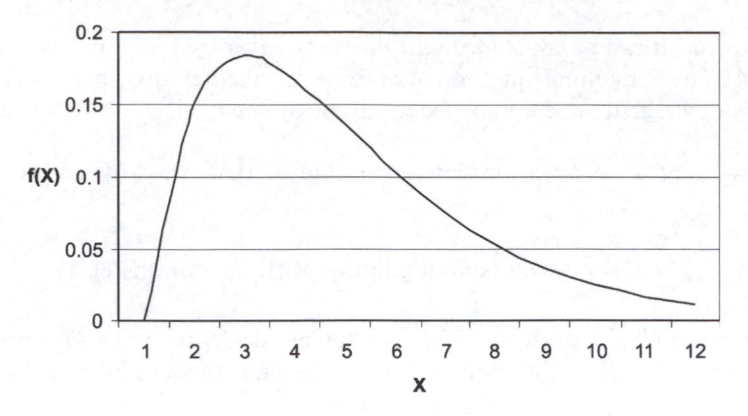

Figure 4.4. Continous PDF.

They are defined as follows:

1. The probability density function (PDF) may take any value between 0 and 1.0 as illustrated in Figure 4.4.
2. The area under the curve of the PDF is equal to 1.0.

These properties are expressed as follows:

$$0 \leq p(X) \leq 1.0$$

$$\int_{-\infty}^{\infty} p(X) \cdot dX = 1.0$$

The cumulative distribution $F(x)$ expresses the probability that the random variable X is less than or equal to a given value x:

$$F(x) = P(X \leq x) = \int_{-\infty}^{X} f(x) \cdot dx$$

Conversely the probability of observing a higher value is given by following expression:

$$P(X > x) = 1.0 - F(x)$$

The nature of a cumulative distribution is indicated in Figure 4.5. By definition the function approaches 1.0 asymptotically.

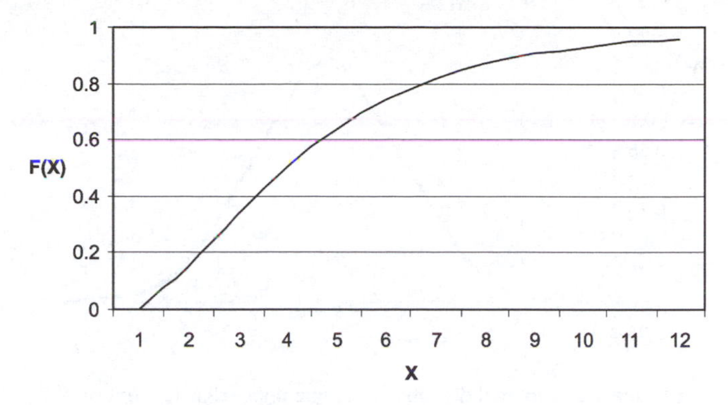

Figure 4.5. Cumulative probability distribution.

4.4.1 The Normal Distribution

Certain risk and engineering problems apply the normal or Gaussian distribution. The variable X is said to be normally distributed with mean μ and variance σ^2 and is written $N(\mu, \sigma^2)$. The probability density function is given by:

$$f(x) = \frac{1}{\sqrt{2\pi\sigma^2}} e^{(x-\mu)^2/(2\sigma^2)}$$

By introducing the standardized variable:

$$Z = \frac{X - \mu}{\sigma}$$

the PDF takes a simpler form:

$$g(z) = \frac{1}{\sqrt{2\pi}} e^{-z^2/2}$$

The further implication of this definition is that variable Z is also normally distributed with the parameters $N(0, 1)$. The PDF is bell-shaped as shown in Figure 4.6.

Example

Problem

The maintenance of the steering system of a vessel involves testing the voltage in a critical circuit. The voltmeter is supposed to read 0 volts in a specific circuit if the system is OK. The reading of the voltmeter can be expressed by z. Past experience has shown that the readings have a mean value of 0 volts and a standard deviation of 1 volt when the system is

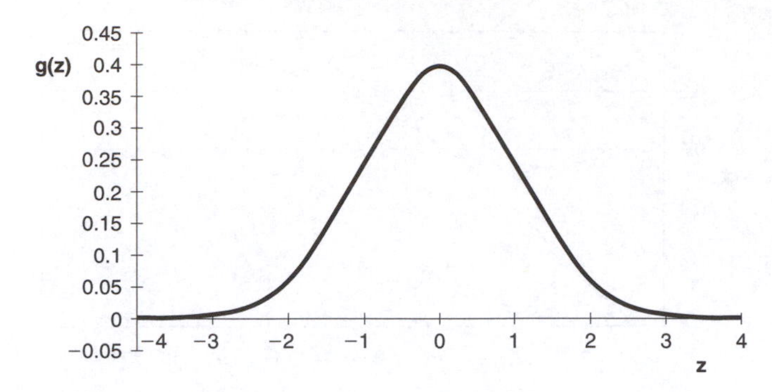

Figure 4.6. Normal distribution, probability density function.

in an acceptable condition. Find the probability that the reading of the voltmeter will show a value between 0 and 1.43 volts.

Solution

The problem can be expressed as follows: $P(0 < z < 1.43)$. This represents the area under the curve from 0 to 1.43.

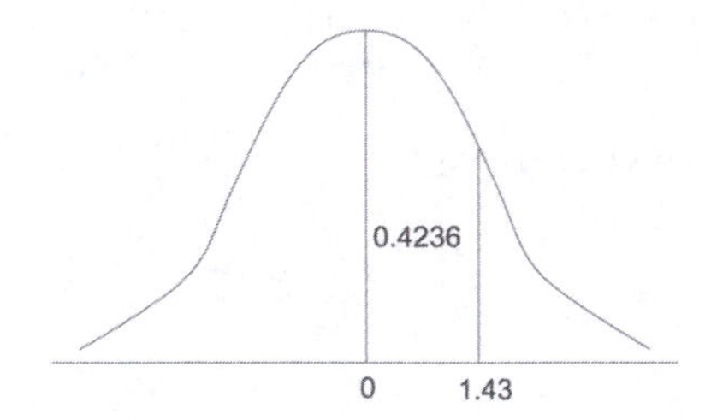

By looking up a table for the normal distribution, one may read the value of the area under the curve from zero to a given value z. The table below indicates the layout. The first two digits are found in the left-hand column (1.4), and the third digit is found by scanning across the top row (0.03). The value of the area is found in the crossing of the row and column.

Z	0.0	0.01	. . .	0.03	. . .
0.0					
0.1					
. . .					
1.4				0.4236	
. . .					

4.4.2 Poisson Approximation

As discussed earlier, the number of accidental events in a given time period may be described by the Poisson distribution. However, it was also pointed out that this distribution becomes increasingly symmetrical for greater values of the parameter λ. This fact can be utilized in computations. An approach might be to substitute the Poisson with the normal distribution.

Let us take following situation. A shipping company has experienced 20 serious occupational accidents on average per year for the last ten years. For the last year, however, 29 serious accidents were reported. The question again is whether this indicates a higher risk level. Let us assume that the annual number of serious accidents is Poisson distributed with $\lambda = 20$. The probability of having at least 29 observations (X) is given by looking up a table:

$$P(X \geq 29) = 1 - P(X < 28) = 1 - 0.966 = 0.034$$

It can, however, be shown that the Poisson distribution is increasingly well approximated by the normal distribution for increasing values of λ:

$$P\left(\frac{X - \lambda}{\sqrt{\lambda}} \leq z\right) \rightarrow G(z)$$

The mean is given by: $\mu = \lambda$
and the standard deviation by: $\sigma = \lambda^{1/2}$

Let us apply this approximation to the example above:

$$P(X \geq 29) = 1 - P(X < 29) = 1 - \phi\left(\frac{X - \lambda}{\sqrt{\lambda}}\right) = 1 - \phi\left(\frac{29 - 0.5 - 20}{\sqrt{20}}\right)$$
$$= 1 - \phi(1.9) = 1 - 0.9713 = 0.029$$

We see that this approximation gives a somewhat smaller value but still outside the confidence interval corresponding to a significance level of $\alpha = 0.05$.

4.4.3 Estimating the Mean of a Normal Distribution

Given n observations drawn from a normal distribution with unknown mean μ and unknown standard deviation σ, we have a distribution of uncertainty for the true mean given by the Student-t distribution:

$$\mu = t(n - 1)(\widehat{\sigma}/\sqrt{n}) + \overline{X}$$

or

$$\mu = t(n-1)\left(\frac{s}{\sqrt{n-1}}\right) + \overline{X}$$

where $t(n-1)$ is the *Student-t distribution* with $(n-1)$ degrees of freedom, s is the sample standard deviation and $\widehat{\sigma}$ is the unbiased single point estimate of the true standard deviation. The Student-t distribution is symmetric and unimodal about zero. The distribution is somewhat flatter than the N distribution. For larger values of n ($n > 30$) the expression can be approximated by:

$$\mu \approx \text{Normal}(0, 1)\left(\frac{s}{\sqrt{n-1}}\right) + \overline{X}$$
$$\approx \text{Normal}\left(\overline{X}, \frac{s}{\sqrt{n}}\right)$$

4.4.4 Monitoring Accident and Loss Numbers

The annual figures for losses and serious accidents for Norwegian vessels are reported regularly. Table 4.1 shows a set of data for a 10-year period (1983–92). The number of total losses varied between 6 and 24 with a mean value of 15.4. The loss number has in other words varied by a factor of 4 within this period, and this is reflected by a high value for the standard deviation (6.3). See the plot in Figure 4.7.

Table 4.1. Losses and serious accidents, 1983–92

Year	Total losses (No.)	Serious accidents (No.)	Fleet size (No.)
1983	22	211	4782
1984	10	195	4762
1985	13	190	4643
1986	12	205	4444
1987	24	156	4364
1988	23	196	4600
1989	19	169	4750
1990	15	189	4839
1991	6	177	5000
1992	10	186	4545
Mean	15.4	187.4	4672.9
St. dev.	6.3	16.5	191.7

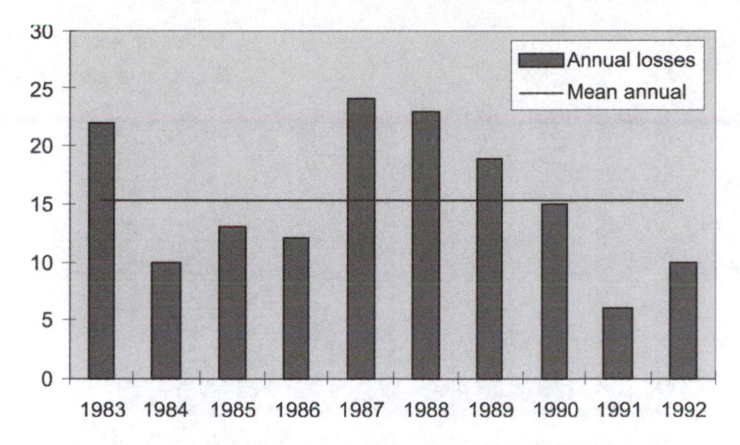

Figure 4.7. Annual number of total losses.

One may then question what is the uncertainty related to the estimated mean loss rate on the basis of the given data. In accordance with the model outlined in the previous paragraph, we have:

$$\mu = \pm t(n-1)\left(\frac{s}{\sqrt{n-1}}\right) + \overline{X} = \pm t(10-1)\left(\frac{6.3}{\sqrt{10-1}}\right) + 15.4 = \pm t(9) \cdot 2.1 + 15.4$$

Looking up a table for the Student-t distribution with 9 degrees of freedom and CDF value $F(t) = 0.95$ gives $t = 1.833$. This gives the following maximum and minimum values for the mean:

$$\mu_{\max} = 1.833 \cdot 2.1 + 15.4 = 19.2; \qquad \mu_{\min} = -1.833 \cdot 2.1 + 15.4 = 11.6$$

The 90% confidence interval for μ (0.05–0.95) is therefore:

$$\mu = 11.6 - 19.2 \text{ (losses/year)}$$

This shows that the uncertainty related to the mean loss number is considerable and that one should be cautious about drawing any conclusion about changes in risk level from single observations of loss numbers.

Observing the trend for losses in Figure 4.7, it might be tempting to postulate some kind of cyclical character for the period. The period started with a high value in 1983, then showed reduced frequency until 1987–88 when there was another peak, before the number again started to decrease. However, one should keep in mind that the absolute number of annual losses is fairly small and therefore does not give a firm basis for any such conclusion about trends. This is supported by the serious accident data shown in Figure 4.8. We see that the cyclical tendency is less pronounced for the annual accident

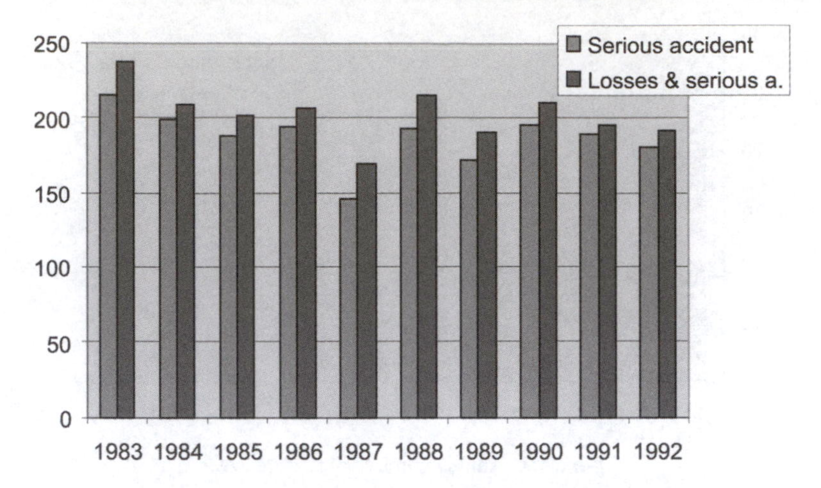

Figure 4.8. Serious accidents and total losses, 1983–92.

figures. The data are to a certain degree giving an opposite message as we have a minimum in 1987 and secondly a downward trend for the whole period.

The average number of serious accidents and losses is 203 per year. These numbers been adjusted for variation in exposed fleet size.

In the same manner as for losses, we may check the uncertainty related to the mean accident number:

$$\mu = \pm\, t(n-1)\left(\frac{s}{\sqrt{n-1}}\right) + \overline{X} = \pm\, t(10-1)\left(\frac{18.1}{\sqrt{10-1}}\right) + 203 = \pm\, t(9)\cdot 6.0 + 203$$

Applying the Student-t distribution with 9 degrees of freedom and CDF value $F(t) = 0.95$, we had already found $t = 1.833$. This gives the following maximum and minimum values:

$$\mu_{\max} = 1.833 \cdot 6.0 + 203 = 214; \qquad \mu_{\min} = 1.833 \cdot 6.0 + 203 = 192$$

The 90% confidence interval for μ (0.05–0.95) is:

$$\mu = 192 - 214 \text{ accidents and losses/year}$$

It is evident that the relative uncertainty for the annual accident numbers is considerably smaller.

4.4.5 Analysis of Time Series

In the discussion of the time series data in the previous section, the matter of trends or cycles was commented on briefly. We will look further into that problem here. The data for losses and serious accidents are shown in a line diagram in Figure 4.9. Although

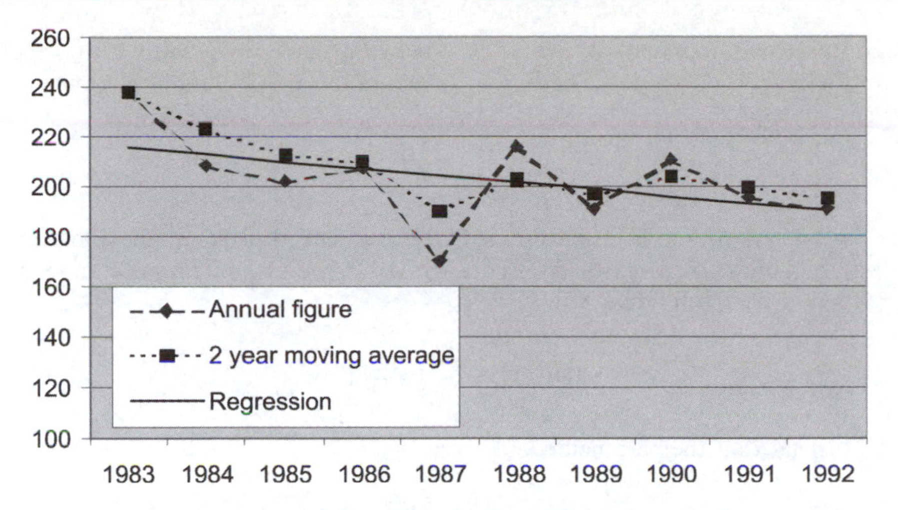

Figure 4.9. Losses and serious accidents.

the curve for annual figures shows some fluctuation, there is an indication of a weak downward trend.

One way to clarify possible trends is to apply so-called moving averages. A two-period moving average is computed as follows:

$$Y^*_{t,2} = (Y_t + Y_{t-1})/2$$

The general expression for the n-period moving averages is:

$$Y^*_{t,N} = (Y_t + Y_{t-1} + \ldots + Y_{t-N+1})/N$$

The curve for the two-period moving averages is shown for the accident data in Figure 4.9. It is clear that this technique removes some of the 'noise' and makes the trend more visible.

However, in order to get a firmer idea of the presence of a trend, application of regression analysis might be a better approach. A linear regression model expresses the stochastic variable Y as a function of X:

$$Y = \beta_0 + \beta_1 \cdot X + \varepsilon$$

where:

β_0 = intercept parameter
β_1 = slope parameter
ε = random error

The random error ε represents the difference between the true value of Y and the value given by the regression model. The basis for estimation of the model parameters is the following expression:

$$\widehat{Y} = b_0 + b_1 \cdot X$$

The parameters b_0 and b_1 are estimated by the so-called least squares method which minimizes the sum of squares of difference (SSD) of the estimated value of Y and the measured value, or the residual SSD:

$$\text{SSD(res)} = \sum \left(\widehat{Y} - Y_m\right)^2$$

It can be proved that the parameters are given by:

$$b_1 = \frac{\sum (X_m - \overline{X})(Y_m - \overline{Y})}{\sum (X_m - \overline{X})^2}$$

$$b_0 = \overline{Y} - b_1 \cdot \overline{X}$$

A simpler way to compute this parameters is to apply the *Solver function* in the Microsoft Excel spreadsheet to find values for b_0 and b_1 that minimize the expression for SSD(res). By applying the least squares method, the following linear model is estimated for the total number of losses and serious accidents per year:

$$N_{\text{L\&SA}} = 215.8 + 2.87\ (1983 - \text{YEAR})$$

Some of the computations are shown in Table 4.2. By introducing the linear model, the standard deviation relative to the regression line was reduced somewhat in relation to the original value:

$$\sigma_{\text{total}} = 18.1 \quad \text{was reduced to} \quad \sigma_{\text{res}} = 15.9$$

But the values also show that a considerable part of the variation is not accounted for by the linear model. Another way of expressing the goodness of fit of the model is to take the fraction between SS described by the model and the total SS:

$$R^2 = \frac{\sum (\widehat{Y} - \overline{Y})^2}{\sum (Y_m - \overline{Y})^2} = \frac{\text{SS(regr)}}{\text{SS(total)}}$$

This so-called coefficient of determination is for the present case:

$$R^2 = 680.9/2959.6 = 0.230 = 23\%$$

which confirms our first assessment of the correlation.

Table 4.2. Regression analysis by least squares method

Year	All accidents (No.)	Regression	SS(res)	SS(regr)	SS(total)
1983	238	215.8	489.1	167.1	1228.1
1984	209	212.9	17.8	101.1	34.1
1985	202	210.1	68.6	51.6	1.2
1986	207	207.2	0.1	18.6	16.6
1987	170	204.3	1199.3	2.1	1101.9
1988	216	201.4	210.1	2.1	170.6
1989	191	198.6	60.6	18.6	146.3
1990	211	195.7	225.5	51.6	61.4
1991	195	192.8	6.6	101.1	56.1
1992	191	190.0	0.9	167.1	143.4
Mean	202.9	SS	2278.6	680.9	2959.6
St. dev.	18.1	St. dev.	15.9	8.7	18.1

Table 4.3. Analysis of variance computations

Source	Sum of squares	Degrees of freedom	Mean square
Due to regression	$b_1 \sum (X_m - \overline{X})(Y_m - \overline{Y})$	1	MS(regr)/1 = SSD(regr)/1
Residual	$\sum (Y_m - \widehat{Y}_m)^2$	$N-2$	MS(res) = SSD(res)/($N-2$)
Total corrected for mean	$\sum (Y_m - \overline{Y})^2$	$N-1$	

A third approach would be to test the significance of the slope b_1 in the regression model. One may test whether the coefficient is equal to zero or in other words that the model *does not* explain the variation in accident rate:

$$H_0: \quad \beta_1 = 0$$

against the alternative: $H_1: \quad \beta_1 \neq 0$

This test can be accomplished by applying following F (Fisher) statistic based on the mean sum of squares (MS):

$$F_{\text{calc}} = \frac{\text{MS(regr)}}{\text{MS(res)}}$$

We have the following analysis of variance calculation sheet (Table 4.3).

Table 4.4. ANOVA case

Source	Sum of squares	Degrees of freedom	Mean square
Due to regression	680.9	1	680.9
Residual	2278.6	$10-2=8$	284.8
Total corrected for mean	2959.6	$10-1=9$	

The actual computations for the present case can partly be based on the SSD in Table 4.3 and are summarized in Table 4.4. F_{calc} is computed as follows:

$$F_{calc} = 680.9/(2278.6/8) = 2.39$$

The test criteria for F_{calc} are taken from a F tabulation for a specified significance level α and (ν_1, ν_2) degrees of freedom. Looking up the table, we have:

Assuming: $\alpha = 0.05$
We get: $F_{TAB}(0.05, 1, 8) = 5.32 > F_{calc} = 2.39$

It can be concluded that F_{calc} is within the confidence range consistent with the H_0 hypothesis. The linear model should in other words be rejected as the b_1 coefficient is not significantly different from zero. We should therefore stick to the simple 'constant level' model:

$$N_{L\&SA} = 202.9 \text{ (accidents and losses/year)}$$

4.5 CONSEQUENCE ESTIMATION

4.5.1 Distribution Characteristics

The most often used risk parameters are the accident frequency and the measure of consequence. In this chapter we shall focus on the second parameter which has certain important characteristics:

- The consequences of an accident may take different forms such as human injury and loss (fatality), environmental pollution, material and economic losses.
- Accident statistics are mainly based on high-frequency events with minor consequences.
- As risk managers we are more concerned with low-frequency and large-consequence events.
- Uncritical use of accident statistics may therefore give a misleading picture of the worst-case scenario.

Case

The fatality rate in the Norwegian offshore sector in the 1980s was as follows:

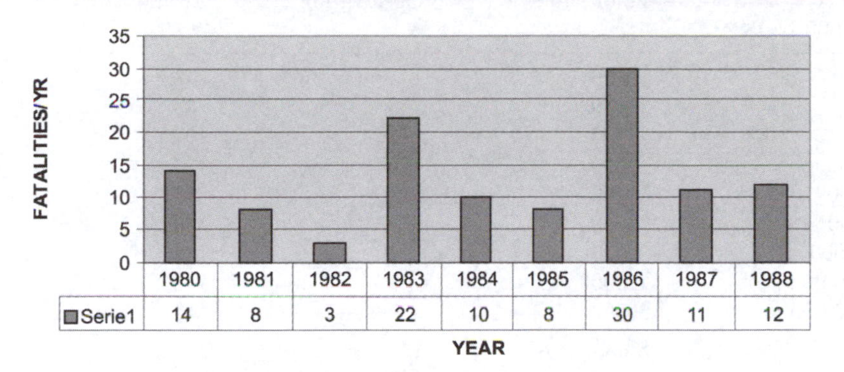

The following can be stated about the annual number of fatalities for this period:

Average number:	13.1 fatalities/year
Minimum number:	3 fatalities/year
Maximum number:	30 fatalities/year

However, the tragic fact was that in 1989 we had 119 fatalities in one single accident. One may then ask whether the statistic figures from the previous period could say anything about the probability of a catastrophe of this magnitude. The immediate answer might be to say 'no', as the mean fatality number was 13.1. Even the largest fatality number in the period was 30, which was less than 1/3 of the accident in 1989. However, if we could establish the distributional characteristics of the fatality, the chances might be brighter.

4.5.2 Fitting a Non-parametric Distribution

Rather than estimating the parameters of a known distribution, one may generate an empirical distribution directly on the basis of the observed data. Let us take data for the economic loss as a result of ship accidents as a case to demonstrate the approach (Table 4.5).

A non-parametric or empirical distribution is established as follows:

1. Select ranges for the loss variable (column 1).
2. Estimate average point value for each range (column 2).
3. List the number of observations in each range N_i (column 3). The sum of observations is given below (ΣN).

Table 4.5. Economic loss in accidents

1 Range of X (loss in 1000 NOK)	2 Point value: X	3 Observations N	4 Accumulated N	5 CDF
1–100	20	48	48	0.32432
100–200	120	35	83	0.56081
200–500	300	24	107	0.72297
500–1000	600	16	123	0.83108
1000–2000	1200	10	133	0.89865
2000–5000	3000	7	140	0.94595
5000–10,000	6000	4	144	0.97297
10,000–20,000	12000	2	146	0.98649
20,000–50,000	30000	1	147	0.99324
	Sum	147		
	Sum + 1	148		

4. Compute the accumulated number as follows:

$$AN_{i+1} = AN_i + N_{i+1}$$

5. Compute the 'artificial' CDF value in following manner:

$$F(x) = AN_{i+1} / \left(\sum N+1\right)$$

The 'trick' of adding 1 to ΣN reflects the fact that CDF approaches the value 1.0 asymptotically.

The result is shown in the right-most column. The distribution is plotted in Figure 4.10 with a logarithmic scale for the abscissa. It can be concluded by observation that the curve fits the data reasonably well.

4.5.3 The Log-Normal Distribution

Certain consequence parameters, such as the number of lives lost or the size of an oil spill, seem to follow a very skewed distributions. Stated simply it means that:

- Accidents with minor or lesser consequences represent the majority of the total number of events.
- However, a limited number of accidents lead to great or catastrophic consequences.

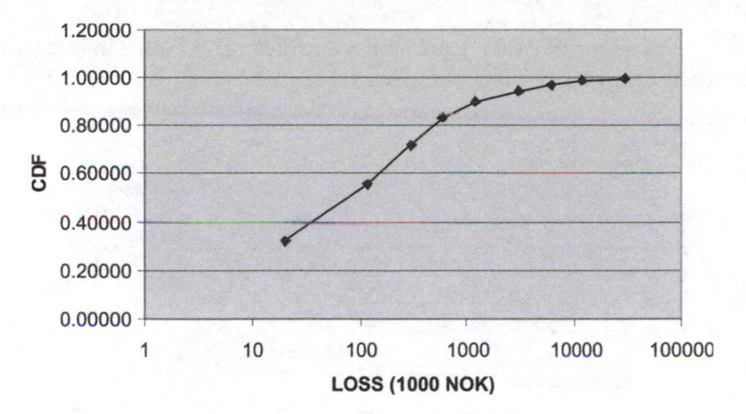

Figure 4.10. Economic loss per accident.

The log-normal (LN) distribution has properties that makes it suitable for describing consequence phenomena. If the random variable $\ln X$ is normally distributed, $N(\mu_1, \sigma_1)$, then the variable X is said to be log-normally distributed, $LN(\mu, \sigma)$. The PDF can be expressed as follows:

$$f(x) = \frac{1}{\sqrt{2\pi}\sigma_1} \frac{1}{x} e^{-((\ln x - \mu_1)^2 / 2\sigma_1^2)}$$

where

$$\mu_1 = \ln\left[\frac{\mu^2}{\sqrt{\sigma^2 + \mu^2}}\right]$$

$$\sigma_1 = \sqrt{\ln\left[\frac{\sigma^2 + \mu^2}{\mu^2}\right]}$$

The expected value and variance are given by:

$$E(X) = e^{(\mu_1 + \sigma_1/2)}$$

$$\mathrm{Var}\, X = e^{2\mu_1 + \sigma_1}(e^{\sigma_1^2} - 1)$$

Example

It has been pointed out that the log-normal distribution gives a good representation of variables that extend from zero to + infinity. Another observation is that it models well variables that are a product of other stochastic variables. The figure below shows the PDF and the CDF for the normally distributed variable $\ln X$, given by $N(10, 2)$.

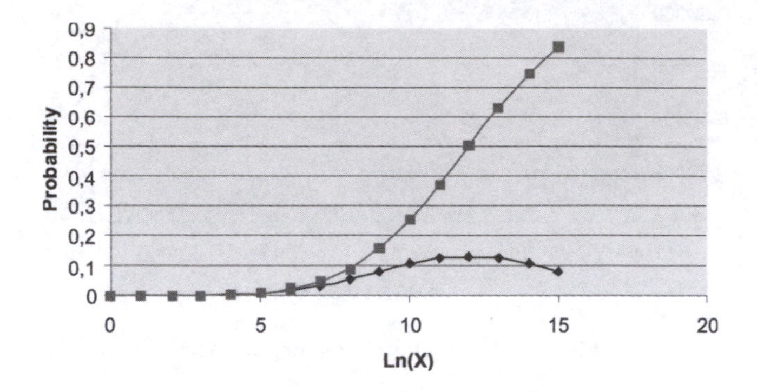

The corresponding CDF for X, which is log-normally distributed, is shown in the diagram below. It is evident that distribution models a variable that may take large values.

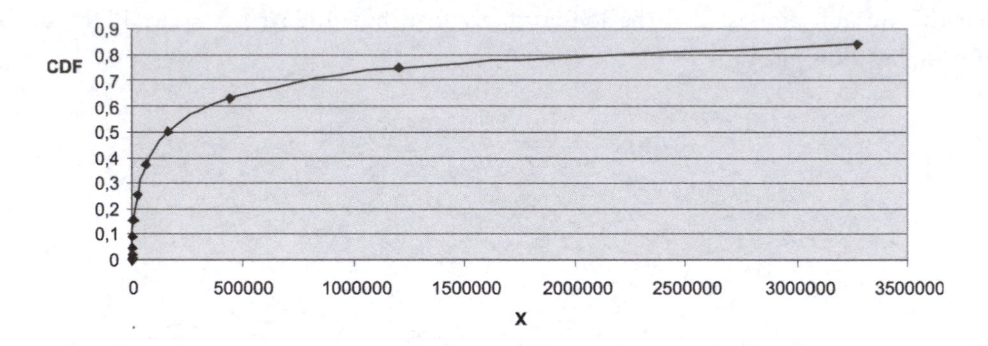

4.5.4 Fitting a Parametric Distribution to Observed Data

Vose (2000) has given some basic rules for deciding whether to apply a theoretical distribution when we are going to model a stochastic variable. Some key points are:

- Does the theoretical range of the variable match that of the fitted distribution?
- Does the distribution reflect the characteristics of the observed variable?

In order to illustrate the practical approach, we will use a set of data for cargo oil outflow as a result of ship accident (see Table 4.6).

It has been proposed that oil outflow volume may be described by a log-normal distribution because:

- The distribution range is positive numbers.
- It is highly skewed.

Table 4.6. Oil outflow distribution based on 22 ship accidents

Outflow size (tons)	No. of observations
10–100	9
100–500	8
500–1000	2
1000–5000	1
5000–10,000	1
10,000–50,000	1

Table 4.7. Excel datasheet: Estimation of log-normal distribution

Range: X	X	Observations	Observed PDF	Observed CDF	Estimated CDF	Estimated PDF	Squared diff PDF
10–100	20	9	0.3913	0.3913	0.41628	0.41628	0.0006
100–500	200	8	0.3478	0.7391	0.69874	0.28246	0.0016
500–1000	600	2	0.0870	0.8261	0.80789	0.10915	0.0003
1000–5000	2000	1	0.0435	0.8696	0.89490	0.08701	0.0006
5000–10,000	6000	1	0.0435	0.9130	0.94546	0.05056	0.0011
10,000–50,000	20000	1	0.0435	0.9565	0.97644	0.03098	0.0004
							0.0047
Mean	4803.3	Sum	22				
St. dev.	7771.3	Sum+1	23				
μ_1	3.66						
σ_1	3.14						

- The outflow may be seen as a product of a number of failures: accident, load condition and penetration of hull barrier.

In the following paragraph we will give a stepwise description of the approach applied. The numerical computations were done with Excel and are summarized in Table 4.7.

The approach is as follows:

1. List the ranges for observed outflow amount in tonnes.
2. Select subjectively a point value X within each range.
3. List the number of observations N for each range.
4. The observed PDF value is computed as follows:

The total number of observations:	$\Sigma N = 22$
PDF value:	$f(x) = N/(\Sigma N + 1)$

5. The observed cumulative value:

$$F_i(x) = f_i(x) + f_{i-1}(x), \quad \text{where } f_0(x) = 0$$

The theoretical distribution function is estimated by means of the Solver function in the Excel spreadsheet:
- Recall that the variable X is $LN(\mu, \sigma)$ distributed if $\ln(X)$ is $N(\mu_1, \sigma_1)$ distributed.
- The first step is to select a set of arbitrary values for μ_1 and σ_1.
- These values are entered into the function that is found under the Excel function menu. The function returns the CDF value $F(x)$.

6. The estimated PDF values are simply computed by applying the following formula:

$$f_i(x) = F_i(x) - F_{i-1}(x)$$

7. The theoretical distribution is obtained by first computing the sum of squared deviations between observed and estimated CDF values. These are shown in the right-most column.
8. The final step is to apply the Solver function, which is a search algorithm:
   - Minimize: sum of squares of deviations of PDF
   - By selecting optimum values for μ_1 and σ_1
9. The solution found by Solver was:

$$\mu_1 = 3.66 \quad \text{and} \quad \sigma_1 = 3.14$$

The theoretical distribution function is plotted in Figure 4.11.

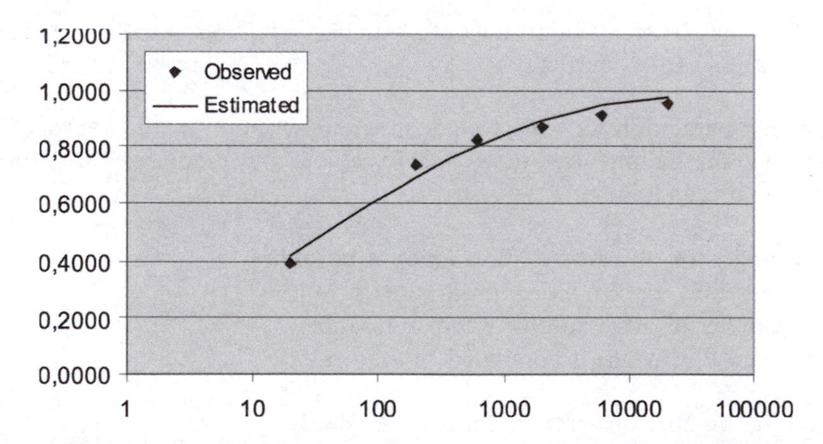

Figure 4.11. Oil outflow from ship accidents given by a log-normal distribution.

4.5.5 Estimating a Worst-Case Scenario

As pointed out earlier, the risk manager is not primarily concerned about the 'average' accident but rather the worst-case scenario. With the previous case in mind, the problem may be stated as follows: What is the risk of having a ship accident leading to an oil outflow of at least 100,000 tons?

Let us assume that the frequency of accidents leading to oil spill has been studied for a certain operation and estimated to be:

$$\theta_A = 6 \text{ accidents/year}$$

The probability of having a spill greater than 100,000 tons is:

$$P(S \geq 100{,}000) = 1 - F(S < 100{,}000)$$

Using the CDF in Figure 4.11 we get $F(S < 100{,}000) = 0.9937$, or

$$1 - F(S < 100{,}000) = 0.00626$$

The *return period* is defined as the *average time* between events of a certain magnitude, and may be written:

$$T_R = \frac{1}{\theta_A \cdot P(S \geq 100{,}000)} = \frac{1}{\theta_A \cdot [1 - F(S < 100{,}000)]}$$

which gives the following estimate:

$$T_R = 1/(6 \cdot 0.00626) = 26.6 \text{ years}$$

It may, however, be questioned whether this estimate is sufficiently precise.

Another way of stating the risk of this catastrophic scenario is to ask what is the probability of having this event in any given year? This may be answered in the following way:

1. Taking a conservative view: What is the maximum number of accidents in one year? Assuming a Poisson distribution and a CDF value $F(N_A) = 0.95$, we obtain, by looking up a table:

$$N_A = 10 \quad (\text{Exact value: } F(10) = 1 - 0.0413 = 0.9587)$$

2. The next question is: What is the probability that one out of these 10 accidents will lead to a spill greater than 100,000 tons? This can be seen as a *binomial* situation:

$$p(x) = \left\{\frac{n!}{x!(n-x)!}\right\} p^x(1-p)^{n-x}$$

$$p(1) = \left\{\frac{10!}{1! \cdot 9!}\right\} \cdot 0.00626^1 \cdot (1-0.00626)^9 = 0.059$$

   The risk of the catastrophic scenario on an annual basis is 6%. In other words this is a situation that is fairly probable or at least far from being improbable! This conclusion may therefore lead to an improvement in the operation.

3. The last point was not quite correct as there is a remote probability that even more than one accident may lead to a spill of at least 100,000 tons. The probability that all 10 accidents give a catastrophic spill can be written:

$$p(10) = 1 - \left\{\frac{10!}{0! \cdot 10!}\right\} 0.0026^0 \cdot (1-0.0026)^{10} = 1 - 0.9387 = 0.061$$

   The result is almost identical for the simple reason that having more than one catastrophic spill is a very remote outcome.

Given the probability of 0.06 for this disaster scenario, we may estimate the return period:

$$T_R = 1/(0.06) = 16.7 \text{ years}$$

It can be concluded that this estimate gives a much lower return period than the first one (26 years).

4.5.6 Extreme Value Estimation

In many situations the risk manager is, as already pointed out, more concerned with the worst-case situation rather than the average loss number. It may then be more feasible to focus on the extreme values in each observation period rather than using the whole set of data.

Table 4.8 reports the most serious single accident measured by the number of fatalities for the offshore sector in the years 1973–80. It can be observed that the variation is quite large and is best illustrated by the last two years where the number went from 1 fatality to 123 fatalities as a result of the *Alexander L. Kielland* loss.

It is possible to estimate a so-called extreme value distribution on the basis of such a sample set. The approach is basically the same as described in the preceding section with a few modifications.

Table 4.8. Maximum number of fatalities per accident: the Norwegian offshore sector, 1973–80

Accident	Year	Fatalities: X
Helicopter emergency landing	1973	4
Diving bell	1974	2
Alpha capsule	1975	3
Deep Sea Driller	1976	6
Helicopter crash	1977	12
Helicopter crash	1978	18
Unspecified	1979	1
Alexander L. Kielland capsize	1980	123

Table 4.9. Excel sheet: estimating extreme value distribution

Fatalities: X	Rank: N	Observed CDF	Observed PDF	Estimated PDF	Estimated CDF	Sum of squared deviations
1	1	0.1111	0.1111	0.1341	0.1341	0.0005
2	2	0.2222	0.1111	0.1194	0.2535	0.0010
3	3	0.3333	0.1111	0.0896	0.3430	0.0001
4	4	0.4444	0.1111	0.0699	0.4129	0.0010
6	5	0.5556	0.1111	0.1028	0.5157	0.0016
12	6	0.6667	0.1111	0.1697	0.6855	0.0004
18	7	0.7778	0.1111	0.0857	0.7711	0.0000
123	8	0.8889	0.1111	0.2046	0.9757	0.0075
$N+1$	9				Sum:	0.0121

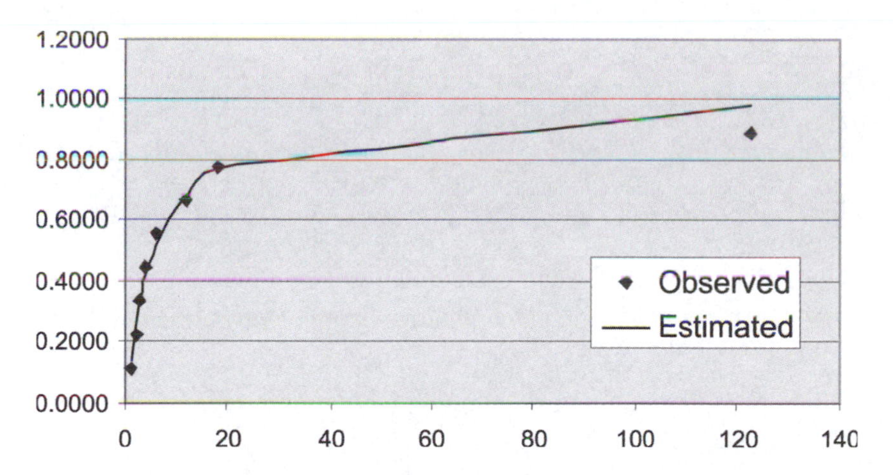

Figure 4.12. Maximum number of fatalities per offshore accident.

The first step is to order the observation by increasing loss magnitude given by the fatality number. As we only have single observations for each year, the probability distribution is approximated by ordering the observations with the same value $f(x) = 0.11$. This figure is obtained by dividing 1 by $(N + 1) = 9$ where N is the number of observations. In this case also it was decided to apply the log-normal distribution, and the model was fitted to the data with Excel's Solver as described in an earlier section (see Table 4.9).

The result is plotted in Figure 4.12. It can be seen that the model estimates the fatality number pretty well with the exception of the largest one. It is clear that even this model underestimates the probability of this event. This is, however, not unexpected bearing in mind that the available data cover a fairly short period.

The fact that the *Alexander L. Kielland* number (123 fatalities) lies below the model curve may give same weight to the suspicion that the safety control deteriorated somewhat during the 1970s.

REFERENCES AND SUGGESTED READINGS

Beyer, W. H., 1968, *Handbook of Tables for Probability and Statistics*. The Chemical Rubber Co., Cleveland, Ohio.

Brilon, W., 1973, Konfidenzintervalle von Unfallzahlen. *Accident Analysis and Prevention*, Vol. 5, 321–341. Pergamon Press, Oxford, UK.

Evans, M. et al., 2000, *Statistical Distributions*. Wiley-Interscience, New York.

Juran, J. M. and Gryna, F. M., 1988, *Juran's Quality Control Handbook*, chapter 23: Basic statistical methods. McGraw-Hill, New York.

Lillestøl, J., 1997, *Probability and Statistics* [In Norwegian: *Sannsynlighetsregning og statistikk Med anvendelser*]. Cappelen Akademisk Forlag, Oslo.

Vose, D., 2000, *Risk Analysis: A Quantitative Guide*. John Wiley, Chichester.

5

DECISIONS IN OPERATION

Experience is a good teacher, but can sometimes be very expensive.
(Norwegian proverb)

5.1 INTRODUCTION

As a part of the management and operation of the company, certain preventive and consequence-reducing measures are proposed for implementation. In this chapter we will outline how such measures or controls can be assessed with due consideration of uncertainty factors. As already discussed, accident data are subject to different forms of uncertainty: measurement problems, limited data, external effects, and unknown mechanisms in processes and accident development.

Many of the examples given are related to occupational safety and manpower training.

5.2 WORK ACCIDENT MEASUREMENT

5.2.1 Accident Frequency Rate

The frequency of work accidents is given in terms of AFR, which is a measure related to a standardized exposure (S):

$$\text{AFR} = \frac{\text{DI}}{\text{EMP} \cdot \text{AH}} S$$

where:

$$\begin{aligned} \text{DI} &= \text{number of injuries or work accidents per year} \\ \text{EMP} &= \text{number of employees} \\ \text{AH} &= \text{average annual hours work per employee} \\ &= 40 \cdot 50 = 2000 \text{ (hours/year)} \\ S &= 200{,}000 \text{ worker-hours/year} \end{aligned}$$

This means that the accident frequency is scaled relative to 200,000 worker-hours per year. It is therefore necessary to include the actual worker-hours which are expressed by the term (EMP · AH). The scaling factor is based on a standardized company with the following operation:

$$S = 100 \text{ employees} \cdot 40 \text{ hours/week} \cdot 50 \text{ weeks/year}$$
$$= 200{,}000 \text{ worker-hours}$$

It should be mentioned that other scaling factors are also used, such as 100,000 and 1 million work-hours. It should also be kept in mind that the term lost time incidence (LTI) rate is used instead of AFR.

Figure 5.1 shows how the LTI rate dropped in Texaco during an improvement programme on work safety (Wills et al., 1996). The LTI parameter can also be applied to measure the effect of specific safety measures. Schlumberger Anadrill (Aitken, 1996) correlated the accident frequency against the number of risk reports handed in per employee as shown in Figure 5.2. Better incidence reporting seems to contribute to fewer accidents.

Example

A company has reported the following accident figures for two departments:

Department	A	B
EMP = employees	6	30
DI = number of injuries/year	40	250

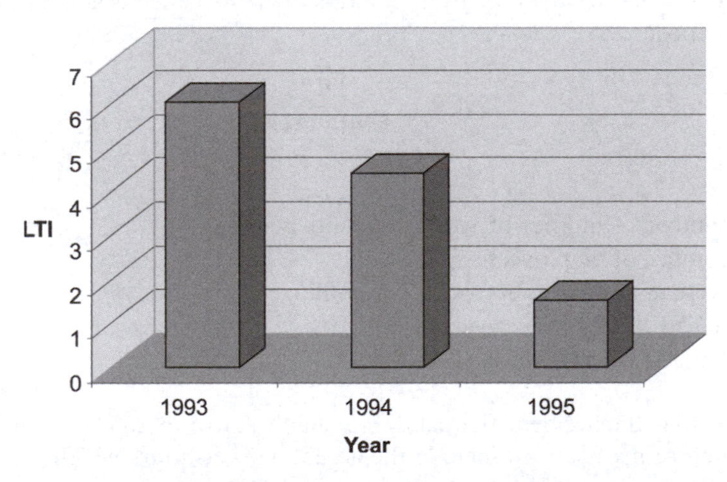

Figure 5.1. Lost time incidence rate (LTI) at Texaco Inc. during a three-year improvement plan: LTI per 100,000 work-hours. (Adapted from Wills et al., 1996.)

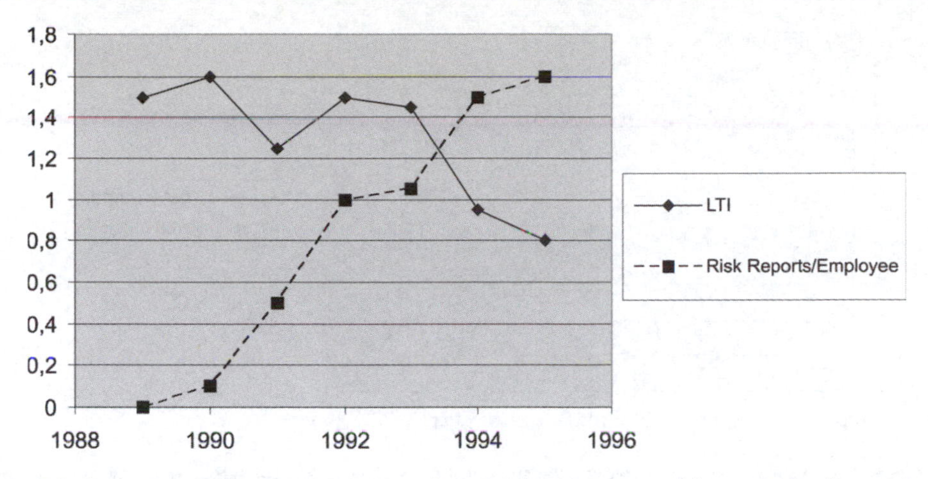

Figure 5.2. Lost time incidence rate versus number of risk reports per employee, Schlumberger Anadrill. (Adapted from Aitken, 1996.)

The following values for the accident frequency rate can be computed:

$$\text{AFR}_\text{A} = \frac{6}{40 \cdot 2000} 200{,}000 = 15$$

$$\text{AFR}_\text{B} = \frac{30}{250 \cdot 2000} 200{,}000 = 12$$

It can be concluded that department B has a lower injury frequency:

$$(15 - 12)100/15 = 20\% \text{ lower}$$

5.2.2 Accident Severity Rate

The accident severity rate (ASR) measures the consequence of a work injury in terms of number of lost work-days on the same basis as for the frequency:

$$\text{ASR} = \frac{\text{LWD}}{\text{EMP} \cdot \text{AH}} S$$

where LWD = accumulated lost work-days per year for 200,000 worker-hours.

Like the frequency rate, the severity rate may also be used to monitor the development of the safety conditions in an operation. Aitken (1996) compared the number of lost work-days against the number of days spent on safety training. As shown in Figure 5.3, the lost work-days went down dramatically during a four-year period when the number of days spent on training was increased significantly.

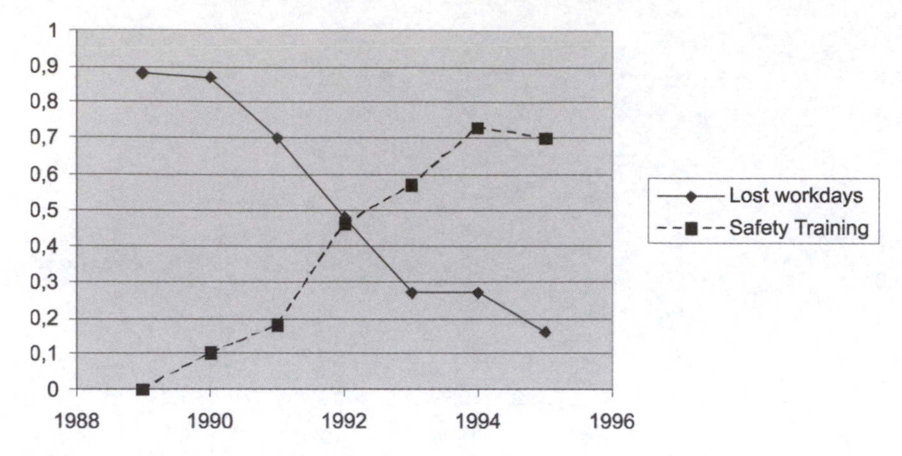

Figure 5.3. Lost work-days per employee versus number of days spent on safety training per employee, Schlumberger Anadrill. (Adapted from Aitken, 1996.)

Example

A company with 25 employees has reported 105 lost work-days for the recent year. The accident severity rate is:

$$\text{ASR} = \frac{105}{25 \cdot 2000} 200{,}000 = 420$$

5.2.3 Frequency-Severity Indicator

In risk analysis we are often confronted with the matter of selecting a single measure in order to rank different alternatives from a safety point of view. For so-called concept-related risk (the risk of ships), we often use the following measure:

$$R = p \cdot C$$

where p is the probability of an accident and C is the expected outcome. An alternative measure for work-related accidents is the so-called frequency-severity indicator:

$$\text{FSI} = \sqrt{\frac{\text{AFR} \cdot \text{ASR}}{1000}}$$

This criterion has something in common with the one above apart from the square sign. A rational argument for squaring the expression is the fact that our risk aversion is not linear with the numerical severity of an accident. An accident taking 20 lives is not necessarily twice as serious as one leading to 10 fatalities.

Example

A shipping manager has analysed the safety performance of the crew members he is handling for two periods. The result was as follows:

Period	1991–95	1996–2000
AFR	13.5	10.2
ASR	205	220

The question is whether the risk level has gone down or not. The following computations can be made:

$$1991\text{–}95: \qquad FSI = \sqrt{\frac{13.5 \cdot 205}{1000}} = 1.66$$

$$1996\text{–}2000: \qquad FSI = \sqrt{\frac{10.2 \cdot 220}{1000}} = 1.50$$

In other words, it can be concluded that the worker risk level has gone down during the 10-year period.

5.3 SAFETY COMPETENCE: CORRELATION ANALYSIS

An important aspect of any safety programme is to continuously assess attitudes and competence among the crew or employees. There are different sources that may be used for such an assessment:

- Examination scores
- Inspection and evaluation of work behaviour
- Questionnaire study
- Assessment of personnel by their supervisors

In order to cross-check this kind of information, one may perform correlations on the data from such studies. Let us take the following situation: a company has invested in a safety awareness and training programme and has later done an evaluation of the competence of the workforce. This leaves us with two sets of data:

1. Training program examination score (Score).
2. Safety rating by supervisor (Rating).

The assessment data on the competence for the crew of a vessel are shown in Table 5.1. Both sets were based on a ranking scale from 1 (low) to 10 (high). It can be seen that the mean Score is 7.5, which is somewhat higher than the mean Rating value of 7.1.

Table 5.1. Assessment of safety programme

Crew member	Rating (X)	Score (Y)
1	4	5
2	9	8
3	7	9
4	9	8
5	3	4
6	7	8
7	8	8
8	5	7
9	10	8
10	6	5
11	8	9
12	8	7
13	6	7
14	9	10
15	8	10
Mean	7.1	7.5
St. dev.	2.0	1.8

The standard deviation for the exam scores (1.8) is slightly smaller than for the ratings (2.0). One possible interpretation of these observations might be that:

- The exam scores are overestimating the safety competence.
- The rating approach is better at differentiating the competence among the individual crew members.

The results are also plotted in a scatter diagram as shown in Figure 5.4. Although we can see a certain relation between score and rating, there is also considerable scatter of the data.

A more precise measure of the relationship between the two assessment parameters is the correlation coefficient:

$$R_{XY} = \frac{S_{XY}}{S_X \cdot S_Y}$$

where:

$$S_{XY} = \frac{1}{N}\sum (X_i - \overline{X})(Y_i - \overline{Y})$$

$$S_X = \frac{1}{N}\sum (X_i - \overline{X})^2$$

$$S_Y = \frac{1}{N}\sum (Y_i - \overline{Y})^2$$

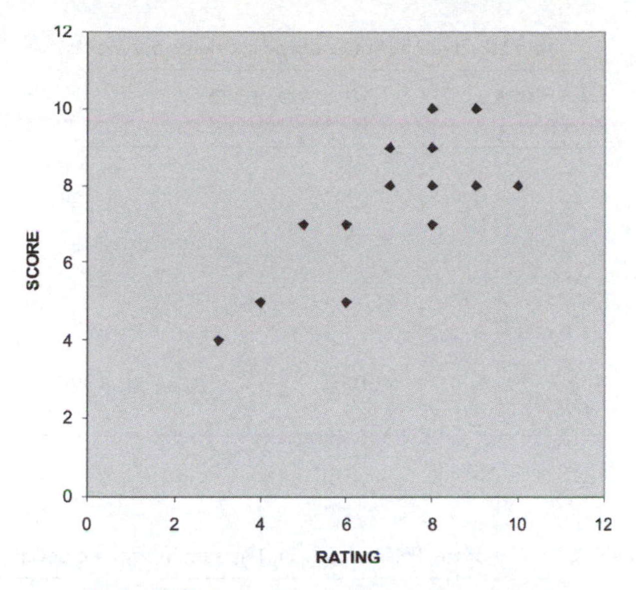

Figure 5.4. Scatter diagram: score versus rating.

Numerical computations by means of Excel gives:

$$S_X = 2.00; \qquad S_Y = 1.77; \qquad S_{XY} = 2.53$$

which gives:

$$R_{XY} = 0.717$$

The fact that the correlation coefficient may take values from -1 to $+1$ indicates that we have a fairly good relationship between exam scores and supervisor ratings. The safety management might therefore decide to use the examination method only, as this is more efficient and time-saving. However, some effort should be directed towards improvement of the examination programme in order to improve the differentiation between the candidates.

5.4 TESTING OF A DISTRIBUTION MODEL

The previous chapter spent considerable effort on the estimation of distribution models for empirical safety data. Here we shall look a little closer at how to test whether a potential model is appropriate for the data set at hand.

Let us look at following case given by ReVelle and Stephenson (1995). A company has kept records on the number of lost-time accidents (LTA) per week for a period of

Table 5.2. Number of lost-time accidents per week

LTA/Week X	Observed weeks N	$X \cdot N$
0	45	0
1	29	29
2	17	34
3	9	27
4		
5		
6		
Sum:	100	90
Mean:		0.9

Table 5.3. Testing of the Poisson distribution: sum of squared deviations

LTA/week X	Poisson PDF	Poisson PDF corrected	Expected weeks	Observed weeks	$(O - E)^2/E$
0	0.4066	0.4066	40.66	45	0.4639
1	0.3659	0.3659	36.59	29	1.5749
2	0.1647	0.1647	16.47	17	0.0173
3	0.0494	0.0628	6.28	9	1.1767
4	0.0111				
5	0.0020			Sum:	3.2328
6	0.0003				
Part sum	0.0628				

100 weeks. Table 5.2 shows that the number of accidents per week has varied between 0 and 3 with dominance on the lower values. The mean number was 0.9 accidents per week. This indicates that the Poisson distribution might be appropriate to describe the frequency of LTA.

In order to test the feasibility of the Poisson distribution we will compare what this model would give with observed values. Table 5.3 summarizes the computations.

The first step is to compute the PDF for $\lambda = 0.9$. Statistical experience says that any value should not be lower than 0.05, and this requires that the distribution is truncated by grouping the values from $X = 3$ to 6 together, which adds up to 0.0628 (see the shaded area in the table).

The next step is to compute the estimated number of weeks on the basis of a 100 weeks observation period.

It can now be proved that the sum of the relative difference between number of observed weeks (O) and estimated weeks (E) is chi-square distributed:

$$\chi^2_{\text{calc}} = \sum_N \frac{(O - E)^2}{E}$$

The distribution applies for positive values and has the parameter ν, which denotes the number of degrees of freedom and is given by:

$$\nu = (N - 1) - 1$$

where N is the number of observations and the second -1 is the consequence of introducing an estimate for λ in the computations.

If our hypothesis that the number of weeks is given by the Poisson distribution is true (H_0), the calculated value of the chi-square criterion should be less than the critical value. We are then able to test the assumption of a Poisson distribution as follows:

from Table 5.3: $\chi^2_{\text{calc}} = 3.2328$
degrees of freedom: $\nu = (4 - 1) - 1 = 2$
assuming significance level: $\alpha = 0.95$
tabulated value (from handbook): $\chi^2_{2,0.95} = 5.99$

It can be concluded that the Poisson distribution is *valid* as the calculated value (3.23) is less than the tabulated value.

5.5 CHOOSING AMONG ALTERNATIVE TRAINING PROGRAMS

The chi-square test can also be useful for testing other models. Let us take the following case described by ReVelle and Stephenson (1995). A company has tried out training programmes of different duration: 1, 3, 5 and 10 days. The attending crew members were subject to a rating by their supervisors 6 months after the training session. The supervisor used the following ranking: excellent, good or poor.

The result of the assessment is shown in the upper part of Table 5.4. Observation of the data may support the suspicion that there is no clear relationship between course duration and rating. It is interesting to note the low number of 'excellent' ratings for the participants in the 10-day program.

Against this background, it may be interesting to test the following null-hypothesis:

$$H_0 = \text{No correlation between Duration and Rating}$$

Table 5.4. Analysis of training effectiveness

	Excellent	Good	Poor	Sum
(a) Observed rating				
1 day	6	12	0	18
3 days	12	25	6	43
5 days	14	31	12	57
10 days	2	23	7	32
Sum	34	91	25	150
(b) Estimated rating				
1 day	4.08	10.92	3	18
3 days	9.75	26.09	7.17	43
5 days	12.92	34.58	9.50	57
10 days	7.25	19.41	5.33	32
Sum	34	91	25	150
(c) Relatively squared difference				
1 day	0.9035	0.1068	3.0000	4.0103
3 days	0.5209	0.0453	0.1899	0.7561
5 days	0.0903	0.3706	0.6579	1.1188
10 days	3.8048	0.6626	0.5208	4.9883
				10.8736

Given this hypothesis, the distribution of the number of crew members would follow this computational rule:

$$\text{Cell}\{\text{Row}_i, \text{Column}_j\} = \frac{\text{Sum}(\text{Row}_i) \cdot \text{Sum}(\text{Column}_j)}{\text{Sum}(\text{Rows} \,\&\, \text{Columns})}$$

which expresses the assumption of independence by the fact that the number in each cell is only determined by the column and row sums. Applying this rule to each cell in the table, we get the expected result shown in the middle part of Table 5.4.

Based on the upper and middle parts of the table, we are now in a position to calculate the chi-square value as outlined in the previous chapter:

$$\chi^2_{\text{calc}} = \sum_N \frac{(\text{O} - \text{E})^2}{\text{E}}$$

The result of for each cell is shown in the lower part of Table 5.4 and the sum is:

$$\chi^2_{\text{calc}} = 10.8736$$

The number of degrees of freedom is given by following formula:

$$\begin{aligned}\nu &= (\text{Number of rows} - 1)(\text{Number of columns} - 1)\\ &= (4-1)(3-1)\\ &= 6\end{aligned}$$

Assuming a significance level $\alpha = 0.99$, we find the following tabulated value: $\chi^2_{6,0.99} = 16.8$.

It can be concluded that the 'no relationship' hypothesis holds as the calculated value is less than the tabulated value. This means that the variation in ratings is not more than would be expected under the null-hypothesis H_0. Or in other words, it is not possible to explain the variation in rating by the course duration.

It should, however, be pointed out that this conclusion is based on a very high value for the significance level: $\alpha = 0.99$. This reflects our concern of *not rejecting a true* H_0. If we decided to be more open to the alternative hypothesis that there is a relationship between course duration and rating, we might have set the significance level somewhat lower:

$$\alpha = 0.95$$

$$\chi^2_{6,0.95} = 12.6$$

This result did not, however, change our conclusion as the tabulated value still is higher than the one calculated.

REFERENCES

Aitken, J. D. et al., 1996, Committed HSE management vs. TQM: Is there any difference? *International Conference on Health, Safety and Environment*, Paper SPE 35760, 9–12 June, New Orleans, Louisiana.

ReVelle, J. B. and Stephenson, J., 1995, *Safety Training Methods: Practical Solutions for the Next Millennium*. John Wiley, New York.

Wills, T. L. et al., 1996, The use of integrated management systems assessments for continous improvement of EHS programs. *International Conference on Health, Safety and Environment*, Paper SPE 35887, 9–12 June, New Orleans, Louisiana.

PART III
RISK ANALYSIS

6

TRAFFIC-BASED MODELS

Shipwreching on land is worse than at sea — the sea at least has a strand.
(Johan Falkberget, Norwegian author)

6.1 INTRODUCTION

Earlier, it has been shown how to estimate the probability of an accident based on historical accident numbers. The simplest and most intuitively correct manner is to base accident frequency estimates on exposure criteria such as vessel-years (i.e. number of vessels at risk per year). However, such a statistical approach may only describe the mean risk of a large number of ships and not reflect variation in technical standards, environmental conditions and traffic density. In certain instances the analysis of risk will be undertaken for specific fleets or for certain waters or fairways. This demands another method than the statistical approach based on fleet-year exposure. In this chapter it will be shown how the probability of an impact-type accident can be estimated for a specified seaway. By an impact-type accident we mean collision, grounding, stranding or allision (above-water object impact).

6.2 BASIC THEORY

It has earlier been shown that the expected number of ship accidents per unit of time in a specified fairway may be estimated by the following equation:

$$C = \lambda \cdot N$$

where:

C = Expected number of accidents in seaway per time-unit
λ = Number of accidents per vessel-passage of seaway
N = Number of passages per time unit

A voyage may for computational reasons be defined as the passing of a sequence of fairway sections. As a simplification, it is further assumed that the navigational and topological characteristics are relatively constant within each section of the fairway.

Consequently, the traffic density and other environmental conditions can be assumed to be relatively unchanged within each section. Previously it was shown that phenomena with a small chance of occurring have an expected frequency (events per unit of time) that is equal to the probability of realization. This assumption holds, for instance, for the Poisson model. The expected number of impact accidents within the *m*th fairway section can then be expressed as follows:

$$C_m = \lambda_m \cdot N = P(C)_m \cdot N$$

where:

C_m = Expected number of impact accidents per time-unit within the *m*th fairway section
$P(C)_m$ = Probability of impact accident when passing the *m*th fairway section

By referring to the potential accident type with index u, the expected number of accidents of type u within section m of the fairway may be expressed as:

$$C_{m,u} = \lambda_{m,u} \cdot N = P(C)_{m,u} \cdot N$$

Hence the estimated total number of accidents per time-unit for the whole voyage, C_T, may be expressed as follows:

$$C_T = \sum_m \sum_u \lambda_{m,u} \cdot N_m = \sum_m \sum_u P(C)_{m,u} \cdot N_m \tag{6.1}$$

where:

$P(C)_{m,u}$ = Probability of impact accident type u per passage of fairway section m
N_m = Number of passing ships per time unit

The expected accident frequency is in other words calculated by summing over all fairway sections and all accident types. How the fairway is split up into sections will to a certain degree be a subjective matter, but should as already mentioned take the traffic and topography into consideration. It will obviously be a compromise between computational efficiency and a need for homogeneous conditions within each section. A very simplified fairway representation is shown in Figure 6.1 and consists of the following three sections:

A. Fairway with traffic in the same and head-on direction.
B. Crossing traffic in each direction.
C. Fairway with an obstacle (shoal) and traffic in both the same and the head-on directions.

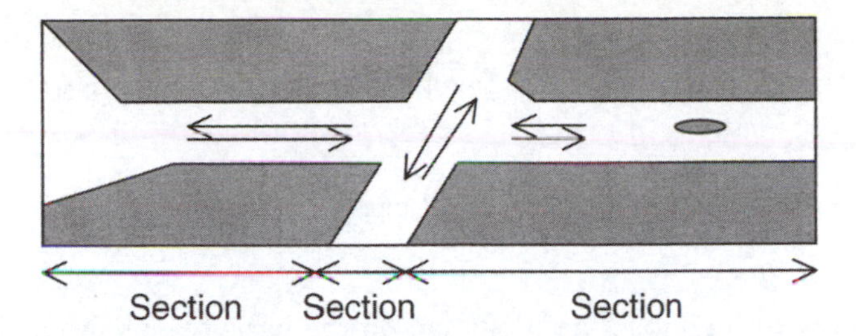

Figure 6.1. Selection of sections for a general traffic fairway.

Table 6.1. Potential accident situations

Section	Potential accident situations
A	1. Collision with ships on the same course 2. Head-on collision 3. Stranding
B	4. Collision with crossing traffic
C	5. Collision with ships on the same course 6. Head-on collision 7. Stranding 8. Grounding on shoal in fairway

This generates a total of eight different accident situations as shown in Table 6.1. The different impact accidents will now be treated in detail in the following sections. Modelling of each accident type will require different approaches.

6.3 A GENERAL MODEL OF IMPACT ACCIDENTS

The models that will be proposed for estimation of impact accident frequency are all based on following premises:

1. The vessel has an *opportunity* to be put at risk.
2. The vessel will be subject to an *incident* that puts it at risk.
3. The vessel is unable to handle the incident and will thereby have an *accident*.

The first requirement means that the vessel is underway or sailing. The second requirement is that for some unspecified reason the vessel has lost control and thereby is subject to an incident. Thirdly, the incident may lead to an accident in the case where the situation is not corrected in due time (see Figure 6.2). The most relevant parameter for the first condition is the duration of the operation or voyage. The second condition can be expressed by a probability of having an incident, whereas the third condition is given by a conditional probability of having an accident given to be in an incident situation.

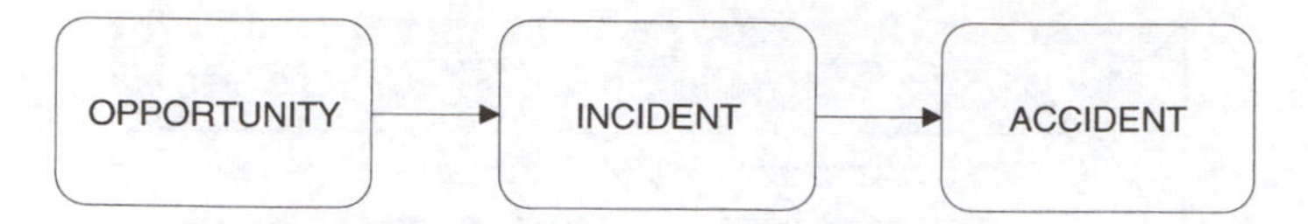

Figure 6.2. Conceptual impact accident model.

The probability of an impact accident can on the basis of this model be expressed by the product of two probabilities that reflect the transitions from normal operation state (opportunity) to the accident state:

$$P(A) = P(C) \cdot P(I|C) \tag{6.2}$$

where:

$P(A)$ = Probability of an impact accident per passage
$P(C)$ = Probability of losing vessel control per passage
$P(I|C)$ = Conditional probability of having an accident given loss of vessel control (incident)

The probability of losing control, $P(A)$, is assumed to have a constant value that is independent of time and reflects the overall operational standard of the vessel or group of vessels. The conditional probability of having an impact accident after losing control is in reality a function of the vessel's ability to handle emergencies. Rather than trying to estimate this directly, one may either compute the number of accidents relative to the number of incidents, or assess the probability on the basis of the traffic or fairway condition. It is fairly evident that the risk of an accident is greater the more dense the traffic or narrow a fairway. In the following sections, different approaches will be shown for estimating the conditional probability of having an impact accident $P(I|C)$.

The equation will for convenience be written as follows:

$$P_a = P_c \cdot P_i \tag{6.3}$$

where the indexes denote accident (a), loss of control (c) and impact (i).

6.4 GROUNDING AND STRANDING MODELS

A ship moving in a restricted seaway without any other traffic is subject to stranding and grounding hazards. The coastal zones, shoals, rocks and islands are basically stationary objects relative to the vessel. The estimation of the probability that an incident will lead to an accident will be based on certain assumptions of how the vessel moves in the critical phase. As the first step we will model this aspect and subsequently look at the probability of losing vessel control.

6.4.1 Grounding

The grounding scenario is based on a straight fairway section as shown in Figure 6.3. Assume that control of a ship is lost owing to failure in the navigation system due to either technical or human factors or both. The ship's lateral position within the fairway width is assumed to be random at the time when control is lost. The distance of the fairway section is denoted D, and let us further assume that the vessel is positioned randomly anywhere along this track (longitudinally). In the critical (incident) phase it is a simplification, assuming that the vessel continues on an unchanged straight course. The situation is shown in Figure 6.3.

The probability that the uncontrolled vessel hits the obstacle is then exclusively dependent on the dimensions of the fairway and the beam of the ship:

$$P_i = \frac{B + d}{W} \tag{6.4}$$

where:

W = Average width of fairway
d = Cross-section of obstacle, e.g. shoal, rock, island, etc.
B = Breadth of vessel

This fairly simple model is based on the assumption that the vessel may have any transverse position in the seaway and that the breadth of the critical corridor is given by the term $c_i = B + d$ as indicated in Figure 6.4. The probability is thereby given by the ratio between these two terms.

In a seaway with a number of obstacles the conditional probability of an impact is given by the union of the cross-section of the obstacles:

$$P_i = \frac{1}{W} \cdot \left[B + \left(d_1 \bigcup d_2 \bigcup \ldots \bigcup d_k\right)\right]$$

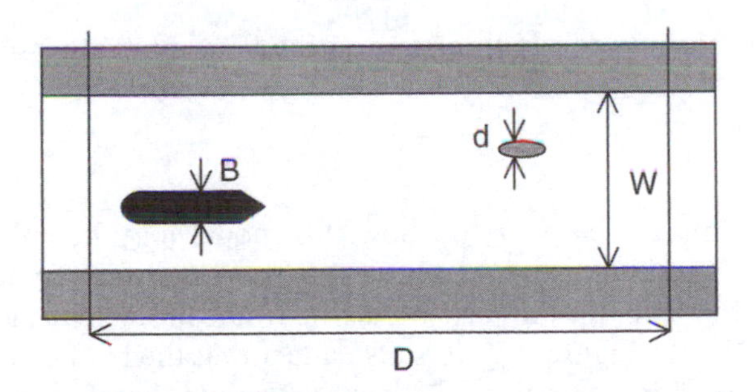

Figure 6.3. Modelling of a grounding accident.

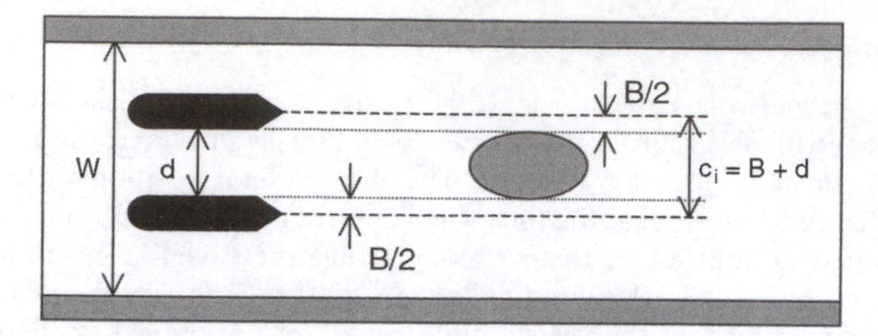

Figure 6.4. Characteristic parameters of the grounding situation.

or simply by assuming no overlap between the obstacles:

$$P_i = \frac{1}{W} \cdot \left[B + \sum_k d_k \right] \tag{6.5}$$

In a fairway with numerous shoals and other hindrances the number of such obstacles may also be expressed in the following manner:

$$K = \rho \cdot D \cdot W$$

where ρ = obstacle density (obstacles/area-unit).

By replacing the sum expression in Eq. (6.5) with the number of obstacles, the following expression is obtained:

$$P_i = \frac{1}{W} \cdot (B + \rho \cdot D \cdot W \cdot d)$$

$$= \frac{B}{W} + \rho \cdot D \cdot d$$

If the ship's beam is considered small relative to the fairway width, we get:

$$P_i = \rho \cdot D \cdot d \tag{6.6}$$

A final comment should be made on how this model may be enhanced in order to improve its validity. Instead of using the physical barriers of a fairway to specify potential lateral positions of the ship, a lane may be defined that more realistically describes the actual maritime traffic (see Figure 6.5). It is also a fact that the traffic density is decreasing as one approaches the shore. A more realistic model would therefore be to apply a traffic distribution model to reflect this.

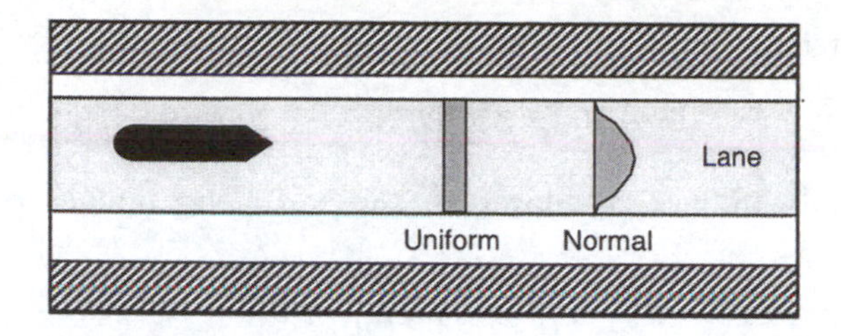

Figure 6.5. Enhanced grounding scenario.

Case Study

Problem

A new oil refinery is under planning. Oil is going to be transported to and from the terminal by tankers through a fjord of width 1 km. There is an island in the middle of the fairway representing a grounding hazard. The width of the island is equal to 100 meters. The planned capacity of the oil refinery requires 6 shipments for export and 3 shipments for import daily. The mean beam of these ships is 20 metres. The risk of grounding has to be quantified in order to compare the risk of oil spill for this and eventually other locations of the refinery.

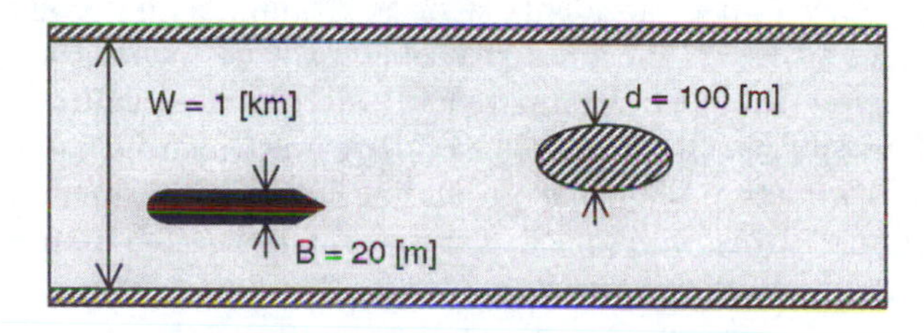

Solution

Estimate the expected number of groundings on the island per year.

Assumptions:
The probability of losing navigational control P_c is equal to $1.4 \cdot 10^{-4}$ per passage of the fairway. The ship's lateral position within the width of the fairway is uniformly distributed. Importing and exporting vessels are respectively leaving and entering the port in ballast condition.

Analysis

The number of ships passing the fairway each year is:

$$N_p = 2 \cdot (6 + 3) \text{ (passages/day)} \cdot 365 \text{ (days/year)} = 6570 \text{ (passages/year)}$$

The impact diameter is:

$$d_i = 20\ (\text{m}) + 100\ (\text{m}) = 120\ (\text{m})$$

The conditional probability of grounding after loss of navigational control is:

$$P_i = \frac{d_i}{W} = \frac{120\ (\text{m})}{1000\ (\text{m})} = 0.120$$

The probability of grounding per passage of the fairway is:

$$P_a = P_c \cdot P_i = 1.4 \cdot 10^{-4} \cdot 0.120 = 1.68 \cdot 10^{-5}$$

Based on the assumptions and the given probability data, a grounding on the island within one year is equal to $1.68 \cdot 10^{-5}$. The average time between groundings is given by the reciprocal value:

$$T = 10^5/1.68 = 59{,}524 \text{ years}$$

In other words, not a very likely event.

Comment

The probability of other impact accidents such as strandings and collisions should also be estimated in order to assess the total risk of impact accidents for this fairway. The total accident frequency should be compared for alternative locations. The conditional probability of an oil spill given an impact accident must also be estimated in order to have the complete risk picture. Whether grounding leads to an oil spill is dependent on a number of factors such as ship speed, cargo containment system (hull design) and weather conditions.

6.4.2 Stranding

Recalling the straight line fairway scenario in the previous section, there is also a risk of stranding. The term stranding is used for the impact with the shoreline in contrast to the impact with individual shoals and islands in the fairway. A random position for the vessel in the seaway is again assumed and as an average value the centre location as shown in Figure 6.6.

The model will be based on the assumption that in the case of loss of control the vessel may continue on any course ahead, e.g. a course within a span of 180°. As can be seen from the figure, the critical angle leading to stranding at both sides is equal to α. It is fair to assume that the length of the fairway D is considerably greater than the width W, or:

$$\left(\frac{W}{D}\right)^2 < 1$$

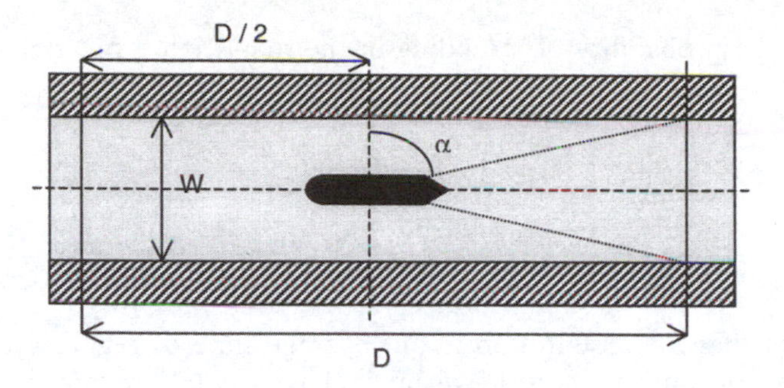

Figure 6.6. Stranding model.

The conditional probability of stranding is given by the ratio of the critical angle to the total angle (expressed for one lateral side):

$$P_i = \frac{\alpha}{\pi/2} = \frac{\arctan((D/2)/(W/2))}{\pi/2}$$

By replacing the arctan term by the following series expansion, we get:

$$P_i = \frac{\pi/2 + \sum_n (-1)^n \cdot (W/D)^{2n-1}/(2n-1)}{\pi/2} = \frac{2}{\pi} \cdot \left[\frac{\pi}{2} - \frac{W}{D} + \left(\frac{W}{D}\right)^2 - \left(\frac{W}{D}\right)^3 \ldots\right]$$

The two first components in the series may be used as an approximation for the whole series expression. The equation then takes the following simplified form:

$$P_i \approx 1 - \frac{2}{\pi} \cdot \frac{W}{D} \tag{6.7}$$

6.4.3 A Comment on the Stranding Model

The estimates from the model rest to a large degree on the relative distance of the fairway section that is studied. Let us take a fairway 10 nm long and of width 0.5 nm. The conditional stranding probability is:

$$P_i \approx 1 - \frac{2}{\pi} \cdot \frac{0.5}{10} = 0.032 \approx 3\%$$

On the other hand one may assume that the maximum time that the vessel will continue without control is 10 minutes. With a speed of 15 knots, this corresponds to a distance of 2.5 nm. It would therefore seem more correct to model two sections, each of a distance of 5 nm. The average distance sailed within a section of 5 nm is one half or 2.5 nm.

The estimate of the probability of stranding in the first section but not stranding in the other section is:

$$P_1 = 1 - \frac{2}{\pi} \cdot \frac{0.5}{5} = 0.064 \approx 6\%$$
$$P_2 = 1 - P_1 = 0.936$$
$$P_{1\text{–}2} = P_1 \cdot P_2 = 0.064 \cdot 0.936 = \mathit{0.060} = 6\%$$

This confirms that the assumption about average time to regain vessel control and thereby selection of fairway section distance has a vital impact on the estimated probability. On the other hand, it should not be forgotten that these models are used primarily for comparing alternatives and that less weight is put on the absolute numbers.

6.5 LOSS OF NAVIGATIONAL CONTROL

In order to calculate the probability of having an impact accident within a fairway, the probability of losing navigational control P_c has to be quantified also. It might be the case that the value of P_c is different for stranding and grounding situations as they represent different navigation tasks. Hence, the value of P_c for stranding and grounding situations should be estimated separately. We have the following general expression based on Eq. (6.3):

$$P_c = \frac{P_a}{P_i} \tag{6.8}$$

The probability of loss of control P_c can be estimated on the basis of observation of traffic, counting of accidents and estimating the geometric probability P_i for a specific fairway. In the following sections it will be shown how this was done in some pioneering studies for Japanese coastal waters.

6.5.1 Japanese Traffic Studies (Fujii, 1982)

Uraga Strait

The Uraga Strait, which is located at the entrance to Tokyo Bay, has several obstacles which make it necessary for any passing ship to change course several times in order to avoid stranding. Roughly estimated, the conditional probability of stranding in case of loss of control could be set to be $P_i = 1.0$.

The number of accidents for ships greater than 300 GRT had been counted for the period from 1966 to 1970 and was in total $N_a = 16$. The corresponding number of ship passages (or movements) in the same period was N_m = 140,000. The loss of control probability can then be estimated for this fairway:

$$P_a = \frac{N_a}{N_m} = \frac{16}{140{,}000} = 1.1 \cdot 10^{-4}$$
$$P_c = \frac{P_a}{P_i} = \frac{1.1 \cdot 10^{-4}}{1.0} = 1.1 \cdot 10^{-4}$$

The probability of losing navigational control per passage of the Uraga Strait may therefore be set to be $1.1 \cdot 10^{-4}$.

An interesting fact was that 15 of the ships involved in accidents were sailing under a foreign flag. However, the foreign flag vessels represented only 50% of the traffic through the fairway. Consequently there was a significantly higher accident risk for foreign vessels compared with Japanese.

The Bisanseto fairway

Ozeishima Island is located in a curved part of the Bisanseto fairway. The probability of impact, given loss of navigational control, P_i, for the fairway was estimated to be 0.25 on the basis of the topological characteristics (Table 6.2). Based on Eq. (6.8), the probability of loss of navigational control for different vessel size categories was estimated in the same manner as for the Uraga Strait.

Naruto Strait

Zakace is the headland in the narrowest part of the Naruto Strait. The hindrance due to Zakace constitutes one-fifth of the fairway width. The geometrical probability of an impact is therefore estimated to be 0.20. The traffic flow and the number of impact accidents are presented in Table 6.3.

Akashi Strait

During the construction work on a bridge over the 4 km wide Akashi Strait, a platform was positioned in the middle of the fairway. There were several ship impacts with the platform during the 70-month construction period. The impact diameter of the platform was 0.2 km. The geometrical probability is calculated according to Eq. (6.4):

$$P_i = \frac{0.2}{4.0} = 0.05$$

Table 6.2. Characteristics of the Bisanseto fairway grounding accidents

Tonnage (GRT)	N_a	N_m	P_a	P_i	P_c
< 100	21	300,000	$0.7 \cdot 10^{-4}$	0.25	$2.8 \cdot 10^{-4}$
100–500	15	180,000	$0.8 \cdot 10^{-4}$	0.25	$3.3 \cdot 10^{-4}$
< 500	6	120,000	$0.5 \cdot 10^{-4}$	0.25	$2.0 \cdot 10^{-4}$

Table 6.3. Characteristics of the Naruto Strait grounding accidents

N_a	N_m	P_a	P_i	P_c
11	730,000	$0.2 \cdot 10^{-4}$	0.20	$0.8 \cdot 10^{-4}$

Table 6.4. Characteristics of the Akashi Strait impact accidents

N_a	N_m	P_a	P_i	P_c
16	2,430,900	$0.07 \cdot 10^{-4}$	0.05	$1.4 \cdot 10^{-4}$

Table 6.4 summarizes the Akashi Strait study findings. The resulting probability of loss of vessel control was estimated to be $1 \cdot 10^{-4}$.

Summary of the Japanese investigations

The investigations presented above show that the probability of losing navigational control varies from $0.8 \cdot 10^{-4}$ to $3.3 \cdot 10^{-4}$. Based on these investigations, the following mean value is proposed:

$$P_c = 2.0 \cdot 10^{-4} \text{ (1/passage)}$$

In certain risk assessment studies it might be necessary to take the effect of sailing distance into consideration. Assuming that the average distance of the critical part of the fairway in the previous studies can be set to 10 nm, the loss of control frequency can be computed as:

$$\mu_c = P_c/D = 2.0 \cdot 10^{-5} \text{ (1/nm)}$$

6.5.2 Alternative Estimates

In order to qualify the results of the Japanese studies, an alternative approach might be tried. The failure frequency of the steering system was estimated in an American investigation (Ewing, 1975) as:

$$\lambda_{ss} = 0.41 \text{ (failures/year)}$$

Assuming 48% of the time at sea, we have 175 sailing days each year and the following hourly frequency:

$$\lambda_{ss} = 0.41/(175 \cdot 24) = 1 \cdot 10^{-4} \text{ (failures/hour)}$$

The relative distribution of factors of causes leading to grounding accidents for Norwegian ships greater than 1599 GRT is shown in Table 6.5 (Kristiansen and Karlsen, 1980). On the basis of this investigation it could be concluded that 1/50 of the

Table 6.5. Distribution of primary causal factors in grounding accidents for Norwegian ships greater than 1599 GRT, 1970–78

Causal factor group	Causal factor	Frequency	
		abs.	%
I. External factors	G. External conditions influencing navigation and auxiliary equipment	8	1.9
	I. Less than adequate markers and buoys	27	6.4
	P. Reduced visibility	53	12.6
	Q. External influences like channel and shallow water effect.	79	18.9
II. Technical failure	A. Failure in ship's technical systems	24	5.7
	C. Serviceability of navigational aids	8	1.9
	D. Remote control of steering and propulsion	3	0.7
	F. Failure in communication equipment	2	0.5
III. Navigation factors	B. Bridge design and arrangement	1	0.2
	F. Error/deficiency in charts or publications	34	8.1
	M. Bridge manning and organization	35	8.4
	O. Internal communicational failure	5	1.2
	X. Inadequate knowledge and experience	4	1.0
IV. Navigation error	R. Failure due to navigation and manoeuvring	49	11.7
	T. Wrong use of the information from buoys and markers	35	8.4
	S. Failure in operation of equipment	10	2.4
	U. Wrong appreciation of traffic information	2	0.5
V. Non-compliance	N. Inadequate coverage of watch	24	5.7
	V. Special human factors	10	2.4
VI. Other ships	H. Fault or deficiency of other ship	—	—
	Y. Navigational error on other ship	6	1.4
Sum		419	

Source: Kristiansen and Karlsen (1980).

accidents were caused by failure of the steering machine. Hence, the total failure rate can be estimated to be:

$$\mu = 50 \cdot 1 \cdot 10^{-4} = 5 \cdot 10^{-3} \text{ (failures/hour)}$$

The American study further estimated that only 5% of the failures led to an impact accident. By assuming that the mean sailing speed is equal to 10 knots, the following estimate of the accident frequency can be made:

$$\mu_c = 0.05 \cdot 5 \cdot 10^{-3}/10 = 2.5 \cdot 10^{-5} \text{ (failures/nm)}$$

Table 6.6. Collision, grounding and impact accidents on the Norwegian coast for the period 1970–78

Fairway segment	Distance (nm)	Traffic N_m (1/year)	Accidents N_a (1/year)	$P_a = N_a/N_m \cdot 10^{-4}$ (1/passage)	$\mu_a = P_a/D \cdot 10^{-5}$ (1/nm)
Oslo fjord	92	28,600	16.9	5.9	0.64
Langesund – Tananger	183	20,900	11.8	5.6	0.31
Tananger – Bergen	101	65,600	30.7	4.7	0.46
Bergen – Stadt	125	50,300	28.2	5.7	0.46
Stadt – Kristiansund	114	28,600	23.3	8.1	0.71
Kristiansund – Rørvik	155	21,600	24.8	11.5	0.74
Rørvik – Støtt	147	8,600	15.8	18.4	1.25
Støtt – Harstad	150	19,500	32.3	16.6	1.10
Harstad – Sørøysund	175	12,500	26.7	26.7	1.22
Sørøysund – Kirkenes	212	7,700	24.1	31.3	1.48
Weighted mean value	129	26,400	—	8.9	0.69
Norwegian coast	1454	263,900	253.2	—	

Source: Kristiansen (1980).

This estimate of the frequency for loss of control compares well with the figure based on the Japanese studies.

The Norwegian study (Kristiansen, 1980) previously referred to also estimated the impact accident rate per distance unit as shown in Table 6.6. The coast was split into ten main fairway segments. The estimated accident frequency given in the right-hand column varied from $0.3 \cdot 10^{-5}$ to $1.5 \cdot 10^{-5}$ accidents per nautical mile, i.e. by a factor of 5. This variation may be explained by different dominating ship types, fairway characteristics and environmental factors. It should also be kept in mind that the characteristic distance (D) does not necessarily reflect the typical traffic pattern, although the majority of vessels are assumed to sail along the coast.

The mean value of $0.69 \cdot 10^{-5}$ accidents/nm combined with an assumed geometrical probability of 25% indicates the following loss of control frequency:

$$\mu_c = P_c = \frac{P_a}{P_i} = \frac{\mu_a}{P_i} = \frac{0.69 \cdot 10^{-5}}{0.25} = 2.8 \cdot 10^{-5} \text{ (failures/nm)}$$

Even this estimate compares well with the previous values.

In certain risk studies it will be of interest to estimate the number of impact accidents within a specified period. We have the following expression for the accident probability per passage (vessel movement):

$$P_a = \frac{N_a}{N_m}$$

or rewritten:

$$N_a = N_m \cdot P_a$$

From the basic definition of the impact accident model we have:

$$P_a = P_c \cdot P_i = \mu_c \cdot D \cdot P_i$$

This gives the following computational expression for grounding:

$$N_a = N_m \cdot \mu_c \cdot D \cdot \left(\frac{B+d}{W}\right)$$

and stranding:

$$N_a = N_p \cdot \mu \cdot D \cdot \left(1 - \frac{2}{\pi} \cdot \frac{W}{D}\right)$$

6.6 COLLISION

In contrast to grounding and stranding, collision represents an impact between two moving objects and not only one relative to a stationary hazard. A collision may also vary in terms of how the vessels are approaching each other: head-on, crossing or overtaking. These situations will be modelled somewhat differently, although the basic approach is the same as for groundings in the sense that the critical impact cross-section is taken into consideration.

6.6.1 Head-on Collisions

A ship is exposed to meeting traffic as outlined in Figure 6.7. The subject ship is exposed to head-on approaching ships within a section of a fairway with distance D and average width W. The modelling approach assumes the own (subject) vessel, denoted by index 1, is approaching a traffic flow of vessels denoted by index 2. We also introduce the relative sailing distance D' expressing the fact that both groups of vessels are moving.

B_1 = Mean beam of meeting ships (m)
v_1 = Mean speed of meeting ships (knots)

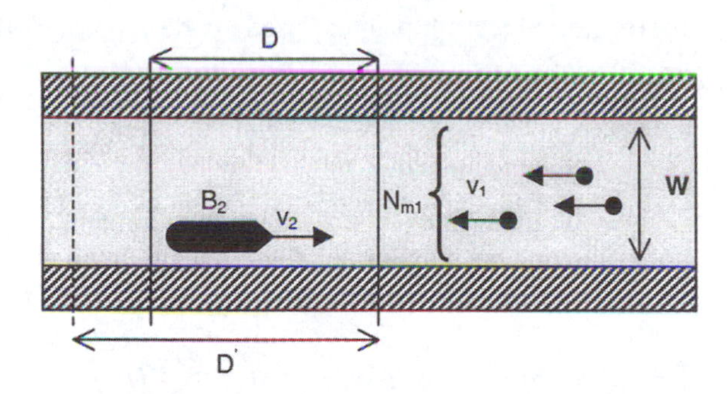

Figure 6.7. Modelling of head-on collision accidents.

B_2 = Beam of subject ship (m)
v_2 = Speed of subject ship (knots)
N_{m1} = Arrival frequency of meeting ships (ships/unit of time)
D' = Relative sailing distance (nm)

The mean number of meeting ships within a square nautical mile of the fairway is referred to as the density of the oncoming ships. This density is calculated as the number of ships entering the fairway within a time period relative to an area characterized by the width of the fairway and the sailed distance of the first meeting ship:

$$\rho_s = \frac{N_{m1} \cdot T}{(v_1 \cdot T) \cdot W} = \frac{N_{m1}}{v_1 \cdot W}$$

where:

ρ_s = Traffic density of meeting ships (ships/nm^2)
T = An arbitrary period of time (hours)

The subject ship (index 2) spends T_2 time units to sail the specified section of the fairway and has a relative speed v to the meeting traffic:

$$T_2 = \frac{D}{v_2}; \qquad v = v_1 + v_2$$

The subject ship sails the distance D' relative to the oncoming ships:

$$D' = v \cdot T_2 = (v_1 + v_2) \cdot \frac{D}{v_2}$$

The impact diameter of a collision is equal to the sum of the exposed ship's beam and the beam of the meeting ships:

$$B = B_1 + B_2$$

Hence the area, A, where the subject ship is exposed to danger of collisions within the fairway is equal to:

$$A = B \cdot D' = (B_1 + B_2) \cdot (v_1 + v_2) \cdot \frac{D}{v_2}$$

The expected number of collisions per passage of the fairway, given that control has been lost, is given by the product of the exposed area and the traffic density:

$$N_i = A \cdot \rho_s = (B_1 + B_2) \cdot (v_1 + v_2) \cdot \frac{D}{v_2} \cdot D \cdot \rho_s$$

or:

$$N_i = \frac{(B_1 + B_2)}{W} \cdot \frac{(v_1 + v_2)}{v_1 \cdot v_2} \cdot D \cdot N_{m1}$$

If the main parameters of the subject ship are equal to the parameters of the meeting ships, the expression is greatly simplified:

$$N_i = 4 \cdot B \cdot D \cdot \rho_s$$

A Head-on Collision Case

Problem

The number of reported head-on collisions for a specific fairway has in recent years been about 14 annually, with little variation. Last year, however, there were 20 reported collisions. Should we assume that this last higher number of collisions is an exception or not? Find the expected number of head-on collisions per year on the basis of the general collision model.

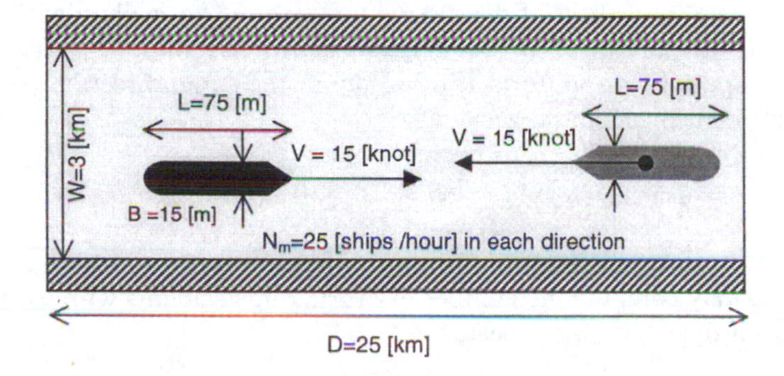

Assumptions

- The lateral position of the ships within the fairway is uniformly distributed.
- The probability of losing control $P_c = 2 \times 10^{-4}$ per passage of the fairway.

Analysis

The traffic density of meeting ships is:

$$\rho = \frac{N_m}{V \cdot W} = \frac{25}{15 \cdot 1852 \cdot 3000} = 3 \cdot 10^{-7} \text{ (ships/m}^2\text{)}$$

The impact probability for oncoming traffic is:

$$N_i = 4 \cdot B \cdot D \cdot \rho_s = 4 \cdot 15 \cdot 25{,}000 \cdot 3 \cdot 10^{-7} = 0.45 \text{ (accidents/incident)}$$

The probability of head-on collision for a single vessel in the fairway is:

$$P_a = N_i \cdot P_c = 0.45 \cdot 2 \cdot 10^{-4} = 9 \cdot 10^{-5} \text{ (accidents/passage)}$$

The expected number of head-on collisions considering that we have the same traffic flow in each direction is:

$$N_a = P_a \cdot N_m = 9 \cdot 10^{-5} \cdot 25 \cdot (24 \cdot 365) = 19.7 \text{ (ships/year)}$$

This estimate indicates that the last year registration of 20 head-on collisions is not exceptional.

6.6.2 Overtaking Collision

In overtaking encounters the vessels involved are sailing in the same direction but at different speeds. The estimation of overtaking collisions is basically identical to head-on collisions apart from the expression for relative speed. Assuming that the subject vessel (subscript 2) is exposed to a uniform traffic flow in the same direction (subscript 1) we have that the number of potential accidents is:

$$N_i = \frac{(B_1 + B_2)}{W} \cdot \frac{(v_1 - v_2)}{v_1 \cdot v_2} \cdot D \cdot N_{m1}$$

Alternatively, we may compute the number of overtaking accidents within a unidirectional traffic flow with a distribution of speed:

$$N_1 = \frac{(B_1 + B_2)}{W} D \cdot N_m \sum f_x \cdot f_y \left(\frac{1}{v_x - v_y} \right)$$

where f_x and f_y denote fractions of the total traffic flow N_m with speed v_x and v_y respectively. The summation is taken over all combinations of different speeds within the traffic flow.

An Overtaking Situation

Problem

A straight fairway of distance 25,000 m and width 3000 m has a co-directional traffic of 25 vessels per hour. Mean breadth of vessels is 15 metres. How many overtaking collisions can be expected on an annual basis?

Traffic observations have shown that the speed distribution is roughly as follows:

Fraction of traffic (%)	30	50	20
Speed (knots)	12	15	18

Solution

The expected number of encounters is given by:

$$N_i = \frac{2B \cdot D \cdot N_m}{W} \sum f_x \cdot f_y \left(\frac{1}{v_x - v_y} \right) = k \sum \cdots$$

The constant term is:

$$k = 2 \cdot 15 \cdot 25{,}000 \cdot 25/3{,}000 = 6.25 \cdot 10^3$$

The summation part in the expression is:

$$\sum \cdots = \begin{bmatrix} 0.30 \cdot 0.50\left(\frac{1}{12} - \frac{1}{15}\right) + 0.30 \cdot 0.20\left(\frac{1}{12} + \frac{1}{18}\right) \\ +0.50 \cdot 0.20\left(\frac{1}{15} + \frac{1}{18}\right) \end{bmatrix} \frac{1}{1852}$$

$$= 2.9 \cdot 10^{-6}$$

This gives:

$$N_i = 6.25 \cdot 10^3 \cdot 2.9 \cdot 10^{-6} = 0.018 \text{ (encounters/passage)}$$

Assuming a probability of loss of navigational control $P_c = 2 \cdot 10^{-4}$, we get the following estimate for the overtaking collision frequency:

$$N_a = 0.018 \cdot 2 \cdot 10^{-4} \cdot 25 \cdot 8760 = 0.79 \text{ (accidents/year)}$$

Comment

Let us make a comparison with the previous head-on collision case. We found then that the two opposing traffic flows generated 19.7 accidents per year. Two similar flows would have generated $0.79 \cdot 2 = 1.6$ overtaking accidents per year. This means that the ratio between head-on and overtaking is: $19.7/1.6 = 12$, which is a substantial difference.

6.6.3 Crossing Collision

A collision between crossing vessels is slightly more complex to analyse. This stems from the fact that the scenario represents two encounter situations. As shown in Figure 6.8, we have that the subject ship (subscript 2) is exposed to crossing ships within a section of a fairway of length D and width W. Making a distinction between the striking and the struck vessel, it is immediately clear that both sets of vessels may have either role.

The description of the traffic situation is based on following nomenclature:

B_1 = Beam of crossing ships (m)
L_1 = Length of crossing ships (m)
v_1 = Speed of crossing ships (knots)
B_2 = Mean beam of subject ship (m)
L_2 = Mean length of subject ship (m)
v_1 = Mean speed of subject ship (knots)
N_{m1} = Arrival frequency of meeting ships (ship/unit of time)

The mean number of crossing ships within a square nautical mile of the fairway is referred to as the density of the crossing ship. Analogous to the meeting situation described in the previous section, the density of the crossing traffic is given by the following equation:

$$\rho_m = \frac{N_{m1} \cdot T}{(v_1 \cdot T) \cdot W} = \frac{N_{m1}}{v_1 \cdot W}$$

where T is an arbitrary selected time period. The subject ship (2) takes T_2 hours to pass the section of the fairway where it is exposed to the crossing traffic:

$$T_2 = \frac{D}{v_2}$$

The relative speed between the vessels is given by vector summation (Figure 6.9).

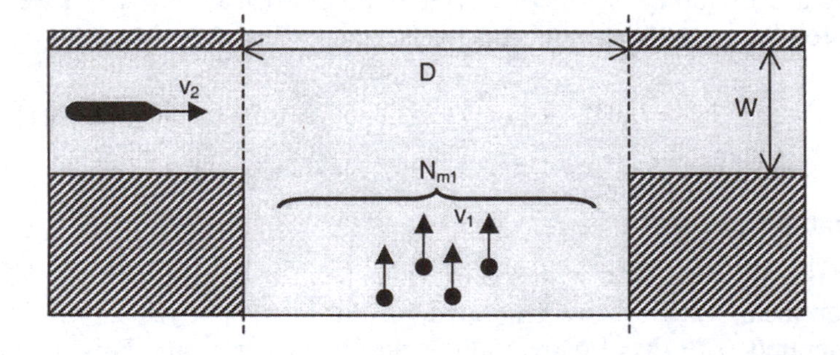

Figure 6.8. Crossing vessels situation.

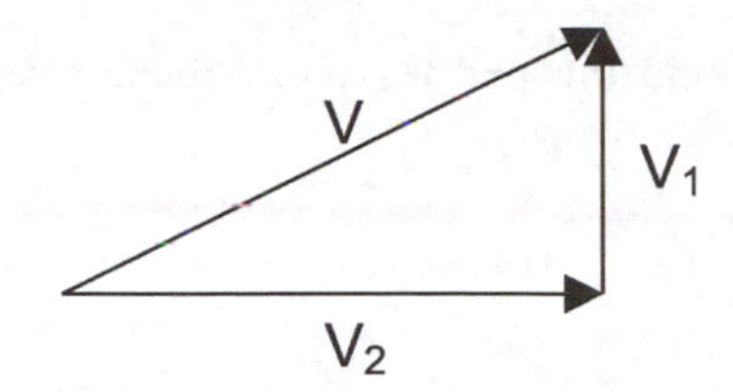

Figure 6.9. Relative speed of crossing vessels.

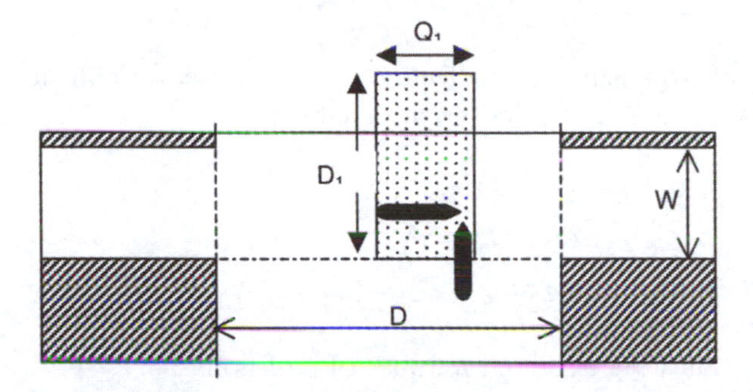

Figure 6.10. Exposed area to collision for situation I.

The crossing ships sail a distance D_1 in the course of the period that the subject ship is exposed in the critical fairway section:

$$D_1 = v_1 \cdot T_2 = v_1 \cdot \frac{D}{v_2}$$

The following development of the model is split into two owing to the fact that we have as already mentioned two collision situations:

1. A crossing vessel hit the subject vessel.
2. The subject vessel hit a crossing vessel.

Let us consider the first situation. The impact diameter of a collision is equal to the sum of the exposed ship's length and the mean beam of the crossing ships:

$$Q_1 = (B_1 + L_2)$$

Hence the area where the subject ship is exposed to the collision hazard (Figure 6.10) is:

$$A_1 = Q_1 \cdot D_1 = (B_1 + L_2) \cdot D \cdot \frac{v_1}{v_2}$$

The expected number of collisions per passage of the fairway is given by the product of the exposed area and the traffic density:

$$\begin{aligned} P_{i1} &= A_1 \cdot \rho_m \\ &= (B_1 + L_2) \cdot D \cdot \frac{v_1}{v_2} \cdot \frac{N_{m1}}{v_1 \cdot D} \\ &= (B_1 + L_2) \cdot \frac{N_{m1}}{v_2} \end{aligned}$$

The same line of argument can be followed for the second collision situation except that the roles of striking and struck ship are changed:

$$\begin{aligned} D_2 &= D \\ Q_2 &= L_1 + B_2 \\ A_2 &= Q_2 \cdot D_2 = (L_1 + B_2) \cdot D \end{aligned}$$

This results in the following expected number of collisions:

$$P_{i2} = A_2 \cdot \rho_m = (L_1 + B_2) \cdot \frac{N_{m1}}{v_1}$$

The total expected number of side collisions is the sum of the two calculated figurers P_{i1} and P_{i2}:

$$\begin{aligned} P_i &= P_{i1} + P_{i2} \\ &= (B_1 + L_2) \cdot \frac{N_{m1}}{v_2} + (L_1 + B_2) \cdot \frac{N_{m1}}{v_1} \\ &= \frac{N_{m1}}{v_1 \cdot v_2} \cdot [(B_1 + L_2) \cdot v_1 + (L_1 + B_2) \cdot v_2] \end{aligned}$$

Assuming that the subject ship and the crossing ships all have identical characteristics, the expression is further simplified and visualizes the basic model, which says that the potential number of encounters is equal to the traffic density times the exposed area:

$$P_i = \frac{N_{m1}}{v} \cdot 2 \cdot (B + L) = \rho_m \cdot 2 \cdot (B + L) \cdot D \tag{6.9}$$

A Crossing Encounter Situation

Problem

Two traffic lanes cross each other at 90°. Seven collisions have been recorded on an annual basis in recent years. The local community is concerned by the considerable amount of oil spilled as a result of these accidents. They demand implementation of new safety measures. The proposal has been challenged by the coast administration, which claims that the safety level is acceptable. Give an assessment of the situation. For simplicity, identical traffic volumes and same ship characteristics can be assumed.

Solution

Assess the safety standard of the vessels navigating the seaway by estimating the probability of loss of control.

Data

Width of fairways: $W = D = 3\,\text{km}$
Ship length: $L = 75\,\text{m}$
Speed: $v = 15$ knots
Traffic: $N = 5$ ships/hour
Ship beam: $B = 15\,\text{m}$

Analysis

Traffic density:

$$\rho = N/(v \cdot D) = 5/(15 \cdot 1852 \cdot 3000) = 6.0 \cdot 10^{-8} \text{ (ships/m}^2\text{)}$$

Impact probability:

$$P_i = 2 \cdot D \cdot (L + B) \cdot \rho = 2 \cdot 3000 \cdot (75 + 15) \cdot 6.0 \cdot 10^{-8} = 0.032 \text{ (1/incident)}$$

On the basis of extensive traffic studies it has been shown that the probability of losing navigational control is $P_c = 5 \cdot 10^{-4}$ (per passage).

The probability of impact accident is:

$$P_a = P_c \cdot P_i = 5 \cdot 10^{-4} \cdot 0.032 = 1.6 \cdot 10^{-5}$$

The *expected* number of collisions per year is therefore:

$$N_a = N_m \cdot P_a = 25 \cdot (24 \cdot 365) \cdot 1.6 \cdot 10^{-5} = 3.5$$

Assessment

The fact that the fairway in question has twice as many collisions as might be expected, indicates that there is a problem related to the crossing of traffic. The local community therefore has a good case in their demand for further investigation of the conditions.

6.7 A GENERAL COLLISION MODEL

In the previous paragraphs two standardized situations that may lead to a collision have been analysed, namely meeting and crossing encounters. However, in some fairways it may be difficult to specify any dominating traffic flows. In such situations it may be more convenient to assume that all courses have the same likelihood.

6.7.1 Basic Model

Figure 6.11 shows a fairway section with characteristic dimension W and randomly distributed traffic. We will now estimate the probability that the entering vessel S_m will collide with the vessels (S_n) in the fairway.

The entering ship has a speed v_m and will be exposed to the traffic for a period of:

$$T_m = \frac{W}{v_m}$$

The probability that any of the S_n ships are present in the fairway is given by:

$$P_n = \frac{W}{v_n} \cdot \frac{1}{\tau}$$

where τ = annual operational time.

Given that the fairway each year is exposed to N ship movements, the expected number of impacts between entering vessel (m) and existing traffic is:

$$N_i = T_m \cdot \sum_{n=1}^{N} P_n \cdot P_{mn} \tag{6.10}$$

where P_{mn} expresses the conditional probability per time unit for the event that S_m collides with S_n. This property is given by the geometry and dimensions related to the general encounter situation depicted in Figure 6.12.

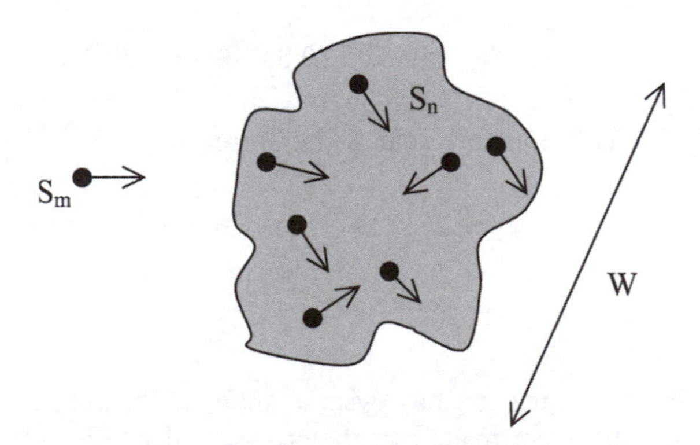

Figure 6.11. Traffic with random courses.

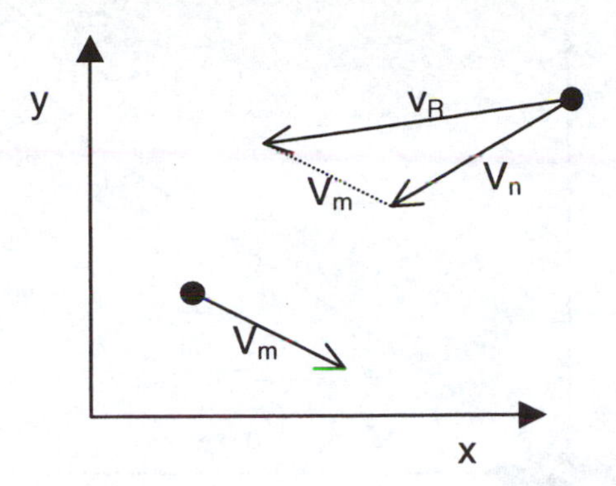

Figure 6.12. General meeting situation.

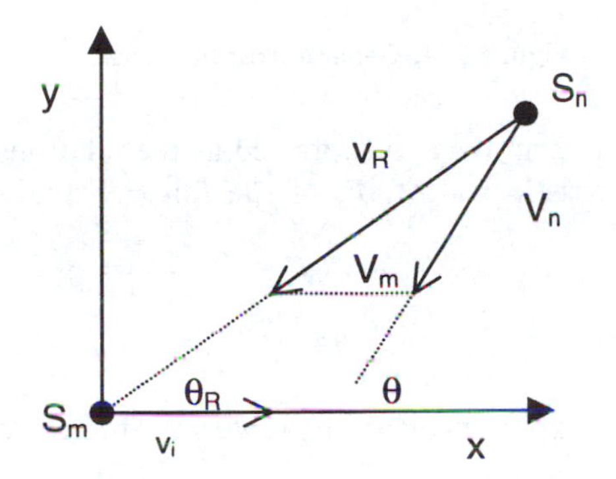

Figure 6.13. Transformed meeting situation.

The relative speed v_R is given by vector addition:

$$\vec{v}_R = \vec{v}_m + \vec{v}_n$$

The general meeting situation may be transformed as shown in Figure 6.13 by aligning the v_m direction along the x-axis with basis in the origin. The angle of crossing between vessels is given by θ and the direction of the relative speed is θ_R in the transformed coordinate system.

The impact diameter (Figure 6.14) is defined as the exposed cross-section *normal* to the direction of the relative speed and given by the vector sum:

$$\vec{d}_i = \vec{d}_{im} + \vec{d}_{in}$$

The impact diameter is a function of the meeting situation parameters:

$$\vec{d}_{im} = d_{im}(\theta_R)$$
$$\vec{d}_{in} = d_{in}(\theta_R, \theta)$$

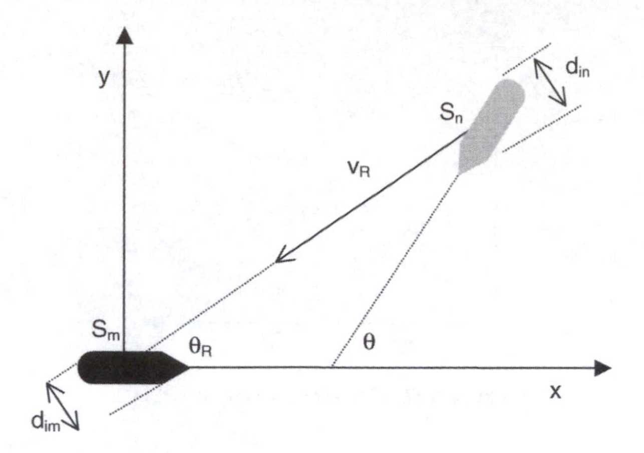

Figure 6.14. General crossing scenario.

The probability of an impact may be expressed as the relationship between the impact diameter and the characteristic width, W, of the fairway analogous to the grounding model (Eq. 6.4):

$$P_i = \frac{\vec{d}_i}{W}$$

The entering ship S_n is exposed to the encountering ship for a period of:

$$T_n = \frac{W}{\vec{v}_R}$$

Hence the impact probability per unit of time is given by:

$$\frac{P_i}{T_n} = \frac{\vec{d}_i \cdot \vec{v}_R}{W^2}$$

As both the vector d_i and the vector v_2 are functions of the meeting situation, the mean value of the impact probability has to be computed by integration over all meeting angles:

$$P_{mn} = \frac{P_i}{T_n} = \frac{1}{\pi} \cdot \int_0^{\pi} \frac{\vec{d}_i \cdot \vec{v}_R}{W^2} \cdot d\theta \tag{6.11}$$

Owing to symmetry it is sufficient to integrate only from 0 to π. We will not show the rest of the development of the model here.

6.7.2 A Model Approximation

Both the integration of the expression above and the summation over all vessels (M) in the seaway can be considerably simplified by assuming equal vessel characteristics (main dimensions and speed). In that case it can be shown that the number of impacts is given by:

$$N_i = \frac{N}{\tau \cdot v} \cdot \left(\frac{4}{\pi} \cdot L + 2 \cdot B \right) \tag{6.12}$$

The expected number of ships in the fairway at any given time is:

$$N_N = \left(\frac{W}{v} \cdot \frac{1}{\tau} \right) \cdot N$$

The traffic density is then:

$$\rho_n = \frac{N_M}{W^2} = \frac{W}{v} \cdot \frac{N}{\tau} \cdot \frac{1}{W^2} = \frac{N}{v \cdot \tau \cdot W} \tag{6.13}$$

which, inserted in Eq. (6.12), gives the following expression:

$$N_i = \rho_n \cdot W \cdot \left(\frac{4}{\pi} \cdot L + 2 \cdot B \right) \tag{6.14}$$

6.7.3 Circular Impact Cross-Section

The integration of the general expression given in Eq. (6.11):

$$P_{mn} = \frac{1}{\pi} \cdot \int_0^{\pi} \frac{\vec{d}_i \cdot \vec{v}_R}{W^2} \cdot d\theta$$

can also be considerably simplified by assuming identical, circular cross-sections for the vessels involved (Figure 6.15):

$$d_{is} = 2 \cdot d_i$$

The size of the impact diameter for each ship is still unknown. The following calculation is performed to calculate the size of the circular impact diameters.

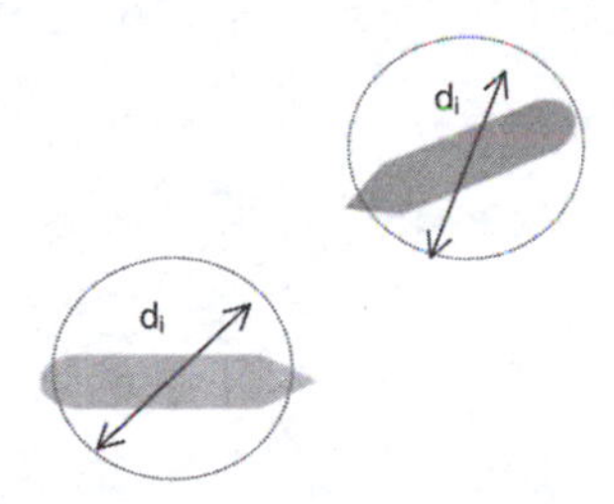

Figure 6.15. Circular impact diameter.

Further, by also assuming equal velocity for all ships, the following expression can be developed:

$$P_{mn} = \frac{8 \cdot d_i \cdot v}{\pi \cdot W^2}$$

Inserting this expression in Eq. (6.10) we obtain, by summing over all vessels (N):

$$N_i = \frac{W}{v \cdot \tau} \cdot \frac{8 \cdot d_i}{\pi}$$

By applying the traffic density expression the expected number of collisions is:

$$N_i = \rho_n \cdot W \cdot \frac{8 \cdot d_i}{\pi}$$

6.7.4 Other Crossing Angles

We have so far analysed perpendicular crossings and random distributed crossing angles. It is possible to show that the general expression for an arbitrary crossing angle is given by:

$$P_i = \frac{d \cdot W \cdot \rho_1}{v_2} \left(\frac{v_2}{\sin\theta} - \frac{v_1}{\tan\theta} + v_1 \right)$$

where θ denotes the crossing angle and d is a circular collision diameter (see Figure 6.16). Assuming identical speed, one gets the following values:

$$\theta = 30^\circ \qquad P_i = d \cdot W \cdot \rho_1 \cdot 1.27$$
$$\theta = 60^\circ \qquad P_i = d \cdot W \cdot \rho_1 \cdot 1.57$$
$$\theta = 90^\circ \qquad P_i = d \cdot W \cdot \rho_1 \cdot 2.00$$

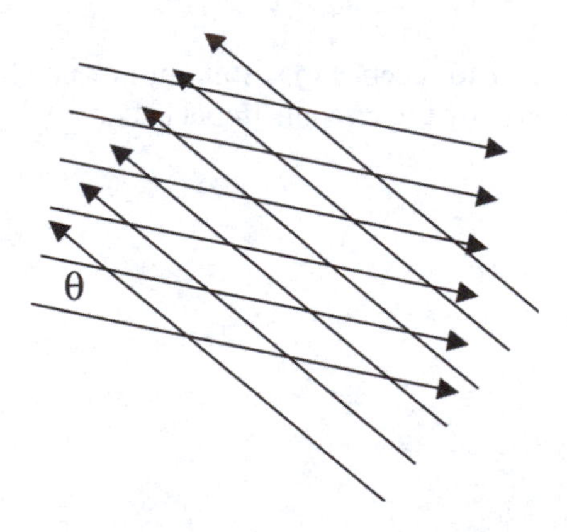

Figure 6.16. Crossing with a relative bearing θ.

Table 6.7. Comparison of P_i for different collision situations

	Head-on	Crossing	Random course	Circular
Basic expression	$4 \cdot D \cdot \rho_m \cdot B$	$2 \cdot D \cdot \rho_m \cdot (B+L)$	$\rho_n \cdot (4 \cdot L/\pi + 2 \cdot B) \cdot D$	$8W \cdot d_i \cdot \rho_n/\pi$
Standardized $L=6B=d_i$, $D=W$	$4 \cdot D \cdot \rho_m \cdot B$	$14 \cdot D \cdot \rho_m \cdot B$	$9.6 \cdot D \cdot \rho_n \cdot B$	$48 \cdot D \cdot B \cdot \rho_n/\pi$
Relative	1	3.5	2.4	3.8

6.7.5 Comparison of Collision Situations

In the previous sections we have shown alternative models for estimating the probability of having a collision given loss of navigational control. As a general conclusion, it is clear that the risk of colliding is a function of the traffic density and the distance of the fairway in question.

The different models can be compared by standardizing some key particulars relating to vessels and fairway:

$$\text{Vessels:} \quad L = 6B \qquad \text{Fairway:} \quad D = W$$

Table 6.7 shows the different models and how they compare by applying the above-mentioned data. The collision estimates vary by almost a factor 4 (or to be precise, 3.8). The low value for head-on encounters can be explained by the fact that the exposed cross-section is at a minimum by $2B$. The high value for the random course and circular impact area can likewise be explained by the large cross-section that is L (or $6B$) reflecting all possible collision angles. The average value of P_i is best represented by the random course model.

One should be careful in drawing any conclusion from this analysis as to the relative risk of different collision forms. It should be kept in mind that the collision frequency for a given seaway is also a function of the dominating traffic flow and the complexity of handling the encounter situation as reflected by P_c. The latter will be discussed for collisions in the next section.

6.8 LOSS OF TRAFFIC CONTROL

The probability of an impact accident has been defined as the product of the likelihood of losing navigational control, P_c, multiplied by the likelihood of having an accident given the loss of control incident, P_i:

$$P_a = P_c \cdot P_i$$

If the exposed ship has N_e passages of the fairway, the expected number of collisions per unit of time is given by:

$$N_a = P_c \cdot P_i \cdot N_e$$

or

$$N_a = P_c \cdot N_i \cdot N_e$$

Because both ships in a meeting situation may lose control, the potential number of collisions is twice as high. By introducing the number of loss of navigational control situations per nautical mile within a fairway, the total number of collision is given by:

$$N_a = 2 \cdot \mu_c \cdot D \cdot P_i \cdot N_e$$

where μ_c = number of loss of control situations per nautical mile, and D = distance of the fairway.

It may also be of interest to estimate the number of collisions that are generated by an observed traffic volume. Assume that a specific fairway is navigated by N_m ships within a given period of time. The number of encountering situations is given by the following expression:

$$\tfrac{1}{2} \cdot N_m^2 \cdot P_i$$

The term $\frac{1}{2}$ is applied as it is computationally irrelevant whether only one or both vessels in an encountering situation lose control. This expression substitutes the $(P_i \cdot N_e)$ term in the previous equation:

$$N_a = \tfrac{1}{2} \cdot \mu_c \cdot D \cdot P_i \cdot N_m^2$$

6.8.1 US Ports

The random course model has been used to perform a collision analysis for harbours in United States. Table 6.8 presents some of the results from this investigation. The subject ships in the analysis were both arriving and departing ships with a displacement greater than 1000 tons. The investigation covered seven ports for the period 1969 to 1974.

The study collected data on vessel characteristics, traffic (N_m) and number of collisions (N_a). By applying the random course collision model previously described, the potential number of impacts for each port was estimated (N_i). It is worth noticing that three of the ports had no reported collisions in the actual period. Based on the total material, the probability of losing traffic control was estimated to be:

$$P_c = 7 \cdot 10^{-4} \text{ (failures/ship-movement)}$$

Table 6.8. Collisions in major US ports, 1969–74

Port	N_m	N_a	P_i	μ_c
Los Angeles	16,900	1	158.76	0.000525
Long Beach	9,800	0	53.96	0.00
Boston	7,700	0	28.99	0.00
New York	23,400	3	325.51	0.000768
Tampa	8,200	0	41.59	0.00
Mississippi	14,100	1	121.45	0.000686
Galveston	12,300	2	89.69	0.00186

The mean sailing distance for these ports was 35 nautical miles. Hence the failure frequency becomes:

$$\mu_c = 7 \cdot 10^{-4}/35 = 2.0 \cdot 10^{-5} \text{ (failures/nm)}$$

This estimate is identical with our earlier estimate for failures leading to stranding and grounding.

6.8.2 Japanese Fairways

Extensive theoretical and empirical traffic studies have been performed for some Japanese fairways (Fujii, 1982). The main characteristics for the fairway and vessels are summarized in Table 6.9. The mean value for ship length indicates that it was a question about coastal traffic.

Based on traffic and accident observations, the probability of loss of navigational control leading to collision was estimated for each fairway. A distinction was made between head-on and overtaking traffic (see Table 6.10). It should be noted that there was some variation in failure rates for the fairways. This may easily be explained by possible differences in environmental conditions and topographic factors. The results indicate that head-on encounters are more difficult to handle than overtaking encounters. This is reasonable, taking the time to respond into consideration. Head-on vessels will typically close a separation of 5 nm (9 km) in 10 minutes, whereas overtaking vessels will still be 4.5 nm away after the same duration, assuming a relative speed of 3 knots (i.e. 1/10 of the relative speed of head-on vessels). As can be seen from the mean estimates, the failure rate for head-on collisions is twice as high as for overtaking collisions.

Table 6.9. Japanese traffic studies: fairway and vessel characteristics

Fairway	D (km)	W (km)	N_m (ships/hour)	V (knots)	L (m)
Uraga Strait	25	3	25	20	50
Akashi Channel	10	3	50	15	32
Kanmon Sound	15	0.6	38	14	20

Table 6.10. Loss of traffic control in Japanese fairways

Fairway	Overtaking traffic μ_c (failures/nm)	Head-on traffic μ_c (failures/nm)
Uraga Strait	$1.3 \cdot 10^{-5}$	$3.9 \cdot 10^{-5}$
Akashi Channel	$1.3 \cdot 10^{-5}$	$2.6 \cdot 10^{-5}$
Kanmon Sound	$0.86 \cdot 10^{-5}$	$2.2 \cdot 10^{-5}$
Mean values	$1.5 \cdot 10^{-5}$	$2.9 \cdot 10^{-5}$

6.8.3 Dover Strait

The probability of collisions in the Dover Strait has also been the subject of extensive investigations (Lewison, 1978). After the implementation of a traffic separation scheme (TSS) in Dover, the risk of head-on collision was reduced. However, there has been an increased stranding frequency which might be explained by the fact that TSS separates the main traffic flows and thereby presses some of the traffic nearer to the shore. The cross-channel ferries between England and the Continent also contribute to Dover's traffic pattern.

The traffic in Dover has been monitored for extensive periods and the number of collisions recorded. In order to estimate the collision risk, an encounter or incident was defined as vessels passing each other within a distance of less than 0.5 nm. That corresponds to our model with random distributed traffic and a circular impact cross-section ($2 \cdot 0.5 = 1$ nm). By analysing radar picture recordings, the number of meetings or encounters could be counted. The following collision probabilities were estimated (P_c):

Head-on traffic: $2.7 \cdot 10^{-5}$ collisions/encounter

Overtaking traffic: $1.4 \cdot 10^{-5}$ collisions/encounter

Crossing traffic: $1.3 \cdot 10^{-5}$ collisions/encounter

The study also indicates that head-on encounters have twice as high a collision risk as other situations. The likelihood of loss of navigational control is about equal for crossing and overtaking traffic.

6.8.4 Summary: Control Failure

Based on the investigations that have been examined in the previous section, separate estimates for the collision failure frequency are proposed in Table 6.11. It is proposed that the failure frequency is twice as high for head-on traffic as for the other two forms. But it should also be kept in mind that there is a considerable uncertainty associated with these estimates and the variation may in fact be greater than the difference proposed. As mentioned earlier, the failure frequency may vary with fairway, traffic pattern, environmental conditions and vessel navigation performance.

Table 6.11. Traffic navigation failure

Encounter situation	μ (failures/nm)
Overtaking vessels	$1.5 \cdot 10^{-5}$
Crossing traffic	$1.5 \cdot 10^{-5}$
Head-on traffic	$3.0 \cdot 10^{-5}$

6.9 VISIBILITY

Accident statistics have revealed that a relatively large proportion of impact accidents, and especially collisions, occur in poor visibility. This is not surprising, considering that navigation is dependent on radar and other electronic aids and without the support of direct visual observation of fairway and traffic. It has on the other hand been suggested that technological developments, for instance the introduction of ARPA, have led to reduced prudence by the navigator.

As a part of the traffic studies in the Dover Strait, the effect of visibility was also studied. It was concluded that the visibility factor was quite large and even greater than the effect of the particular encounter situation itself (Lewison, 1978). A traffic separation scheme (TSS) was implemented in Dover Strait in 1977. The effect of visibility was studied before and after implementation of the TSS. Visibility may be defined in various ways, but in the present investigation three classes were applied: clear, mist/fog and thick/dense, as specified in Table 6.12.

The number of collisions per encounter before and after the implementation of TSS is shown in Figure 6.17. There was a certain reduction of collisions in reduced visibility conditions, but on the other hand an increase in clear weather.

Apart from the before and after effect of TSS, it could be concluded that the relative collision risk for the different ranges of visibility remained fairly constant. The development

Table 6.12. Visibility range

Clear	Mist/fog	Thick/dense
Greater than 4 km	200 m–4 km	Less than 200 m

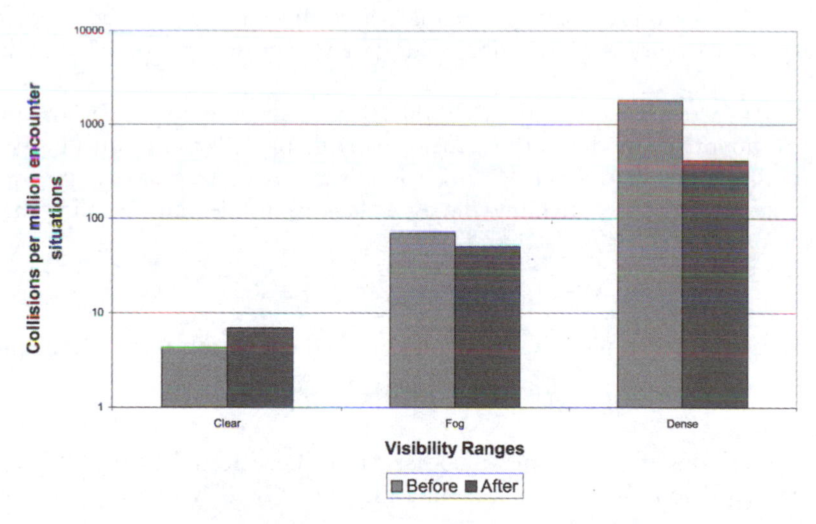

Figure 6.17. Collision accidents before and after implementation of traffic separation distributed on ranges of visibility.

Table 6.13. FCRI for the Dover Strait

Visibility (k)	Clear	Mist/fog	Thick/dense
Relative visibility incidence (VI_k)	0.9457	0.0446	0.0097
Collision probability (P_k)	$6 \cdot 10^{-6}$	$60 \cdot 10^{-6}$	$1800 \cdot 10^{-6}$

of the visibility effect model was therefore based on data for the whole study period. The Fog Collision Risk Index (FCRI) gives the number of collisions per encounter as a function of the relative incidence of the visibility ranges:

$$\text{FCRI} = (P_1 \cdot VI_1 + P_2 \cdot VI_2 + P_3 \cdot VI_3)$$

where:

$$\begin{aligned} P_k &= \text{Probability of collision per million encounters} \\ VI_k &= \text{Fraction of time that the visibility is in the range } k \\ k &= \text{Visibility range: 1, Clear; 2, Fog; 3, Dense} \end{aligned}$$

The estimated parameters of the model are shown in Table 6.13. The data show the dramatic effect of reduced visibility on the collision risk. Although the relative frequency of visibility 'Thick/dense' is less than 1%, the probability increases by a factor of 1800. The resulting value for Dover Strait was:

$$\text{FCRI} = 25.8 \cdot 10^{-6} \text{ (collisions/encounter within 0.5 nm)}$$

The contribution of 'Thick/dense' on this figure is 68% ($0.0097 \cdot 1800 = 17.5$), e.g. without the presence of this visibility condition the probability had been in the order of FCRI $= 8$ or one-third of the actual value. It can therefore be concluded that though marginal visibility is mostly observed, its effect on navigational safety is seldom dramatic.

The fact that there are limited studies of the effect of visibility for other fairways makes it tempting to adapt the model in our general models for collision frequency estimation. What is essential from the model described is the relative effect of visibility ranges on the collision failure rate. By dividing the P_k values by 6, we can visualize the relative importance:

$$P_1^* = 1; \qquad P_1^* = 10; \qquad P_1^* = 300$$

We can then rewrite the model above as follows for the collision failure frequency:

$$\mu = k(1 \cdot VI_1 + 10 \cdot VI_2 + 300 \cdot VI_3)$$

For head-on collisions in Dover the value has earlier been found to be $2.7 \cdot 10^{-5}$ which, inserted in the equation above, and assuming the same visibility frequencies gives:

$$k = 0.63 \cdot 10^{-5}$$

The model head-on collision in Dover can then be written:

$$\mu = 0.63 \cdot 10^{-5}(1 \cdot VI_1 + 10 \cdot VI_2 + 300 \cdot VI_3) \text{ (failures/nm)}$$

The physical meaning of the constant k is the failure frequency corresponding to 100% clear visibility.

A Traffic Separation Case

Problem

A busy harbour is entered through a narrow channel. The channel, however, represents a significant collision risk. It is suggested to implement a traffic separation scheme (TSS) in a channel. Assess the effect of the TSS.

Facts

The channel is 2 km wide. The traffic distribution in each direction after traffic separation is assumed to follow a normal distribution with the peak traffic (mean) 700 m from the west side bank and variance = 62,500 in the south direction and with a peak at 1300 m in the opposite direction and the same variance. The traffic in each direction is 14.4 ships/hour. The mean ship beam is 15 m and mean speed 7 m/sec.

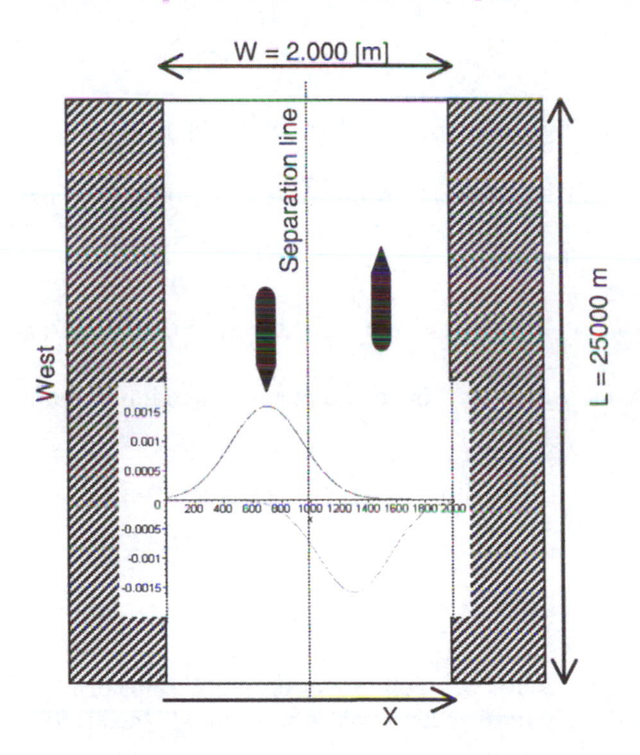

Assumptions

There is no significant crossing traffic. The collision navigation failure rate is equal to $\mu_c = 2.5 \cdot 10^{-4}$ failures/hour.

Experience from other fairways where traffic separation has been implemented shows that some ships will violate the separation scheme and therefore still represent a risk of head-on collision. It is assumed that violators (rogues) constitute about 10% of the traffic.

Analysis of present situation

At present the traffic in each direction is uniformly distributed over the whole channel width. The traffic density is given by:

$$\rho_s = \frac{N_m}{v \cdot W} = \frac{14.4}{3600} \frac{1}{7 \cdot 2000} = 0.286 \cdot 10^{-6} \text{ (ships/m}^2\text{)}$$

The expected number of collisions given loss of control is:

$$N_i = 4 \cdot B \cdot D \cdot \rho_s = 4 \cdot 15 \cdot 25{,}000 \cdot 0.286 \cdot 10^{-6} = 0.43 \text{ (1/passage)}$$

The necessary time for a ship to pass the fairway is:

$$T = D/v = 25{,}000/7 = 3571 \text{ sec}$$

The probability of losing navigational control within the fairway is:

$$P_c = 2.5 \cdot 10^{-4} \cdot (3571/3600) = 2.48 \cdot 10^{-4} \text{ (failures/passage)}$$

The probability of collisions is:

$$P_a = P_i \cdot P_c = 0.43 \cdot 2.48 \cdot 10^{-4} = 1.06 \cdot 10^{-4} \text{ (collisions/passage)}$$

As we have the same traffic flow in each direction, the annual number of collisions is estimated to be:

$$N_a = P_a \cdot N_m = 1.06 \cdot 10^{-4} \cdot 14.4 \cdot 24 \cdot 365 = 13.4 \text{ (collisions per year)}$$

Analysis of the effect of TSS

Presently the traffic has a uniform distribution across the width of the fairway. This means that a ship in the course of a period T meets $N_m \cdot T$ ships or $N_m \cdot T/W$ ships per unit of the channel width. If traffic separation is introduced, the traffic will have another distribution function across the channel. The non-dimensional expressions for the traffic flow are:

$$f_{\text{south}}(x) = \frac{1}{250 \cdot \sqrt{2 \cdot \pi}} \cdot e^{-1/2 \cdot (x-700/250)^2}$$

$$f_{\text{north}}(x) = \frac{1}{250 \cdot \sqrt{2 \cdot \pi}} \cdot e^{-1/2 \cdot (x-1300/250)^2}$$

Owing to symmetry it is only necessary to analyse one direction. Let us look at the violating ships going in the south direction. These ships occupy roughly the section from 700 m to 1000 m. By integrating the function above for this range we get:

$$F_1 = \int_{700}^{1000} f_{\text{south}}(x) \cdot dx = 0.236$$

The violating flow is:

$$N_1 = N_m \cdot F_1 = 14.4 \cdot 0.236/3600 = 9.4 \cdot 10^{-4} \text{ ships/sec}$$

The traffic density of the violating flow is:

$$\rho = 9.4 \cdot 10^{-4}/(7 \cdot 300) = 4.5 \cdot 10^{-7} \text{ ships/m}^2$$

The probability of encounter is:

$$P_i = 4B \cdot D \cdot \rho_s = 4 \cdot 15 \cdot 25{,}000 \cdot 4.5 \cdot 10^{-7} = 0.27 \text{ (1/passage)}$$

The non-violating traffic in the critical section can be estimated as above:

$$F_2 = \int_{700}^{1000} f_{\text{north}}(x) \cdot dx = 0.264$$

The exposed traffic flow is:

$$N_2 = 14.4 \cdot 0.264 \cdot 8760 = 33{,}302 \text{ ships/year}$$

The expected number of head-on collisions due to violating vessels is:

$$N_a = P_i \cdot P_c \cdot N_2 = 0.27 \cdot 2.48 \cdot 10^{-4} \cdot 33{,}302 = 5.23 \text{ collisions/year}$$

Owing to symmetry the actual number of collisions will be twice as high, namely 4.5 collisions/year.

This shows that the effect of introducing TSS is quite dramatic. The frequency of head-on collisions is reduced by a factor of 3:

$$13.4/4.5 = 3$$

REFERENCES

Ewing, L. E., 1975, *Reliability Analysis of Vessel Steering System*. Technical Report TR 958-198-111-1, General Electric Company, Arlington, VA. (NTIS AD-A015 821).

Fujii, Y., 1982, Recent trends in traffic accidents in Japanese waters. *Journal of Navigation*, Vol. 35(1), 90–99.

Kristiansen, S. and Karlsen, J. E., 1980, *Analysis of Causal Factors and Situation Dependent Factors*. Report No. 80-1144, Det norske Veritas, Høvik, Norway.

Kristiansen, S., 1980, *Risk of Sailing the Norwegian Coast – Tank Ship Accidents* [In Norwegian: *Risiko ved beseiling av norskekysten – tankskipsulykker*]. Report No. 80-1142, Det norske Veritas, Høvik, Norway.

Lewison, G. R. G., 1978, The risk of encounter leading to a collision. *Journal of Navigation*, Vol. 31(3), 384–407.

7

DAMAGE ESTIMATION

If there is a possibility of several things going wrong, the one that will cause the most damage will be the one to go wrong
("Murphy's Third Corollary")

7.1 INTRODUCTION

In the previous chapter methodologies for estimating the probability of having an impact accident were described for various types of accidents. The possible resulting damage of the impact accident is, however, still not identified. This chapter describes qualitatively the likely accident characteristics and also the techniques for quantifying the likely damage caused by the accidents.

There are a number of factors that affect the damage extent in a collision:

1. Structural characteristics of vessels involved.
2. The mass of the vessels involved.
3. Speed and relative course.
4. Location of damage.
5. Deformation mechanisms.

In the same manner, damage as a result of grounding will be governed by:

1. Speed and mass of the vessel.
2. Sea floor characteristics.
3. Frictional forces of hull against sea floor.
4. Initiation of local damage (denting, rupture, etc.).
5. Bottom–vessel interaction (lifting of vessel).
6. Deformation mechanisms.

7.2 SURVEY OF DAMAGE DATA

7.2.1 Centre of Damage

The Ship Hydrodynamics Laboratory in Otaniemi did their first studies of accident damage in the 1970s. Kostilainen (1971) made a survey of tanker accidents in the Baltic area for the period 1960–69. Subsequently a larger study covering tankers, general and bulk carriers was undertaken by Kostilainen and Hyvärinen (1976).

In Figure 7.1 the distribution of the longitudinal centre of the damage for grounding accidents is shown on the basis of the two studies mentioned previously. It is not surprising that the foreship is most exposed, and it can be seen that 80% of the damage is located from the bow and 60% of the length aft. In the same manner the damage centre is shown for collisions in Figure 7.2. It is important to keep in mind that collision involves at least two vessels and that the striking vessel will be subject to bow damage. This explains the fact that almost 50% of damage cases are located in the bow area (90–100% of L). If one excludes the striking ship cases, it can be stated that the struck ship is most exposed at both ends and less in the midship area.

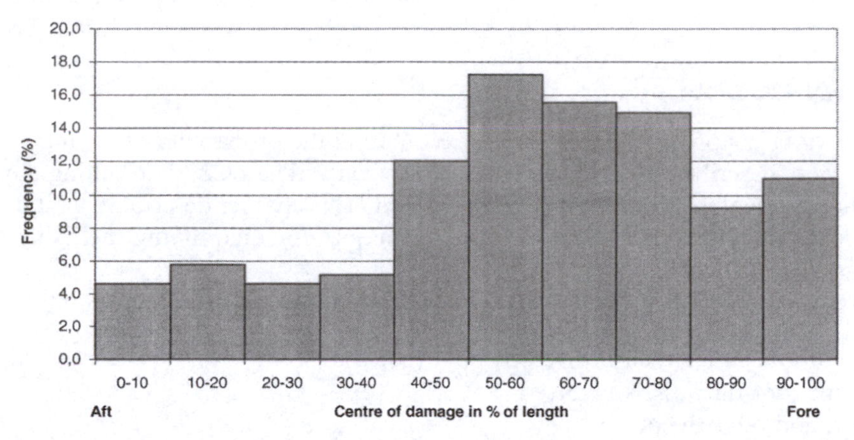

Figure 7.1. Distribution of longitudinal centre of grounding damage, 174 accidents in the Baltic area, 1960–69. (Sources: Kostilainen, 1971; Kostilainen and Hyvärinen, 1976.)

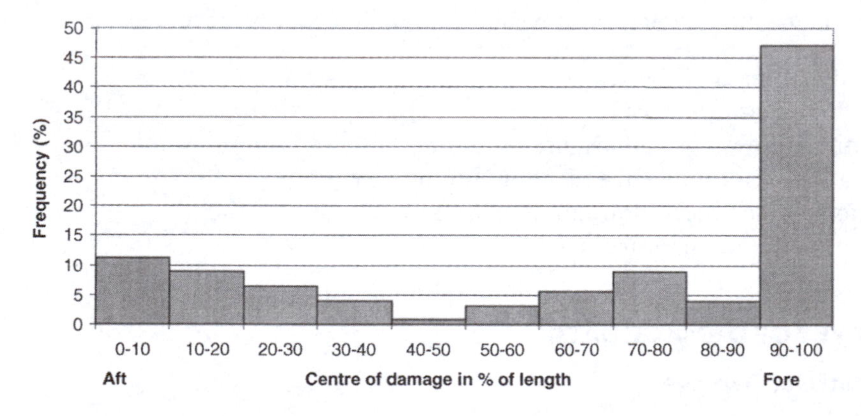

Figure 7.2. Distribution of longitudinal centre of collision damage, 125 accidents in the Baltic area. (Source: Kostilainen, 1971; Kostilainen and Hyvärinen, 1976.)

Alexandrov (1970) did a similar study in connection with an analysis of the effectiveness of life-saving systems. It was based on a greater database that had been prepared by an IMO Working Group and consisted of 485 collisions and 159 grounding reports (IMO, 1993). The striking vessels were also excluded in this study. The distribution is shown in Figure 7.3. It can be concluded that 74% of the vessels were hit between the bow and 60% of the length aft. This compares well with the previously cited studies.

Finally, the mean values for all accidents were taken together for 10% sections in order to cancel out some of the randomness in the material. The result is shown in Figure 7.4. Seen in perspective, these studies give a mixed picture of the probable damage location.

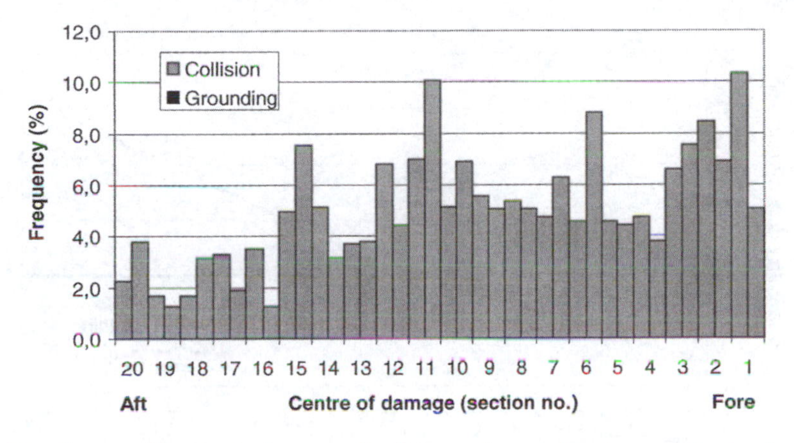

Figure 7.3. Distribution of longitudinal centre of collision and grounding damage. Section length given in 5% of L_{pp}. (Source: Alexandrov, 1970.)

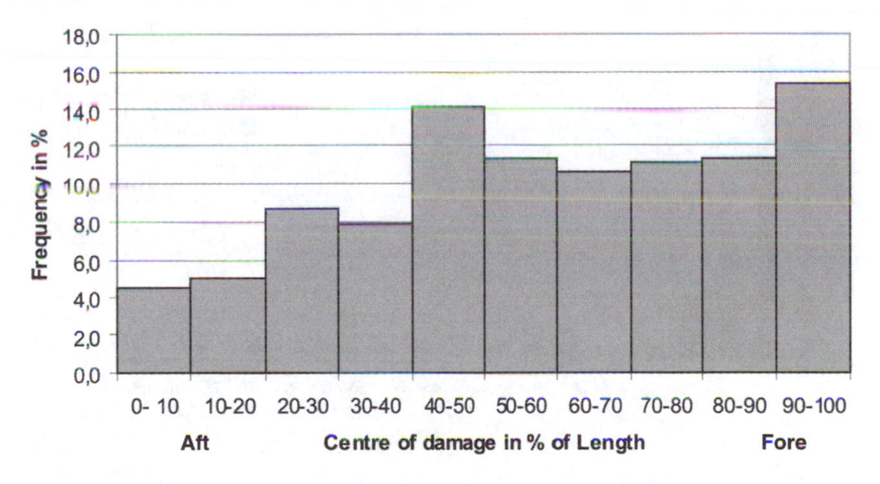

Figure 7.4. Distribution of longitudinal centre of impact damage. Section length given in 10% of L_{pp}. (Source: adapted from data of Alexandrov, 1970.)

This may explain why IMO has chosen simpler models in their regulation for damage and spill estimation for tankers (IMO, 1996). As shown in Figure 7.5, a constant distribution for collision is assumed, which means that all locations have the same probability. For grounding the probability has a maximum at the fore end and decreases moving aft.

7.2.2 Length of Damage

The studies from Baltic waters (Kostilainen, 1971; Kostilainen and Hyvärinen, 1976) referred to earlier also analysed the length of the damage as shown in Figure 7.6.

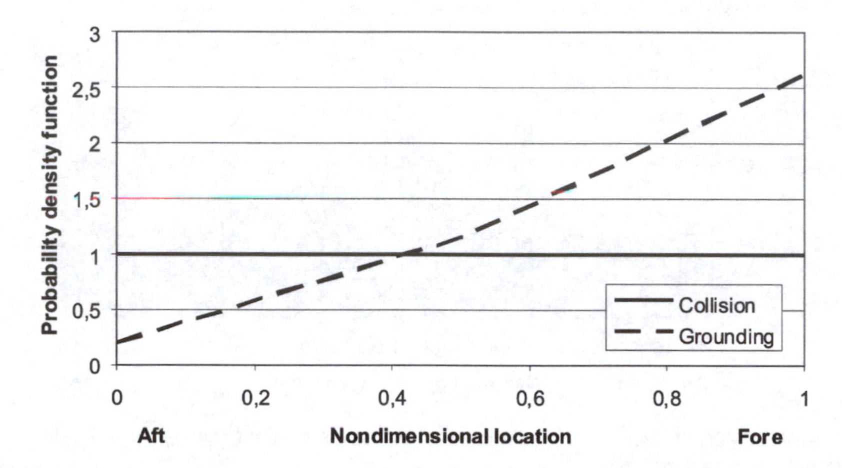

Figure 7.5. Simplified probability density functions (PDF) for damage location. (Source: IMO, 1996.)

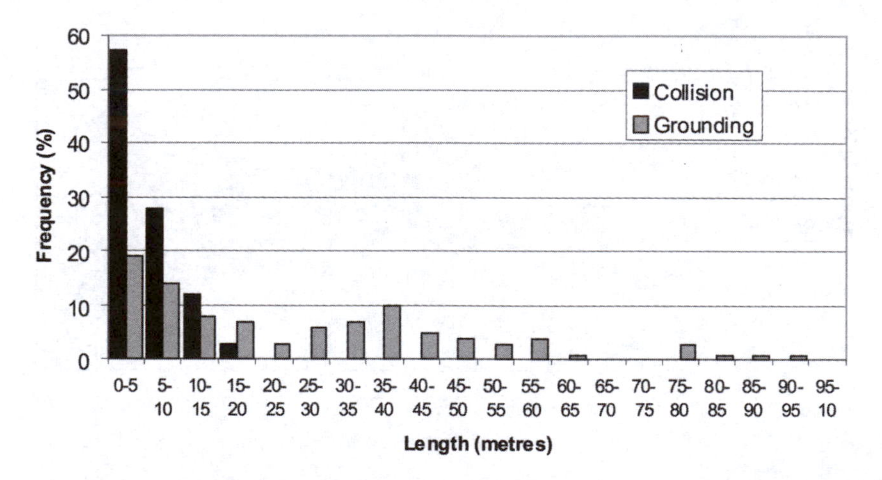

Figure 7.6. Histogram of longitudinal extent of impact damage, impact accidents in the Baltic area, 1960–69. (Sources: Kostilainen, 1971; Kostilainen and Hyvärinen, 1976.)

The longitudinal extent of collision damage was fairly limited and within the range of 20 m. In contrast, a grounding damage length might be much longer as seen from the figure. This might be explained by the fact that grounding damage develops in the longitudinal direction whereas many collisions involve being struck in the transverse direction.

Alexandrov (1970) also studied damage length and the results are shown in Figure 7.7. He found that the length followed more or less the same distribution apart from a few observations of extreme damage for groundings. This is in clear contrast to the previous survey result. The mean damage length was 7.2 m and 6.4 m for collision and grounding respectively.

The MARPOL spill risk model (IMO, 1996) applies more simplified distributions shown in Figure 7.8 as estimates for the damage length. However, it confirms that collision damage is shorter than grounding damage in extent.

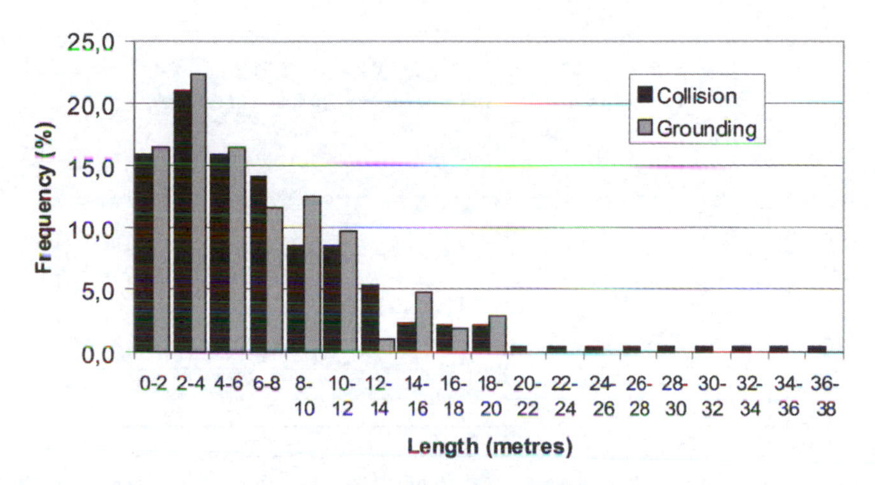

Figure 7.7. Histogram of longitudinal extent of impact damage. (Source: Alexandrov, 1970.)

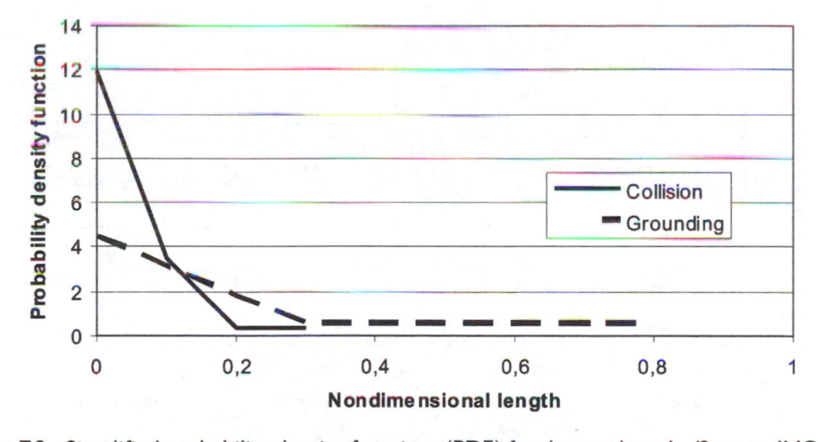

Figure 7.8. Simplified probability density functions (PDF) for damage length. (Source: IMO, 1996.)

7.2.3 Damage Penetration

The extent of penetration has also been analysed in a number of studies. We will here restrict ourselves to refer the recommended design criteria given by IMO (1996) for the estimation of tanker cargo spillage (see Figure 7.9). It can be seen that the probability of

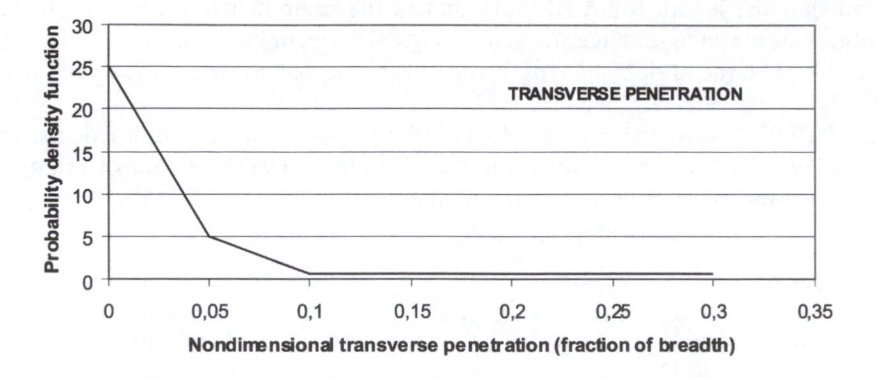

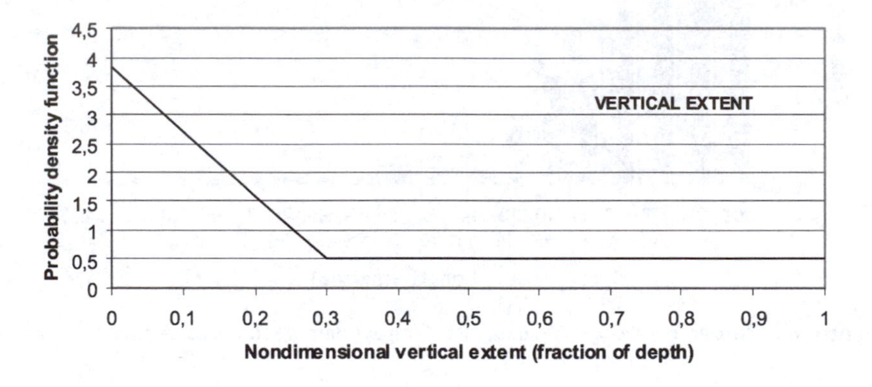

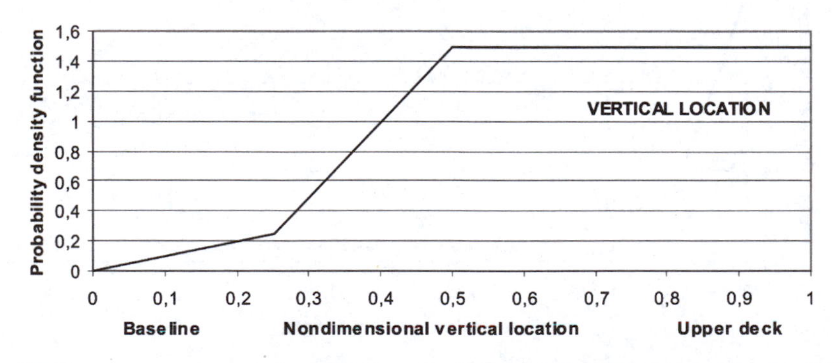

Figure 7.9. Transverse and vertical penetration of collision damage for tankers. (Source: IMO, 1996.)

having a sideways penetration greater than 10% of the ship breadth is fairly low and 75% of the cases have less than 5% penetration. The side of the hull is somewhat more exposed and the cumulative probability of having a damage less than 75% of the ship's depth is 75%. The mean location is 67% of the depth above the keel. This further means that damage will be restricted to an area above the waterline and be fairly limited.

The estimation of vertical and lateral extent of grounding damage for tankers has also been analysed by IMO (1996) and is outlined by probability density functions in Figure 7.10. It can be shown that 78% of the groundings have a vertical penetration less than 10% of the depth. The lateral extent of the damage is obviously greater than for collisions as the bottom area is most exposed. The mean transverse extension is 50% of the ship's breadth. The transverse location follows a uniform distribution which means that all locations have the same probability.

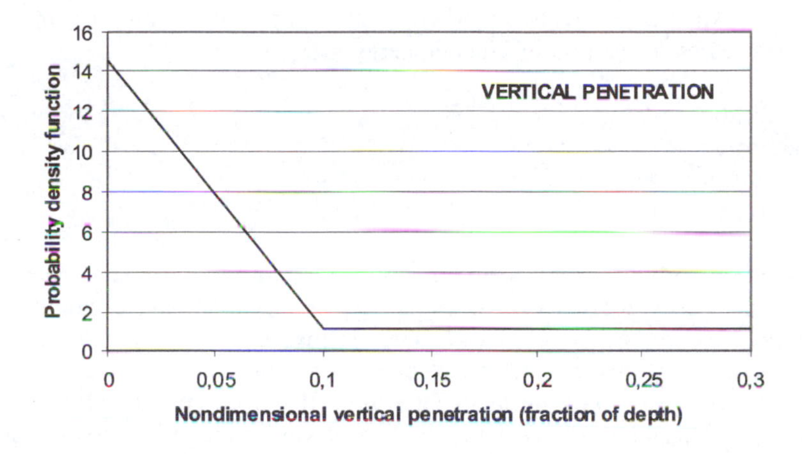

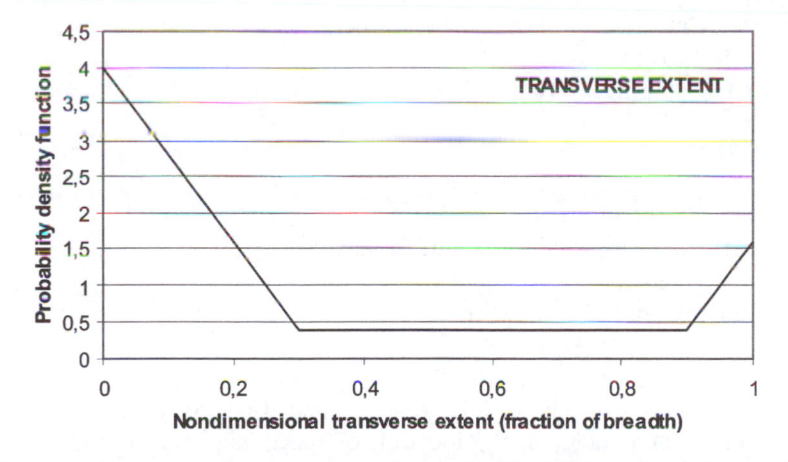

Figure 7.10. Vertical and transverse penetration of grounding damage for tankers. (Source: IMO, 1996.)

7.3 ESTIMATION OF IMPACT ENERGY

7.3.1 Energy Transformation

A critical factor in the modelling of contact damage is the associated impact energy. General considerations of the loss of kinetic energy and how this energy is transferred to plastic deformation of the ship's hull are the basic task of this section. Before the details of grounding and collision accidents are presented, some general parameters for the calculation of impact damage are outlined.

The energy transformation of an impact is dependent on some characteristics of the accident. If the impact forces act through the mass centre of the ship, a central impact occurs. Then the impact does not result in rotation of the ship. If the impact forces do not act through the mass centre of the ship, the impact is described as an eccentric impact. Then some of the kinetic energy in the ship's initial impact direction is transformed to rotation energy of the ship's hull.

The loss of kinetic energy is absorbed in the ship's hull construction. The basic absorption processes of the loss of kinetic energy are:

- Global vibrations
- Local vibrations
- Elastic deformation
- Plastic deformation

In this kind of study the events are generally divided into high-impact energy accidents and low-impact energy accidents. Low-impact energy accidents are accidents where the hull stays intact. Hence, the loss of kinetic energy is absorbed by the membrane strength of the hull. The damage caused by such accidents is minor in relation to high-impact energy accidents. Low-energy impact accidents are not, therefore, discussed further in this chapter. In high-energy impact accidents the loss of kinetic energy is mostly absorbed by plastic deformations of the hull, bulkheads and decks. The membrane strength of the hull has inconsiderable effects. The loss of kinetic energy in a given direction is given by:

$$\Delta\vec{E}_K = \tfrac{1}{2}\cdot\vec{m}_b\cdot\vec{v}_b^2 - \tfrac{1}{2}\cdot\vec{m}_a\cdot\vec{v}_a^2$$

where:

ΔE_K = Loss of kinetic energy
m = Total mass (mass of ship + added mass)
v = Impact velocity
b = Immediately before impact
a = Immediately after impact

The lost kinetic energy is mostly transformed to plastic deformations. These deformations are concentrated in the location of the impact and the impact resistance of the ship's hull.

7.3.2 Energy Transfer in Collision

The collision extent is characterized by several parameters. The main parameters affecting the damage extent are:

- Structural characteristic of struck ship.
- Structural characteristic of striking ship.
- Mass of struck ship at the time of collision.
- Mass of striking ship at the time of collision.
- Speed of struck ship at the time of collision.
- Speed of striking ship at the time of collision.
- Relative course between the striking and struck ship.
- Location of damage relative to ship's length.

In order to quantify the loss of kinetic energy in a collision, a close to right-angled collision where the struck ship is hit at about the centre is considered (Figure 7.11). In this collision situation it may be assumed for simplicity that any yaw movement of the struck ship is not taken into consideration. Minorsky (1959) has proposed that the speed of the striking ship perpendicular to the struck ship's centreline is expressed as follows based on the principles of conservation of momentum:

$$m_1 \cdot v_1 \cdot \sin\alpha = (m_1 + (1 + C_h) \cdot m_2) \cdot v$$

where:

m_1 = The striking ship's mass
v_1 = The striking ship's speed
m_2 = The struck ship's mass
C_h = Added mass coefficient of the struck ship
v = Joint speed perpendicular to struck ship after collision

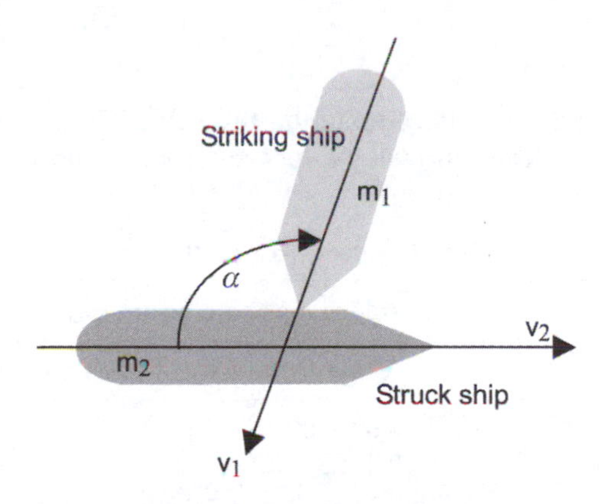

Figure 7.11. Collision scenario.

If we assume that the ship only loses kinetic energy in a right angle to the struck ship, the loss in kinetic energy can be estimated by:

$$\Delta E_k = \tfrac{1}{2} \cdot m_1 \cdot (v_1 \cdot \sin\alpha)^2 - \tfrac{1}{2} \cdot (m_1 + (1 + C_h) \cdot m_2) \cdot v^2$$

The common speed of the two ships after the collision, v, is at a right angle to the struck ship and can be found by the equation of motion. Inserted into the equation above, the loss of kinetic energy is given by:

$$E_I = \tfrac{1}{2} \cdot m_1 \cdot (v_1 \cdot \sin\alpha)^2 - \tfrac{1}{2} \cdot \frac{m_1^2 \cdot (v_1 \cdot \sin\alpha)^2}{m_1 + m_2(1 + C_h)}$$

The resulting expression is as follows:

$$E_I = \frac{m_1 \cdot m_2(1 + C_h)}{2(m_1 + m_2(1 + C_h))} \cdot (v_1 \cdot \sin\alpha)^2 \tag{7.1}$$

In this equation, only the added mass of the struck ship is considered. In a collision there are two ships involved. Hence, the impact energy will be distributed on these ships. The transferred energy is dependent on the total mass and in a central impact the following energy is transferred to the struck ship:

$$E_{t2} = E_I \cdot \left(\frac{1}{1 + m_1/m_2}\right) \tag{7.2}$$

where E_{t2} = energy transferred to struck ship and E_I = lost kinetic energy.

When the ship is retarded, both the mass of the ship and some of the surrounding water of the ship is retarded. Hence the loss of kinetic energy is not only related to the mass of the ship but also to the virtual added mass. The added mass is a function of magnitude, duration and the direction of the retardation. This aspect is discussed by Minorsky (1959), who proposed the following value for the struck ship:

$$C_h = 0.4$$

Zhang (1999) has made a reassessment of the added mass on the basis of recent investigations and proposes the following values given in Table 7.1.

Table 7.1. Added mass coefficient

Motion mode	Added mass coefficient (C_h)	
	Range	Proposed
Surge	0.02–0.07	0.05
Sway	0.4–1.3	0.85
Yaw	0.21	0.21

7.3.3 Estimation of Collision Energy

Zhang (1999) has in his Ph.D. thesis given an extensive analysis of collision energy and associated hull damage. He has developed numerical models with a complexity beyond the scope of our discussion here, but it might be interesting to assess the feasibility of simplified models with reference to his data. Let us take a look at a case involving a collision between two similar container ships with a displacement of 25,500 tonnes and speed 4.5 m/s. Zhang (1999) studied the effect on loss of kinetic energy for different meeting angles and impact points along the length of the hull of the struck vessel. Figure 7.12 gives a summary of the computational results. For collision angles in the range from 60° to 120° the energy loss has a maximum when hitting the midship section. The maximum energy loss might be in the order of 60% of the total kinetic energy. For collision angles of 120° and higher, one approaches a head-on collision that gives the highest energy loss. For a collision angle of 150° a maximum energy loss is also found for impacts at the bow and represents in this case nearly 80% of the kinetic energy.

A Numerical Example

Problem

We have a collision situation for two similar supply ships with a displacement of 4000 tonnes and speed 4.5 m/s. The striking ship hits the other at midship and an angle of 90°. How much energy is absorbed by the struck vessel?

Solution

With the given collision configuration the struck ship will be subject to sway motion and we can then assume an added mass coefficient of 0.85.

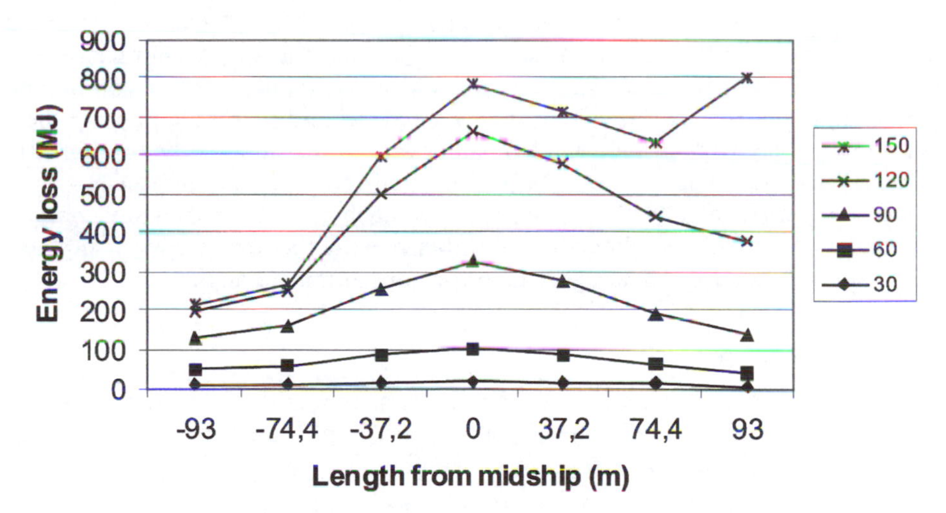

Figure 7.12. Energy loss in collision between two similar container ships. Displacement, 25,500 tonnes; speed, 4.5 m/s; friction coefficient, 0.6; collision angle, 30° to 150°. (Source: adapted from Zhang, 1999.)

By applying Eq. (7.1) we get the following estimate for lost kinetic energy:

$$E_I = \frac{4000 \times 4000 \times (1 + 0.85)}{2(4000 \times 4000 \times (1 + 0.85))} \times (4.5 \times \sin 90°)^2$$
$$= 38.9\,\text{MJ}$$

Numerical simulation by Zhang (1999) gave 35.3 MJ, which represents a difference of only 10%. (See also Figure 7.12.) The result should also be seen in the light of the uncertainty related to the effect of added mass.

As pointed out earlier, the model of Minorsky (1959) was based on the simplification of just taking the impact normal to the longitudinal axis of the struck ship into consideration. It is therefore at risk of underestimating the impact energy. In the following section we will adjust Minorsky's model somewhat for three basic collision scenarios, namely hitting the midship section at angles of 60°, 90° and 120°.

As a case study we use a collision between two container ships each with a displacement of 25,206 tonnes and speed 4.5 m/s. Zhang (1999) has made extensive numerical calculations on the case and will be used as a reference. The computations are summarized in Table 7.2.

First we look at the normal impact ($\alpha = 90°$). The impact energy on to the struck vessel is computed straightforwardly. The second element is the sway motion of the striking vessel due to the fact that the struck vessel is moving normal to the striking vessel during the impact. The struck vessel therefore, so to speak, attacks the bow of the striking vessel. The force takes the friction into consideration. The sway component to the struck vessel is, however, almost twice as large as the yaw component. The total energy lost is 241 MJ, which compares well with the more exact numerical estimate of Zhang (1999) which is 223 MJ.

The second case is the crossing with an angle of 60°. The relative impact speed of the striking vessel is slightly reduced due to the fact that the struck ship has a motion component in the same direction. The yaw component is computed by first computing the component acting normal to the side of the struck vessel and then decomposing the part acting normal to the striking ship. The estimate is exactly the same as the result of Zhang (1999). The lost kinetic energy is only 36% compared to the normal impact.

At a crossing angle of 120° the relative speed is much greater for the sway component due to the opposing motion directions. Combined with the yaw component, the lost kinetic energy is 375 MJ or 56% higher than for normal impact angle.

7.3.4 Collapsed Material

The pioneering work on the analysis of impact damage was done by Minorsky (1959). Based on the analysis of 26 full-scale collision cases, he proposed the following relation between absorbed energy and damaged hull material:

$$E_I = 47.2 \cdot V_C + 32.8 \tag{7.3}$$

Table 7.2. Estimation of lost kinetic energy in collision between two similar container ships of 25,206 tonnes displacement and speed 4.5 m/s: comparison with Zhang (1999, Table 2.7)

Situation	Computation of lost kinetic energy
Sway: $C_h = 0.6$ Yaw: $C_h = 0.21$ Friction coefficient: $\mu = 0.6$	
Angle: $\alpha = 90°$	Sway of struck vessel: $E_I = \frac{25{,}206 \cdot 25{,}206 \cdot (1+0.6)}{2(25{,}206 + 25{,}206 \cdot (1+0.6))} \cdot (4.5 \cdot \sin 90°)^2 = 157\,\text{MJ}$
	Yaw of striking vessel: $E_I = \frac{25{,}206 \cdot 25{,}206 \cdot (1+0.21)}{2(25{,}206 + 25{,}206 \cdot (1+0.21))} \cdot (4.5 \cdot \sin 90°)^2 \cdot 0.6 = 84\,\text{MJ}$
	Total: $E_I = 157 + 84 = 241$ MJ Zhang: $E_I = 223$ MJ
Angle: $\alpha = 60°$	Sway of struck vessel: $E_I = \frac{25{,}206 \cdot 25{,}206 \cdot (1+0.6)}{2(25{,}206 + 25{,}206 \cdot (1+0.6))} \cdot (4.5 - 4.5 \cdot \cos 60°)^2 = 39\,\text{MJ}$
	Yaw of striking vessel: $E_I = \frac{25{,}206 \cdot 25{,}206 \cdot (1+0.21)}{2(25{,}206 + 25{,}206 \cdot (1+0.21))} \cdot (4.5 \cdot \sin 60° \cdot \sin 60°)^2 \cdot 0.6 = 47\,\text{MJ}$
	Total: $E_I = 39 + 47 = 86$ MJ Zhang: $E_I = 86$ MJ
Angle: $\alpha = 120°$	Sway of struck vessel: $E_I = \frac{25{,}206 \cdot 25{,}206 \cdot (1+0.6)}{2(25{,}206 + 25{,}206 \cdot (1+0.6))} \cdot (4.5 + 4.5 \cdot \sin 120°)^2 = 354\,\text{MJ}$
	Yaw of striking vessel: $E_I = \frac{25{,}206 \cdot 25{,}206 \cdot (1+0.21)}{2(25{,}206 + 25{,}206 \cdot (1+0.21))} \cdot (4.5 \cdot \sin 120° \cdot \sin 120°)^2 \cdot 0.6 = 21\,\text{MJ}$
	Total: $E_I = 354 + 21 = 375$ MJ Zhang: $E_I = 338$ MJ

where E_I = absorbed collision impact energy (MJ) and V_C = collapsed material volume of the hull (m^3).

This equation is only valid for high-impact energy accidents (over 50 MJ). The model is not applicable for bow crushing. Depending on which part of the hull is hit and the impact energy, the hull is subject to plastic tension, crushing and folding, and tearing.

Zhang (1999) has analysed the relation between absorbed energy and penetration on the basis of an elaborate numerical model. His estimate are summarized in Table 7.3. Applying regression analysis on these data, the following expression for the penetration was found:

$$L_p = 2.67 \cdot \ln E_I - 1.97 \cdot \ln\left(\frac{m_2}{1000}\right) + 1.66 \tag{7.4}$$

Table 7.3. Absorbed energy as a function of penetration for selected struck ships

L_p (m)	Struck vessel					
	Ro-Ro 15,800 t	Ro-Ro 27,000 t	Ferry 16,073 t, 0 m/s	Ferry 16,073 t, 4 m/s	Tanker 100′ tdw	Tanker 293′ tdw
2	8	10	8	7	37	70
3.5	20	25	20	14	70	150
5	35	45	40	23	108	275
8				50		

Source: Zhang (1999).

Case Study

Problem

Will a segregated ballast tanker (SBT) withstand penetration of the inner cargo tank after collision with a container vessel? The container vessel is at reduced speed whereas the tanker is at standstill.

Data

Container vessel: 186 m length, 25,000 t mass, 4.5 m/s
SBT tanker: 40,000 t dw, 50,000 t mass
Breadth of segregated ballast tanks: 5.5 m

Assumption

The tank vessel is hit at midship with a 90°crossing angle.

Analysis

The tanker is only subject to sway motion.

Estimation of impact energy (see Table 7.2):

$$E_I = \frac{25{,}000 \cdot 50{,}000 \cdot (1 + 0.6)}{2(25{,}000 + 50{,}000 \cdot (1 + 0.6))} \cdot (4.5 \cdot \sin 90)^2 = 193\,\text{MJ}$$

The energy absorbed by the SBT tanker is given by Eq. (7.2):

$$E_{t2} = 193 \cdot \left(\frac{1}{1 + 25{,}000/50{,}000} \right) = 129\,\text{MJ}$$

Estimation of penetration depth is based on Eq. (7.4):

$$L_p = 2.67 \cdot \ln 129 - 1.97 \cdot \ln 50 + 1.66 = 6.9\,\text{m}$$

Conclusion

The estimated penetration depth of 6.9 m indicates that the inner cargo tank will be penetrated even at the reduced speed of 4.5 m/s (8.7 knots) as thc breadth of the tank is only 5.5 m.

7.4 STRANDING AND GROUNDING

7.4.1 Absorbed Energy

The estimation of absorbed energy during grounding is more straightforward compared to the collision scenario due to the fact that all kinetic energy is associated with a single vessel. Absorbed energy is given by the following expression:

$$E_I = \tfrac{1}{2}m \cdot (1 + C_h) \cdot V^2 \tag{7.5}$$

7.4.2 The Grounding Scenario

Generally we make a distinction between drift and powered grounding. Drift grounding seldom immediately results in high-energy impacts. The wave action may, however, break down the hull over time. Powered groundings are related to considerably larger impact velocities and consequently have a higher likelihood of extensive damage.

The general description of a stranding or grounding scenario (Figure 7.13) may be described as:

- Lifting the ship against gravity forces.
- Frictional forces generated by rubbing of the hull against the ground.
- Forces involved in the plastic deformation of hull girders.
- Forces involved in the fracture of bottom plating and structural material.

If the gravity forces are large relative to the impact forces, the ship will not be lifted and the bottom will be plastically deformed. The length of the damage has to be past the collision bulkhead in order to reach the cargo tank section. If the impact forces are large

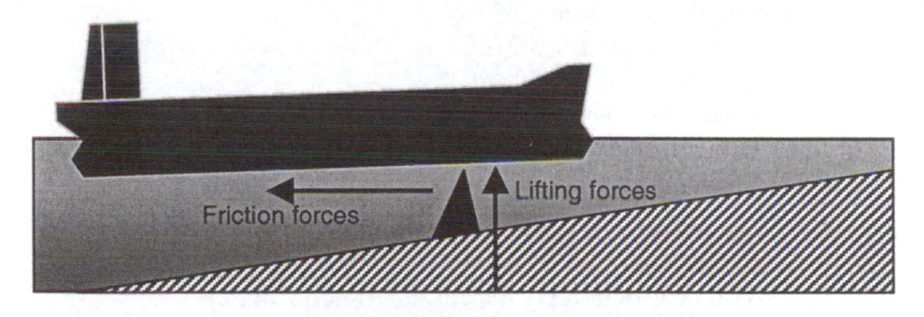

Figure 7.13. Grounding scenario.

relative to the gravity force, the ship is lifted and starts to move over it. The membrane forces may, however, keep the hull intact for some time. As the obstruction moves towards midship the penetration forces will increase. If the membrane forces are large relative to impact forces, the ship may rest or ride over the ground with an intact bottom structure. A more likely scenario, however, is that the impact forces exceed the membrane forces of the hull and the bottom is ruptured. This rupture may occur below the cargo tanks and may result in a cargo spill if the rupture extends to the double bottom.

The stopping distance for the ship hitting a flat bottom is given by (NRC, 1991):

$$x_s = V \cdot \sqrt{\frac{(1 + C_h) \cdot \Delta}{g \cdot \text{TPF} \cdot \sin\alpha \cdot (\mu \cdot \cos\alpha + \sin\alpha)}} \text{ (ft)} \quad (7.6)$$

where:

V = Ship's velocity (knots)
Δ = Ship's displacement (tons)
g = Acceleration of gravity (ft/s^2)
C_h = Hydrodynamic added mass coefficient
α = Inclination of sea floor (degrees)
TPF = Tons per foot immersion
μ = Coefficient of friction (1.2 rocky to 0.4 sand bottom)

This is based on an idealized scenario where the ship stops with the bottom parallel to the seabed and the energy transfer does not include reorientation of the ship.

Example

Problem

A single-hull tanker runs aground with a speed of 11.5 knots. The sea floor is rocky. Estimate the length of the bottom damage.

Data

Vessel data:

Length pp: 304 m = 997 ft
Breadth: 52.4 m = 172 ft
Displacement: 237,000 tons
Water plane area: $A_W = 997 \times 172 \times 0.85 = 145{,}761$ ft^2
TPF = 145,761 × 12/420 = 4165 t/ft
Added mass: 5%

Sea floor:

The sea floor is reasonably even (no protrusions)
Assumed friction coefficient: 0.8 (intermediate)
Elevation 1°

Analysis

The non-dimensional stop length is given by Eq. (7.6):

$$x_s = 11.5 \cdot \sqrt{\frac{(1 + 0.05) \cdot 237{,}000}{32.2 \cdot 4165 \cdot \sin 1^\circ \cdot (0.8 \cdot \cos 1^\circ + \sin 1^\circ)}} = 132\,\text{ft} = 40\,\text{m}$$

Assessment

The relative stopping distance is 132 ft or 13% of the ship's length. This estimate is very sensitive to both to the inclination of the sea floor and the friction coefficient. Both factors may be associated with considerable uncertainty.

7.4.3 Grounding Damage

Zhang (1999) has proposed the following approach for estimating the extent of damage as a result of a grounding. The collapsed hull material is given by:

$$V_C = L_D \cdot B_D \cdot t_{\text{eqv}} \; (m^3)$$

and the absorbed energy is estimated as:

$$E_I = 3.21 \cdot \left(\frac{t_{\text{eqv}}}{B_D}\right)^{0.6} \cdot \sigma \cdot V_C$$

where:

L_D = Length of bottom damage area
B_D = Breadth of damaged area
t_{eqv} = Equivalent bottom plate thickness
σ = Average flow stress = 320 MPa

The equivalent plate thickness takes the contribution of longitudinal stiffeners into consideration. Let us now return to the previous example and estimate the extent of damage during grounding.

Example

Problem

Estimate the length of grounding damage for the single-hull tanker.

Assumptions

Added mass in surge motion: 5%
Bottom plate thickness: 28.5 mm (Zhang, 1999)
Thickness corrected for longitudinal stiffeners: 56.5 mm

Analysis

The grounding impact energy is given by Eq. (7.5):

$$E_I = \tfrac{1}{2} \cdot 237{,}000 \cdot (1 + 0.05) \cdot 5.9^2 = 4331\ \text{MJ}$$

The expected breadth of the damaged area is 17.5% of the breadth based on the distribution given in Figure 7.10, which gives:

$$B_D = 52.4 \times 0.175 = 9.1\ \text{m}$$

We get the following expression for damaged material:

$$V_C = L_D \cdot 9.1 \cdot \frac{56.5}{1000} = L_D \cdot 0.514\ \text{m}^3$$

Finally we get the following expression for absorbed energy:

$$E_I = 4331 = 3.21 \cdot \left(\frac{56.5}{9100}\right)^{0.6} \cdot 320 \cdot L_D \cdot 0.514 = L_D \cdot 25.0$$

The resulting damage is $L_D = 173$ m. Zhang (1999) points out that the actual length was 180 m for a similar vessel involved in a grounding under similar conditions.

Assessment

It is clear that the damage length for grounding on a rock is larger than the stop length on an even and almost flat sea floor. The latter was previously estimated to be 40 m. It is also clear that this estimate is dependent on the assumed breadth of the damaged area, which in this case was based purely on statistical data.

Minorsky's equation cannot be used for grounding accidents because the ship's bottom structure has other damage resistance characteristics than the side ship. A second reason is that the hull deformation in a grounding accident is distributed over a larger part of the hull. An analogous method has, however, been developed by Vaughan (1978). This equation takes the fractured area into account:

$$E_I = 352 \cdot V_C + 126 \cdot A_s\ (\text{MJ})$$

where:

E_I = Absorbed grounding impact energy (tonnes · knots^2)
V_C = Collapsed volume of the hull (m^2, mm)
A_s = Cross-sectional area of indenter (m, mm)

For a single-bottom ship the value of A_s is equal to the product of the bottom plating thickness and the breadth of the indenter. Let us test this model on the previous case. The absorbed energy is given by:

$$E_I = \frac{1}{2} \cdot 237{,}000 \cdot (1 + 0.05) \cdot 45^2 = 22.4 \cdot 10^6 \text{ tonnes} \cdot \text{knots}^2$$

and related to the damage resistance:

$$E_I = 352 \cdot L_D \cdot 9.1 \cdot 56.5 + 126 \cdot 9.1 \cdot 56.5$$

By combining the expressions we get $L_D = 123$ m. This is considerably lower than the previous estimate and casts doubt on the credibility of the model.

7.5 HIGH-SPEED CRAFT (HSC) DAMAGE

The vulnerability of high-speed vehicles has been increasingly focused after a number of serious accidents. It has become apparent that these vessels may be subject to greater damage than is designed for in the IMO HSC Code. The Code is merely assuming damage lengths based on statistics from slow-speed vessels. The Code therefore does not take the higher speed or hull material properties into consideration. Alternative damage models are discussed in a paper by McGee et al. (1999). They refer also to two simplified models. The first one is given by Gallagher (1997), who proposed the following expression for the damage length of HSC after grounding:

$$L_D = \frac{\Delta \cdot V^2}{2.5E \cdot t}$$

where:

S = Damaged length (m)
Δ = Displacement (tonnes)
V = Speed (m/s)
E = Young's modulus (steel 150 to aluminium 75)
t = Plate thickness (mm)

The model is based on the assumption that the vessel has a speed reduction during the accident of 10–15%. The indenting part of the sea floor has a breadth of 0.3 m. It is clear from the expression that the nominator stands for the kinetic energy whereas the denominator represents hull resistance against damage.

Example

Problem

A high-speed craft is grounding with a speed of 45 knots. Estimate the damage length of the aluminium hull.

Data

Displacement of the HSC: 190 tonnes
Speed: 45 knots = 23.2 m/s
Bottom plate thickness: 10 mm

Solution

Damage length:

$$L_D = \frac{190 \cdot 23.2^2}{2.5 \cdot 75 \cdot 10} = 54.5\,\text{m}$$

Assessment

The result indicates that the whole bottom length is opened as the length of the vessel is only 45 m. A more advanced numerical model by McGee et al. (1999) gave a damage length of 40.3 m, indicating that the simplified model may be seen as conservative.

7.6 DAMAGE CONSEQUENCES

An impact accident and the associated puncture of the hull will lead to water ingress and possibly also outflow of oil cargo and bunkers. This will, in the next phase, lead to loss of buoyancy and stability, and ultimately sinking and capsize. It is beyond the scope of this book to analyse the hydrostatic aspects of an impact accident. We will, however, in the following paragraphs look at the spill consequences of a hull penetration (Figure 7.14).

7.6.1 Spill Volume

The maximum oil spill potential is equal to the volume of oil in the penetrated cargo tanks. This volume is generally 98% of the nominal tank volume. Given damage of the side of the

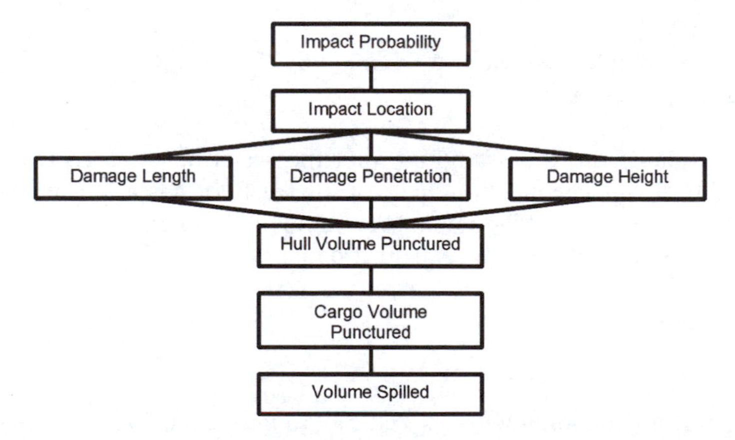

Figure 7.14. General tanker spill analysis.

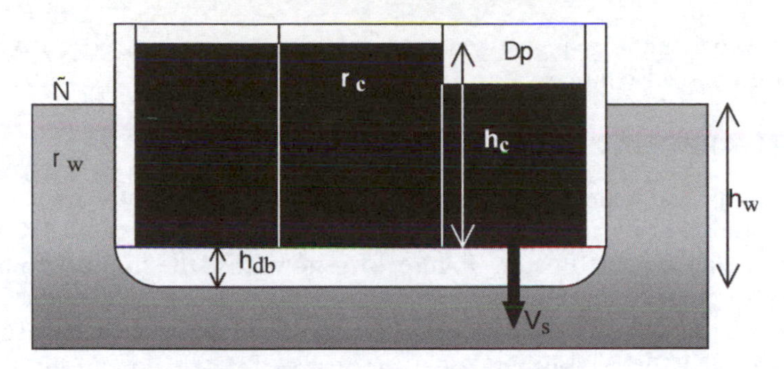

Figure 7.15. Hydrostatic equilibrium.

hull resulting from a collision, it can be assumed that all the oil in the penetrated tanks is released. A bottom damage does, however, introduce another mechanism. As the oil flows through the bottom of the penetrated tank, the pressure at the tank bottom is reduced until a hydrostatic equilibrium pressure is reached. At this stage further oil release is possible only if the draught of the ship is changed.

The spilled volume in the equilibrium state (Figure 7.15) relative to the tank capacity can be expressed as:

$$V_s = 1 - \frac{h_w}{h_c} \cdot \frac{\rho_w}{\rho_c} \ (\%)$$

where:

h_w = Height of water column (≈ draught T)
h_c = Height of cargo column
ρ_w = Water density (seawater = 1.025 t/m^3)
ρ_c = Cargo density (oil ≈ 0.86 t/m^3)

If the penetration is through the double bottom, the initial pressure in the cargo tank and the height of the double bottom has to be considered. The cargo volume spilled is:

$$V_S = 1 - \frac{(\rho_w \cdot (h_w - h_{db}) \cdot g - 100 \cdot \Delta p)}{\rho_c \cdot h_c \cdot g} \ (\%)$$

where:

h_{db} = Height of double bottom
g = Acceleration of gravity = 9.81 m/s^2
Δp = Tank overpressure due to inert gas (≈ 0.05 bar)

By applying the expression to a double-hull tanker design it is found that the spilled cargo volume is in the order of 20% for each penetrated tank section in the case of bottom damage. Tide, waves and mixing of seawater in the remaining cargo oil will over time increase the total spilled volume.

7.7 BOSTON HARBOUR COLLISION RISK STUDY

To illustrate a simplified approach to spill risk estimation, we will show part of a risk study for Boston harbour (Barlow and Lambert, 1979). Liquefied natural gas (LNG) tankers are delivering gas to a harbour at Boston. Figure 7.16 shows that the ship has to pass through a narrow fairway close to densely built-up areas. This, seen in association with the fact that the cargo of a 80,000 m^3 LNG tanker corresponds to 30 nuclear Hiroshima bombs, brought up the need for establishing safety measures for the operation. The objective of this study was to reduce the risk of collision accidents in the harbour area. For this purpose the fairway was divided into four sections.

The main accident phenomena that may lead to release of cargo were:

- Sabotage
- Natural disaster
- Ship-related accidents such as fire
- Ship collisions
- Struck by plane
- Stranding or grounding

As the objective of the investigation was to assess safety measures to reduce risk of collisions, only this accident type was studied. 72 different causes were identified that could contribute to a spill accident. The top section of the fault tree is shown in Figure 7.17.

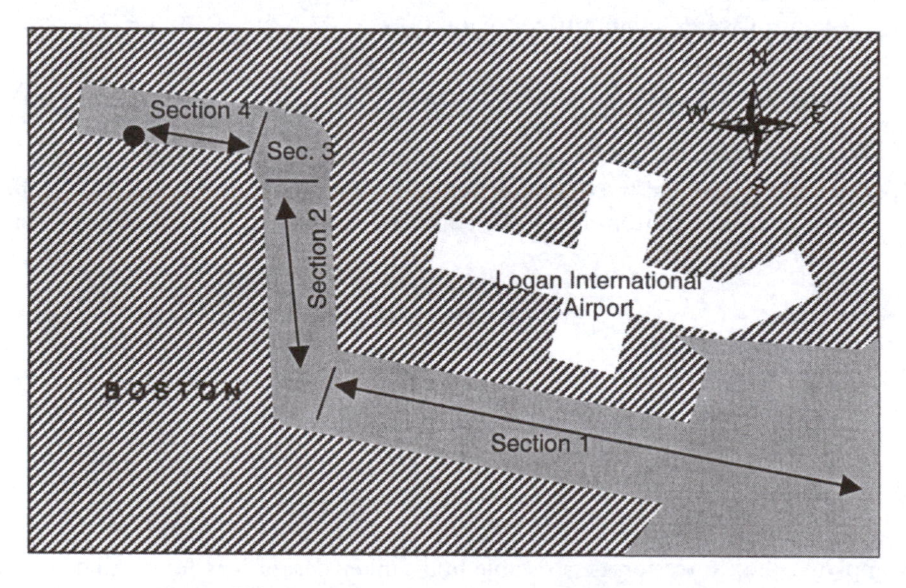

Figure 7.16. Simplified description of Boston harbour fairway.

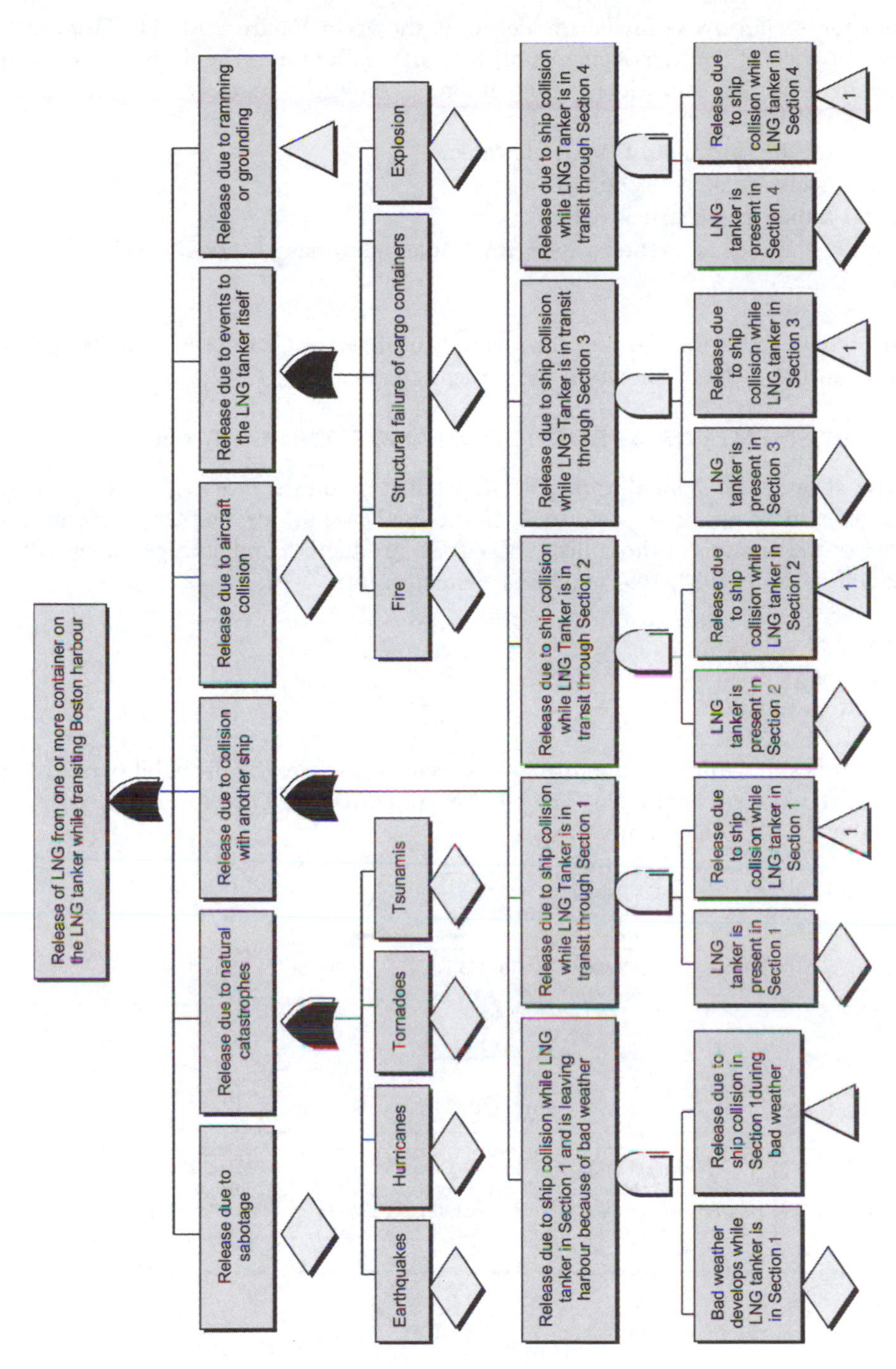

Figure 7.17. Top of a fault tree for estimation of the probability of LNG release in Boston harbour.

The collision scenario was further modelled as shown in Figure 7.18. This approach was applied for each of the four segments of the route leading into the harbour. A cargo spill is dependent on the realization of four different conditions, namely:

1. A ship enters the safe area of the LNG tanker.
2. The two vessels are on collision course.
3. The LNG tanker is the struck vessel.
4. The striking vessel is greater than 1000 tons and has a critical velocity and collision angle.

Items 1 through 3 can be estimated by the methods outlined in Chapter 6, whereas item 4 can be based on the methods shown in the present chapter.

7.8 RISK ASSESSMENT FOR A PRODUCTION AND STORAGE TANKER

For any type of operation related to petroleum production on the Norwegian continental shelf a risk assessment has to be performed. In the final case in this chapter a description will be given of the analysis of the collision risk for a production and storage tanker (PS). The assessment (Dahle, 1988, 1992) involves estimation of:

- Probability of collision
- Collision energy
- Loss of safety function

The assessment is summarized in the form of an event tree where the probability of losing critical, safety functions is computed. The Norwegian petroleum Directorate (NPD) gave a limiting value of 10^{-4} on an annual basis.

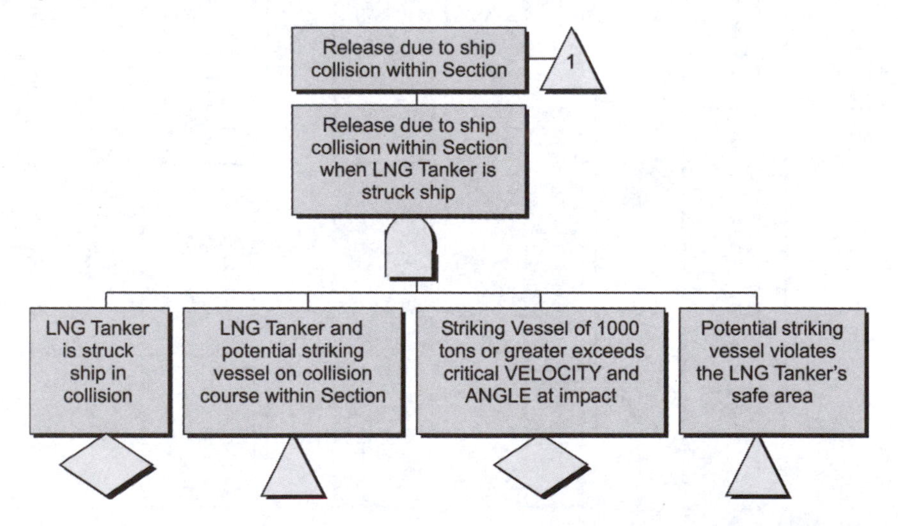

Figure 7.18. Release due to ship collision branch. (Source: Barlow and Lambert, 1979.)

Case Study

Problem

The following example is a part of an actual analysis called 'Risk Assessment of Production and Storage Tankers'. The analysis consists of a traffic modelling part and a damage modelling part in addition to calculations of the effect of implementation of an evasive manoeuvre for the PS in a collision situation. In this presentation the damage modelling part is presented without consideration of an evasive manoeuvre. Based on the results from the traffic modelling, the damage modelling is performed.

Solution

Estimate the collision damage for the production ship (PS). A consequence of a collision with the PS may be a spill of oil from some of its tank sections. Oil may ignite and cause further accident escalation by putting shelter areas, escapeways and structure at risk. The likelihood of these scenarios has to be estimated.

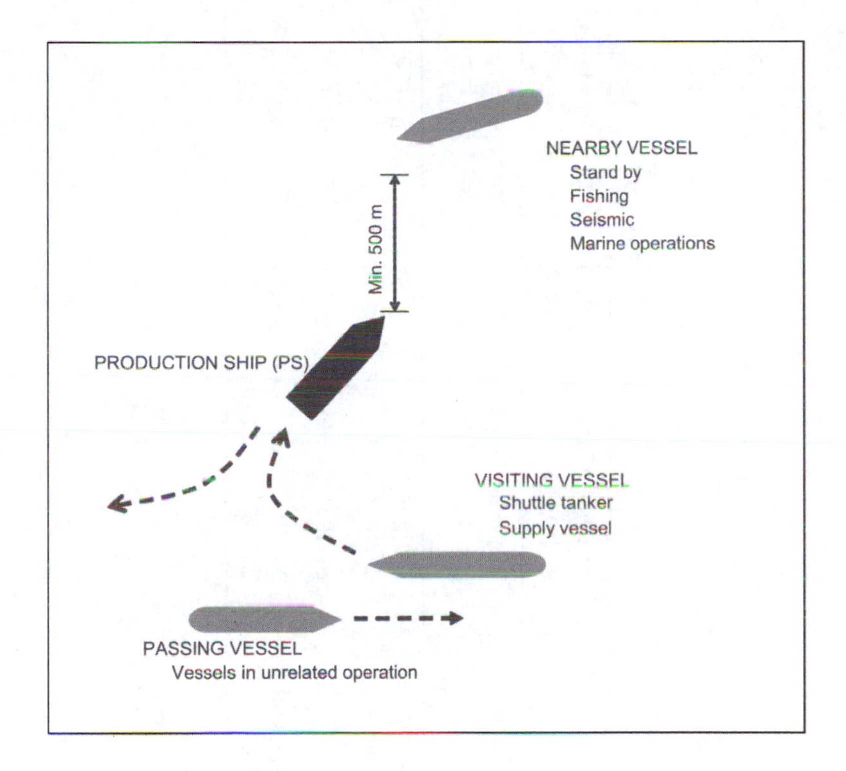

Based on the traffic above, collision probabilities for each traffic segment can be estimated (Table 7.E1).

Based on the computations above, the total probability (random course situations and drifting situations) of collision is estimated to $84.5 \cdot 10^{-4}$.

Table 7.E1

Vessel type		Annual collision probability P_a		Total time within zone per year (hrs)	Conditional probability of impact speed given a collision situation P_i			
		Random	Drifting		Random		Drifting	
					Low	High	Low	High
Visiting	Supply	$15 \cdot 10^{-4}$	$6.2 \cdot 10^{-4}$	55	0.04	0.01	0.1	0
	Shuttle tankers	$30 \cdot 10^{-4}$	$3.1 \cdot 10^{-4}$	110				
Nearby		$26 \cdot 10^{-4}$	$9.9 \cdot 10^{-5}$	1055	0.04	0.01	0.04	0.01
Passing	Non-tankers	$2.7 \cdot 10^{-4}$	$1.6 \cdot 10^{-5}$	108	0.15	0.70	0.04	0.01
	Tanker	$2.9 \cdot 10^{-5}$	$2.1 \cdot 10^{-6}$	11				

Assumptions

For simplicity only two speed ranges are used in the study. Velocities between 3 and 6 knots are considered as *low speed* while 6 to 15.5 knots is considered as *high speed.*

The displacement of the PS is 100,000 t. There are large differences in the size of the shuttle tankers and passing ships. It is assumed that 50% of both the shuttle tankers and the passing tankers are in ballast condition and the displacement set to 70,000 t while the other half are fully loaded with displacement of 150,000 t. 95% of the passing non-tankers are small vessels of 5000 t while the rest of the passing non-tankers are assumed to be merchant vessels with mean displacement 40,000 t.

The added mass coefficient for the striking ship is 0.1 while for the production ship it is 0.7. The probability of having a central collision impact is 0.5 and that of a non-central impact (45°–70°) is 0.25; both situations will lead to hull penetration.

Impact energy

There are two basic collision scenarios. The striking ship may lose propulsion power and drift towards the PS or the striking ship can lose navigational control and continue on a random course. The two collision types have quite different characteristics with respect to impact energy and impact geometry. The bow of the PS is heading against the waves. The wind forces are the dominating load on the drifting vessel. As the angle between the wind and waves is relatively small, a central impact is not likely to occur. Consequently the drifting ship may hit the relatively strong bow of the PS and slide along its side. In this case the impact energy will be low due to the small collision angle and low impact speed. Hence the damage will most likely be minor and local. The drifting collisions are therefore not analysed further.

The probability of having a powered collision is calculated in Table 7.E2.

A ship sailing on a random course will more likely strike the PS in a central impact given the alternative meeting angles. In addition the impact speed will be higher and therefore cause more damage then the drifting vessels. Based on Eq. (7.1) and the assumed added mass coefficients, the fraction of the kinetic energy transferred in plastic deformation of the PS's structure is calculated for all the ships and for both speed categories. The energy transferred to the production ship (Table 7.E3) is calculated by Eq. (7.2).

Structural damage

Based on the collision geometry and the necessary crushed hull volume, the probability of penetrating the hull and cargo tanks is estimated by using Minorsky's equation (7.3) as shown in Table 7.E4. It is assumed that in a central impact about 8.5 m^3 crushed material is necessary to penetrate a cargo tank. Hence the minimum force is 430 MJ. It is further assumed that the energy to penetrate the hull is 40 MJ. This is a crude assumption as the value is outside the valid range for Minorsky's equation. For a non-central impact the energy necessary is 140 MJ.

It can be seen from Table 7.E4 that high-speed impacts only result in cargo tank penetration. Based on these results, an event tree is developed as shown below. It is

Table 7.E2

Vessel type			Col. 1 Fraction within group	Col. 2 Impact speed	Col. 3 Energy (MJ), equation (7.1)	Col. 4 Speed fraction of time within zone	Col. 5 Collision frequency ($P_a \cdot$ Col. 1)	Col. 6 Collision probability (Col. 5 · Col. 4)
Visiting	Supply		1	Low	11	0.04	$15 \cdot 10^{-4}$	$0.6 \cdot 10^{-4}$
				High	44	0.01		$0.15 \cdot 10^{-4}$
	Shuttle tanker	70,000 t	0.5	Low	154	0.02	$30 \cdot 10^{-4}$	$0.6 \cdot 10^{-4}$
				High	616	0.005		$0.15 \cdot 10^{-4}$
		150,000 t	0.5	Low	330	0.02		$0.6 \cdot 10^{-4}$
				High	1320	0.005		$0.15 \cdot 10^{-4}$
Nearby			1	Low	11	0.04	$26 \cdot 10^{-4}$	$1.04 \cdot 10^{-4}$
				High	44	0.01		$0.26 \cdot 10^{-4}$
Passing	Small 5000 t		0.95	Low	11	0.12	$2.7 \cdot 10^{-4}$	$0.32 \cdot 10^{-4}$
				High	44	0.56		$1.51 \cdot 10^{-4}$
	Merchant 40,000 t		0.05	Low	68	0.03		$0.08 \cdot 10^{-4}$
				High	272	0.14		$0.38 \cdot 10^{-4}$
	Tanker 70,000 t		0.5	Low	154	0.075	$2.9 \cdot 10^{-5}$	$0.02 \cdot 10^{-4}$
				High	616	0.035		$0.01 \cdot 10^{-4}$
	Tanker 150,000 t		0.5	Low	330	0.075		$0.02 \cdot 10^{-4}$
				High	1320	0.035		$0.01 \cdot 10^{-4}$
								$5.9 \cdot 10^{-4}$

Table 7.E3. Impact energy transferred to PS in powered collision

Vessel type			Col. 1 M_{ship} (t)	Col. 2 Impact speed	Col. 3 Kin. energy (MJ)	Col. 4 Transferred energy	Col. 5 Collision energy (MJ)
Visiting	Supply		5000	Low	11	0.97	11
				High	44		43
	Shuttle tanker	Ballast	700,000	Low	154	0.71	109
				High	616		437
		Loaded	150,000	Low	330	0.53	175
				High	1320		700
Nearby			5000	Low	11	0.97	11
				High	44		43
Passing	Small		5000	Low	11	0.97	11
				High	44		43
	Merchant		40,000	Low	68	0.81	55
				High	272		220
	Tanker		70,000	Low	154	0.71	109
				High	616		437
	Tanker		150,000	Low	330	0.53	171
				High	1320		700

Table 7.E4

Vessel type			Impact speed	No. of penetrations		P(no penetration)	P(penetration (outside shell))	P(penetration of cargo tank)
				Central ($P=0.5$)	Non-central ($P=0.25$)			
Visiting	Supply		Low	0	0	$0.6 \cdot 10^{-4}$	0	0
			High	1	0	$0.075 \cdot 10^{-4}$	$0.075 \cdot 10^{-4}$	0
	Shuttle tanker	Ballast	Low	1	0	$0.3 \cdot 10^{-4}$	$0.3 \cdot 10^{-4}$	0
			High	2	0	0	$0.075 \cdot 10^{-4}$	$0.075 \cdot 10^{-4}$
		Loaded	Low	1	1	$0.15 \cdot 10^{-4}$	$0.45 \cdot 10^{-4}$	0
			High	2	0	0	$0.075 \cdot 10^{-4}$	$0.075 \cdot 10^{-4}$
Nearby			Low	0	0	$1.04 \cdot 10^{-4}$	0	0
			High	1	0	$0.13 \cdot 10^{-4}$	$0.13 \cdot 10^{-4}$	0
Passing	Small		Low	0	0	$0.32 \cdot 10^{-4}$	0	0
			High	1	1	$0.75 \cdot 10^{-4}$	$0.75 \cdot 10^{-4}$	0
	Merchant		Low	1	0	$0.04 \cdot 10^{-4}$	$0.04 \cdot 10^{-4}$	0
			High	1	1	$0.09 \cdot 10^{-4}$	$0.29 \cdot 10^{-4}$	0
	Tanker	Ballast	Low	1	0	$0.01 \cdot 10^{-4}$	$0.01 \cdot 10^{-4}$	0
			High	2	0	0	$0.005 \cdot 10^{-4}$	$0.005 \cdot 10^{-4}$
		Loaded	Low	1	1	$0.005 \cdot 10^{-4}$	$0.015 \cdot 10^{-4}$	0
			High	2	0	0	$0.005 \cdot 10^{-4}$	$0.005 \cdot 10^{-4}$
					Sum	$3.51 \cdot 10^{-4}$	$2.22 \cdot 10^{-4}$	$0.16 \cdot 10^{-4}$
					Fraction	0.60	0.38	0.02

assumed that there is no loss of safety functions unless spilled oil is ignited. However, if the spilled oil from at least one cargo tank is ignited, the shelter areas and escapeways are assumed to be lost. If the engine room and one cargo tank is penetrated the PS will sink, or in other words there will be loss of all safety functions.

Conclusion

The initial collision probability was $84 \cdot 10^{-4}$. This value gives a return period of approximately 100 years. Of these collisions, 88% are due to random navigation. By eliminating the collisions with insignificant impact energy the risk is reduced by a factor of 10 to a collision probability of $5.9 \cdot 10^{-4}$.

The Norwegian Petroleum Directorate (NPD) has a maximum accepted value for loss of safety functions equal to 10^{-4}. Neither of the three safety functions has a probability greater than this target. Hence the PS meets the requirement with respect to collisions.

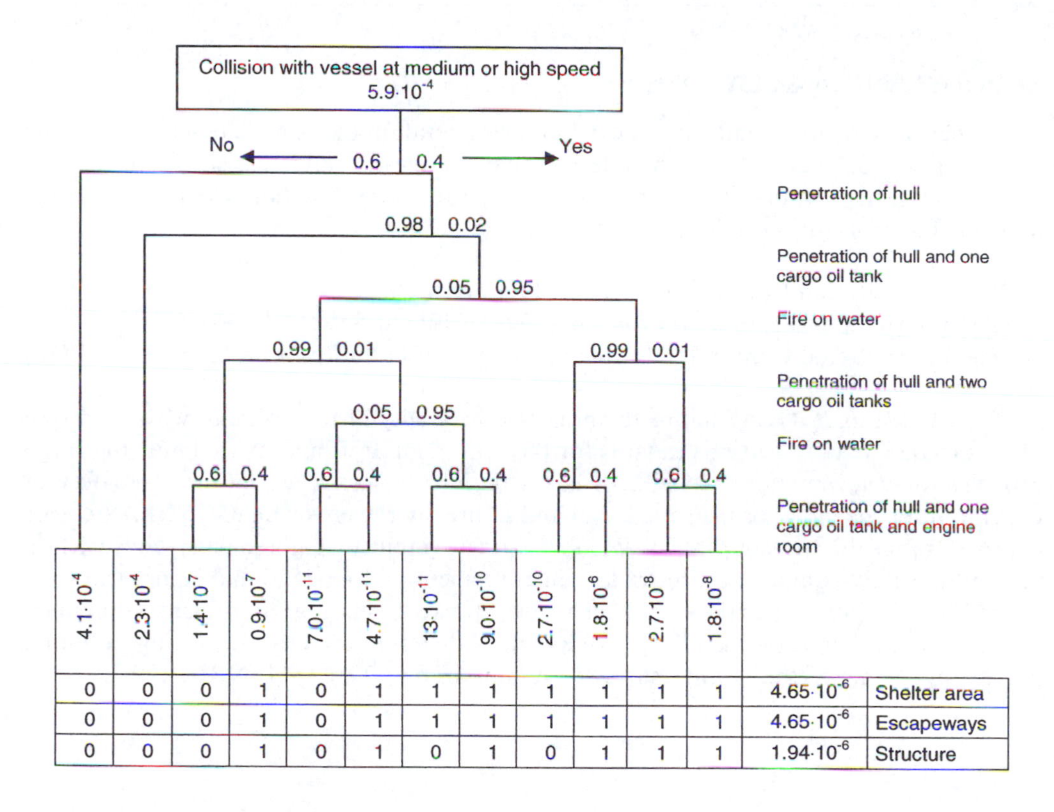

Table 7.4. Elements in MARPOL

Regulation	Description
Segregated ballast tanks (SBT)	Cargo cannot be carried in ballast tanks
Protective location of SBT	SBT are located in ship's side and bottom
Draft and trim requirements	Draft amidships of not less than $2.0 \cdot 0.02 \cdot L$ and trim by stern not greater than $0.015 \cdot L$
Tank size limitations	Maximum tank length 10 m–$0.2 \cdot L$ depending on location. Maximum volume may vary up to 22,500 m^3 for side tanks and 50,000 m^3 for centre tanks
Hypothetical outflow of oil	Maximum hypothetical oil outflow after an assumed damage with specified extents
Subdivision and stability	Subdivision is performed to never give a heel larger than 25° and a final water line above any opening as a consequence of an assumed damage
Crude oil washing (COW)	Cargo tanks are cleaned with oil as washing medium
Inert gas system (IGS)	Lack of oxygen in cargo tanks
Slop tanks	Discharged oil is stored

7.9 RULES AND REGULATIONS

A number of regulations influence the design and operation of a ship. These requirements can roughly be categorized in international, domestic and classification society regulations. This section gives a brief presentation of the international requirements for tankers. The conventions relevant for this chapter are:

- The International Convention on Load Lines (ILLC)
- The International Convention for the Safety of Life at Sea (SOLAS)
- The International Convention for the Prevention of Pollution from Ships (MARPOL)

Both ILLC and SOLAS indirectly influence the ability of the vessel to withstand spill of oil cargo. The ILLC gives standards for the maximum draft and hence limits the cargo capacity for a given ship size. SOLAS has a broader objective: to ensure the safety of the ship itself, its crew, passengers, cargo and indirectly the environment. SOLAS covers a broad range of measures such as subdivisions, stability requirements, construction principles, safety equipment, fire protection equipment and navigational equipment.

MARPOL gives specific requirements for tankers relating to design, equipment and procedures for cargo handling. MARPOL is therefore directly addressing pollution prevention. Some of the key measures are summarized in Table 7.4 (NRC, 1991).

REFERENCES

Alexandrov, M., 1970, Probabilistic approach to the effectiveness of ship life-saving systems. *Transactions of SNAME*, Vol. 14, 391–416.

Barlow, R. E. and Lambert, H. E., 1979, *The Effect of U.S. Coast Guard Rules in Reducing the Probability of LNG Tanker-Ship Collision in Boston Harbour*. May 21, Tera Corporation, Berkeley, CA.

Dahle, E. A., 1992, *Risk Assessment of Production and Storage Tankers*. Det Norske Veritas, Høvik.

Dahle, E. A., 1988, *Damage control on floating platforms* [In Norwegian: *Begrensning av konsekvenser ved skade på flytenede konstruksjoner*]. Lecture Notes, Division of Marine Systems Design, Norwegian University of Science and Technology, Trondheim.

Gallagher, P., 1997, *A Simple Formula for Approximating the Longitudinal Extent of Bottom Damage to High Speed Craft Resulting from Hull Penetrations by Submerged Hazards*. IMO SLF HSC Code Correspondence Group, Document 97-02.

IMO, 1993, *Explanatory Notes for the SOLAS Regulations on Ship Subdivision and Damage Stability of Cargo Ship of 100 Meters and Over*. Resolution A 684(17), Sales number IMO-871E. International Maritime Organization, London.

IMO, 1996, *Interim Guidelines for the Approval of Alternative Methods of Design and Construction of Oil Tankers under Regulation 13 F(5) of Annex I of MARPOL 73/78*, MARPOL 73/78 – 1994 and 1995 Amendments, Sales number IMO-640E. International Maritime Organization, London.

Kostilainen, V., 1971, *Analysis of Casualties of Tankers in the Baltic, Gulf of Finland and Gulf of Bothnia in 1960–1969*. Helsinki University of Technology, Ship Hydrodynamics Laboratory, Report No. 5, Otaniemi.

Kostilainen, V. and Hyvärinen, M., 1976, *Ship Casualties in the Baltic, Gulf of Finland and Gulf of Bothnia in 1971–75*. Helsinki University of Technology, Ship Hydrodynamics Laboratory, Report No. 10, Otaniemi.

McGee, S. P., Troesch, A. and Vlahopoulos, N., 1999, Damage length predictor for high-speed Craft. *Marine Technology*, Vol. 36(4), 203–210.

Minorsky, V. U., 1959, An analysis of ship collisions with reference to protection of nuclear power plants. *Journal of Ship Research*, Vol. 3(2), 1–4.

NRC, 1991, *Tanker Spills Prevention by Design*. National Research Council, National Academy Press, Washington, DC.

Vaughan, H., 1978, Bending and tearing of plate with application to ship-bottom damage. *The Naval Architect*, May, 97–99.

Zhang, S., 1999, The mechanics of ship collisions. Ph.D. Thesis, Department of Naval Architecture and Offshore Engineering, Technical University of Denmark, Lyngby.

8

RISK ANALYSIS TECHNIQUES

The number of rational hypotheses that can explain any given phenomenon is infinite.

(Persig's Postulate)

8.1 INTRODUCTION

The objective of this chapter is to establish a set of tools and techniques that we need to utilize in the process of carrying out a risk analysis and assessment. In order to understand the application, importance and role of these techniques in the context of risk analysis, it is of crucial importance to first gain an understanding of the basic concepts of risk analysis, as well as the underlying components of risk. The first part of this chapter therefore gives a brief introduction to risk analysis and assessment, a concept that is treated in much more detail in later chapters. The second part of the chapter gives some useful basic theory related to system description and structures. Finally, the third and main part of this chapter deals directly with risk assessment techniques. The following five techniques are studied:

- Preliminary Hazard Analysis (PHA)
- Hazard and Operability Studies (HAZOP)
- Failure Mode, Effect and Criticality Analysis (FMECA)
- Fault Tree Analysis (FTA)
- Event Tree Analysis (ETA)

These techniques are utilized in relation to different aspects of risk analysis. The Preliminary Hazard Analysis (PHA) methodology is used to identify possible hazards, i.e. possible events and conditions that may result in any severity. A more extensive hazard identification method is Hazard and Operability Studies (HAZOP), which searches much more systematically for system deviations that may have harmful consequences. The Failure Mode, Effect and Criticality Analysis (FMECA) can be used to identify equipment/system failures and assess them in terms of causes, effects and criticality. The application of an FMECA gives enhanced system understanding as well as an improved basis for quantitative analysis. Fault Tree Analysis (FTA) and Event Tree Analysis (ETA) are the most commonly used methods in terms of establishing the probability of occurrence and the severity of the consequences, for hazards in the context of risk analysis.

8.2 RISK ANALYSIS AND RISK ASSESSMENT

8.2.1 Understanding Risk and Safety

Risk analysis involves analysing a system in terms of its risks. As pointed out in earlier chapters, the concept of risk is central to any discussion of safety. There is a steadily increasing focus on safety in all aspects of life, and in a maritime context risk analysis is nowadays a relatively common investigative and diagnostic element in reviewing system performance with the objective of identifying areas for improvement. Different people tend to understand the term 'safety' differently, and for the sake of this chapter the following definition proposed by Kuo (1997) can be useful: 'Safety is a perceived concept which determines to what extent the management, engineering and operation of a system are free from danger to life, property and the environment.'

As mentioned above, risks and safety are closely linked. But how should we understand the term 'risk'? Risk is a parameter used to evaluate (or judge) the significance of hazards in relation to safety, and as mentioned in the introduction to this chapter, hazards are the possible events and conditions that may result in severity. Risk (R) is normally evaluated as a function of the severity of the possible consequences (C) for a hazard, and the probability of occurrence (P) for that particular hazard:

$$R = f(C, P) \tag{8.1}$$

Both the possible consequences (C) and the probability of occurrence (P) are functions of various parameters, such as human factors, operational factors, management factors, engineering factors and time. It is normal to use the simplest possible relation between C and P, i.e. the product of the two, to calculate the risk (R):

$$R = C \cdot P \tag{8.2}$$

Given this simple equation, we can better understand risk as a concept. For example, a high consequence (C) and a high probability of occurrence (P) for a certain given hazard mean that the risk is high, which will often be considered as intolerable from a safety perspective. On the other hand, a low consequence (C) and a low probability (P) represent a low risk level. A low level of risk will normally be perceived as tolerable in a safety context, but may even be negligible if it is really low. The risk level that results from a high consequence and a low probability, or vice versa, will often be tolerable, but may in extreme cases be either negligible or intolerable. The hazards needing special attention are those where both consequence and probability are significant.

Given this knowledge, estimated risk of hazards can be used to make informed decisions in terms of improving safety. Safety can be improved by reducing the risk, and risks can be reduced by reducing the severity of the consequences, reducing the probability of occurrence, or a combination of the two.

8.2.2 The Risk Analysis and Risk Assessment Process

Risk analysis is the process of calculating the risk for the identified hazards. Experts in this field of study often distinguish between risk analysis and risk assessment. Risk assessment is the process of using the results obtained in the risk analysis (i.e. the risks of hazards) to improve the safety of a system through risk reduction. This involves the introduction of safety measures, also known as risk control options. A principal diagram for the process of risk analysis and risk assessment is illustrated in Figure 8.1.

The first step in the process of risk analysis and risk assessment is to make a problem definition and system description, e.g. to define the vessel and/or the activity whose risks

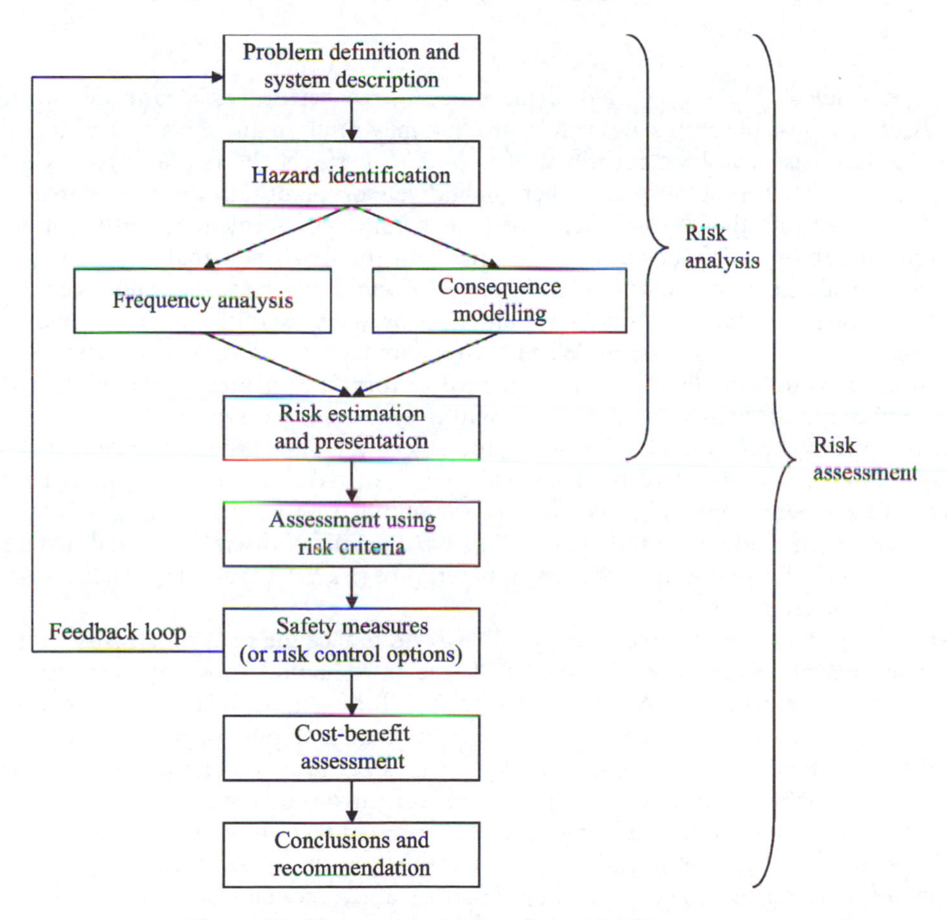

Figure 8.1. The process of risk analysis and risk assessment.

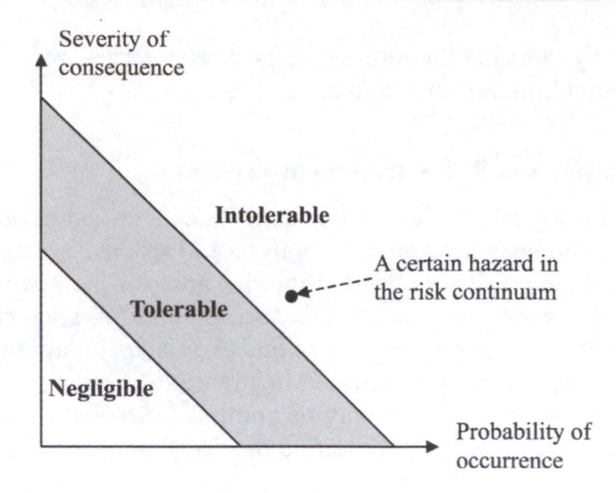

Figure 8.2. Risk presentation using a specific risk acceptance criterion.

are to be studied. The second step of the process is to perform a hazard identification exercise where possible events and conditions that may result in any severity are identified. Once the hazards have been identified, it is time to perform the risk analysis, which is the process of estimating the risks, either qualitatively or quantitatively. First a frequency analysis is used to estimate how likely it is that the different accidents/hazards will occur (i.e. the probability of occurrence). In parallel with the frequency analysis, consequence modelling evaluates the resulting consequences/effects if the hazards really occur. In a maritime context, an accident may have an effect on the vessel, its passengers and crew, the cargo, and/or the environment. When both the frequency and the consequence of each hazard have been estimated, they are combined to form a measures of overall risk. Risk may be presented in many different and complementary forms. Figure 8.2 illustrates the principle of risk presentation using a specific risk acceptance criterion. Figure 8.2 also incorporates an assessment of the hazards in terms of risk, indicating whether they are intolerable (i.e. unacceptable), tolerable (i.e. acceptable) or negligible using continuous risk scales. Often, and particularly in qualitative risk analysis, discrete risk scales are used to assess the relative importance of hazards in terms of risks. An example of such a discrete risk scale is given in Figure 8.3.

In order to make intolerable risks tolerable, or to reduce the risks of hazards to as low a level as reasonably practicable (ALARP), the introduction of safety measures into the system will be necessary. A safety measure may, for example, be the construction and implementation of a marine evacuation system on board a ship. Cost-benefit analysis is a useful tool with regard to assessing safety measures because such an analysis evaluates whether the benefits of such measures justify the costs involved in implementing them. The benefits can be estimated by repeating the risk assessment process with the proposed safety measures in place, thereby introducing an iterative loop into the assessment process as shown in Figure 8.1. Based on the process described above, conclusions may be drawn and recommendations proposed to the shipowner or ship operator, etc.

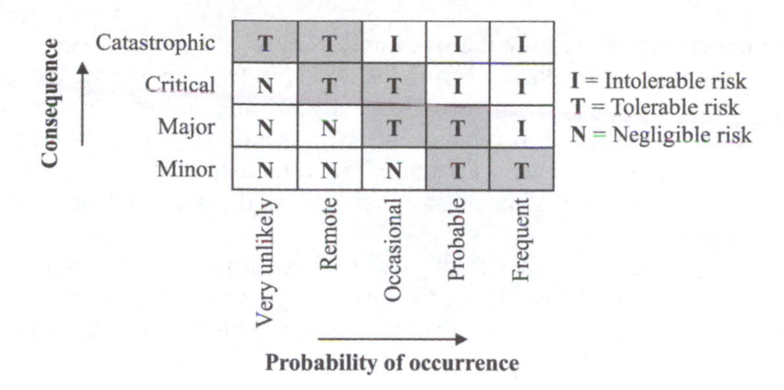

Figure 8.3. Risk presentation using discrete risk scales.

Each of the risk analysis techniques presented later in this chapter can be utilized as tools within the risk analysis and assessment framework presented in Figure 8.1. For example, both Preliminary Hazard Analysis (PHA) and Hazard and Operability Studies (HAZOP) can be used to identify possible hazards. Fault Tree Analysis (FTA) is useful in carrying out the frequency analysis, while Event Tree Analysis (ETA) is a common method used to study possible consequences of hazards.

8.2.3 Analysis Approaches

Risk analysis can be performed quantitatively and/or qualitatively. If any performance is measured with values or by terms in the analysis it is, by definition, a quantitative analysis. A comprehensive and total risk analysis should include the use of both qualitative and quantitative approaches and techniques. The qualitative approach may be included already in the system description phase of the risk analysis process. Both approaches give important and supplementary information about the system.

8.2.4 Required Resources

When performing a risk analysis and risk assessment, several resources are required for a successful result. First, the analyst(s) must have considerable experience with and understanding of the system under consideration, e.g. the operation of a specific oil tanker. This is crucial in terms of identifying the real issue for the analysis, being able to identify and recognize the involved hazards, as well as to establish frequency/probability and consequence models that, as correctly as possible, represent the real world. Substantial knowledge is also necessary in order to be able to make the right simplifying assumptions that keep the complexity of the assessment process within acceptable levels. Such assumptions may include deciding which systems and activities should be included or excluded in the analysis.

Another important resource for the risk analysis and assessment process is statistical data, because these can give an indication of accident frequency and the most likely consequences when a certain hazard occurs. In a maritime context, where the number of serious accidents is quite low due to relatively small ship populations, historical recordings over several decades may be used to establish a statistical basis for risk analyses. The use of statistical data means that risk analysts should be well trained in statistical techniques.

Because of the inherent complexity of most risk assessments, such analyses normally need to combine the work of several people with a wide range of different backgrounds. Therefore, the analysts' teamwork and communication skills are of utmost importance.

8.2.5 Limitations of Risk Analysis

Risk analysis (and assessment) is a powerful tool in obtaining information and increased understanding of a system, its hazards, and the accident mechanisms. This information and understanding makes us able to implement risk control options and thus improve the system's safety. However, one should be aware of the limitations of such analysis, especially in relation to quantitative analysis. The lack of good statistical data due to limited experience is probably the most significant and common limitation in quantitative analysis. This is particularly clear in a maritime context where the number of large-scale accidents is quite low. Lack of statistical data results in huge uncertainties in the outcomes of the analyses, and one should therefore always evaluate these uncertainties and include this evaluation in the decision and recommendation process.

The complexity of most systems makes it necessary to make several simplifying assumptions in order to be capable of performing the analysis. These simplifications also create uncertainties.

A major limitation of traditional risk analyses is that human and organizational factors are usually not given adequate attention. During the last decades it has become a well-established fact that human and organizational factors affect the safety of technically complex systems, conventional ships and other vessels being no exception. These factors materialize themselves as active failures and latent conditions that breach the defences that prevent hazards from becoming severe losses. In technical systems that interact with humans, active human failures are normally considered to be the largest single cause of accidents. Investigations suggest that approximately 60% of all accidents are caused directly by human errors. In addition, some accidents are more indirectly caused by human errors, being a result of so-called organizational factors (e.g. company policies, attitude towards safety, etc.). It is normally easier to take the human and organizational factors into account in qualitative than in quantitative risk analyses.

8.3 BASIC THEORY

8.3.1 System Description

The first step of a risk analysis will normally be to define and describe the system under consideration. This is a step of crucial importance since such a system description is the underlying basis for the risk analysis as a whole. A system may be defined as an orderly arrangement of interrelated components that act and interact to perform a task or function in a particular environment and within a particular period of time. There are often several system levels, and complex systems are generally made up of subsystems in interrelation. Which system levels need consideration depends on the characteristics of the analysis itself. For a risk analysis of a shuttle tanker operation one may, for example, consider the following system levels:

- Offshore loading operations
- Tanker traffic and other ship movements along a coast
- Tanker traffic in a specific fairway
- Unloading operations
- Onboard systems for cargo handling and treatment

It must also be recognized that risk analyses are performed on both existing and planned systems. In a maritime context the system description generally covers the following elements:

- Geographical area: fairway, specific routes or harbours
- Environmental description: sea conditions, meteorological relations, visibility, etc.
- Traffic: transport quantity, frequency/scale of operations
- Vessels: number, capacities, sizes and technical descriptions
- Other activities: surrounding traffic and activities that may introduce hazardous situations

Some terms often used in system descriptions as part of risk analyses are presented in Table 8.1.

Some typical problems often related to the system description step of the risk analysis are the uncertainty of future activities, the complexity of the system and the collection of useful and valid data. These problems introduce the need for simplifications and assumptions. It is very important to clarify, describe and evaluate these problems, because they must be considered when interpreting the results of the analysis.

Table 8.1. System description terms

System description	Specified terms
Functional purpose	Task specification, time period involved, environmental conditions
Component consistency	Identification of subsystems, components and people involved
Functional order	Interrelationship between components and subsystems and the information flow within the system

Example

Problem

The propulsion system of an oil tanker is to be analysed using different risk analysis techniques. In relation to this, a precise basis for the use of these techniques must be produced. Perform a simple system description of the propulsion system.

Solution

The arrangement of the propulsion system is given in Figure 8.4.

Analysis:

Functional purpose: The propulsion system under consideration is the main propulsion system of the vessel. It is used under normal operation and gives the vessel the required speed and manoeuvrability for her whole life period of 24 years. It is allocated 3 days per year for maintenance (off-hire) in addition to the 26 days in harbour. 70% of the remaining sailing time is at full power. In the last 30% only one diesel engine is required in operation. The ship is to sail in ice-free sea conditions and harbours. The time in sea is 336 days per year, and of these 235 days are at full power.

Component consistency: The subsystems included in the analysis are given in Figure 8.4. There are two engineers responsible for the operation and maintenance planning of the system.

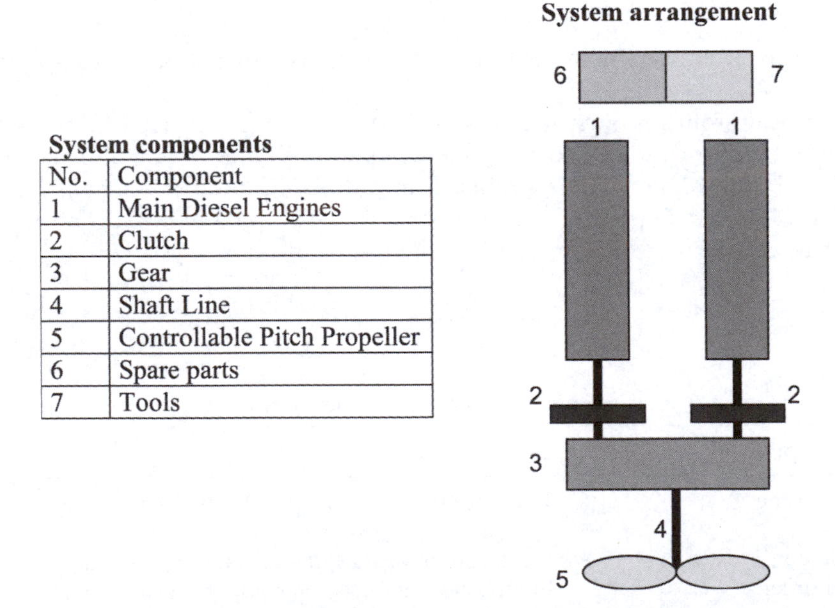

System components

No.	Component
1	Main Diesel Engines
2	Clutch
3	Gear
4	Shaft Line
5	Controllable Pitch Propeller
6	Spare parts
7	Tools

Figure 8.4. Propulsion system arrangement.

Functional order: The two main diesel engines (1) may be operated individually when maximum power is not required. The diesel engines are uncoupled from the gear by the use of a clutch (2). When both diesel engines are in operation both clutches have to be coupled and the total propulsion power is transmitted to the propeller by the gear and shaft line.

Comment:
Generally, a much more comprehensive system definition and description is required as a basis for further risk analysis.

8.3.2 General System Structures

In the quantitative analysis two basic characteristics (or elements) of a system are considered. These are the series structure and the parallel structure. When all components in a system or subsystem have to function in order to allow the system as a whole to function, the components are arranged in a series structure. If, however, only one of the components has to function for the whole system to function, the components are arranged in a parallel structure. If two equal components are in a parallel structure they are redundant. Figure 8.5 illustrates the series and parallel structures using block diagrams.

The probability of structure failure for a series structure and parallel structure is presented below.

Series structure:

$$P_{SF} = P_1 \cdot P_2 \cdot \ldots P_n = \prod_{i=1}^{n} P_i \tag{8.3}$$

Parallel structure:

$$P_{SF} = 1 - (1 - P_1) \cdot (1 - P_2) \cdot \ldots (1 - P_n) = 1 - \prod_{i=1}^{n} (1 - P_i) \tag{8.4}$$

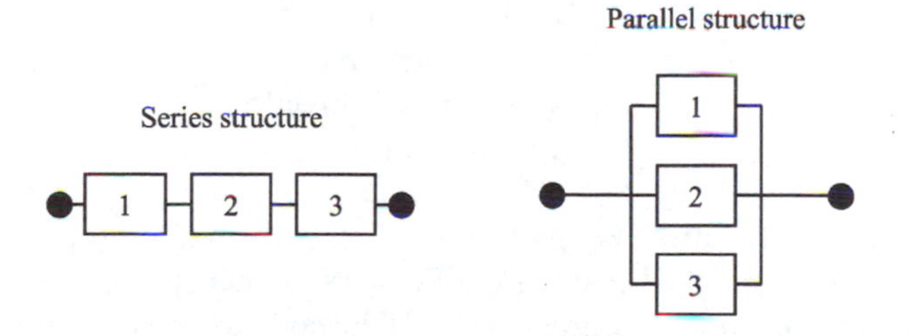

Figure 8.5. General system structures.

where P_{SF} = reliability of structure and P_i = reliability of structure i. The reliability P is defined as the survival probability (of a component or system) and is dependent on the operation time and operational conditions. The failure probability Q is equal to the probability of non-survival $(1 - P)$.

When applying quantitative techniques the validity of the data should be carefully assessed. In general the data are only valid for the environment (i.e. the operational conditions) from which they are collected. Failure data for the same fabricates of a diesel engine in onshore operations may be different from operations offshore in a ship due to different operational conditions/environments. In addition the quality of maintenance greatly influences the failure rate as well as the operational profile. Hence all failure data should be given a validity and sensitivity assessment. The result of such an assessment may be that the data is considered suitable, rejected, or adjusted according to 'expert judgement'.

8.4 PRELIMINARY HAZARD ANALYSIS (PHA)

Systems that are targeted for risk analyses are often quite complex, and the hazards facing the system may not therefore be completely obvious. Hence, after the system description is performed, the next task should be to identify possible hazards. The objective is to identify all possible events and conditions that may result in any severity or harm. A systemized way to identify such hazards is to apply the Preliminary Hazard Analysis (PHA) methodology described in this section.

8.4.1 Principle

The principle (or objective) is to identify hazards that may develop into accidents. This is done by generating situations or processes that are not planned or meant to happen. It is important to identify the hazards as early as possible in the design process in order to implement corrective measures in the design. This is known as proactive risk management/reduction.

8.4.2 Approach

In order to generate the hazardous situations or processes, deviations from the normal operation have to be considered. It may be difficult to get started with this exercise. Some deviations can, however, be established by making use of the cues below:

- More of . . .
- Less of . . .
- Nothing of . . .
- Part of . . .
- Both . . . and . . .
- Another than . . .
- Opposite direction . . .
- Later than . . .

Another approach is to identify parameters related to possible energy transfers. Accidents are often uncontrolled releases or transfers of energy, e.g. as in an uncontrolled fire. By identifying the energy sources, several hazardous events or processes can be established.

Table 8.2. Form applied in PHA

Hazardous element	Trigging event 1	Hazardous condition	Trigging event 2	Potential accident	Effect	Corrective measures
Ship's dependence on buoyancy	Compart-ments not watertight	Potential intake of water is large and uncontrolled	Heavy sea	Ship goes down	Fatalities, environ-mental damage, loss of ship and cargo	Increase number of watertight compartments, avoid heavy sea
...	...	...	...	...	...	...

Source: Henley and Kumamoto (1981).

8.4.3 Elements

The PHA may seem like a very general and non-specific exercise and that is exactly what it is. To facilitate matters for the analyst, it is therefore important to systemize the deviations. There are several ways to do this systemization and the analyst should adapt a system suitable for the system and/or situation he or she is to analyse. Table 8.2 is a general table for identifying the hazards.

Example

Problem

An oil tanker may introduce hazards to personnel, property and the environment. These hazards have to be identified as a basis for further risk analyses. Perform a Preliminary Hazard Analysis (PHA) for the tanker and sketch the accident development using the form presented in Table 8.2 above.

Solution

Figure 8.6 is a diagram of the oil tanker.

Assumptions:
Stability and buoyancy considerations are not treated here. Only kinetic energy and cargo energy are treated further.

Analysis:
Based on the energy considerations a PHA is performed for the tanker's kinetic energy and cargo energy, as shown in Tables 8.3–8.5.

These tables are not exhaustive. Can you, for example, find any other hazardous conditions relating to the kinetic energy of the oil tanker and the cargo's energy?

8.5 HAZARD AND OPERABILITY STUDIES (HAZOP)

A Hazard and Operability Study, popularly known as HAZOP, is a more detailed and comprehensive hazard identification method than the PHA. The basic idea of HAZOP is

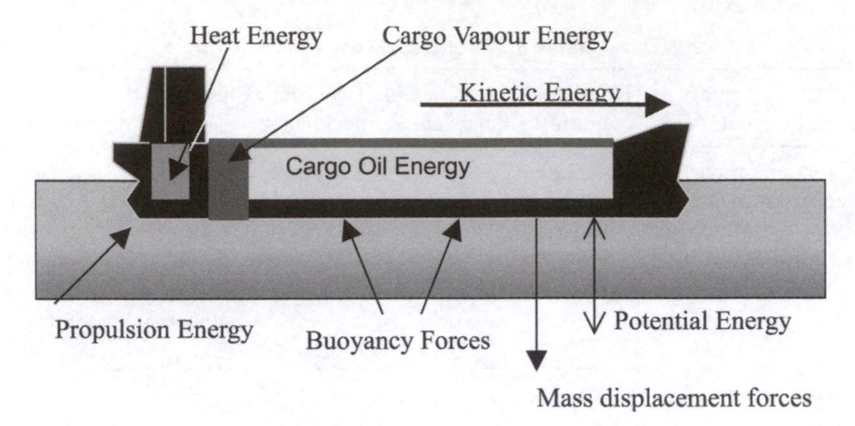

Figure 8.6. The oil tanker under consideration.

Table 8.3. PHA for the tanker's cargo oil

Hazardous element	Trigging event 1	Hazardous condition	Trigging event 2	Potential accident	Effect	Corrective measures
Cargo oil	Rupture of cargo tanks	Cargo oil leaks into the sea	Spill exposes animal life	Spill has consequences for the environment	Environmental damage, hull damage	Increase the rupture resistance of the tanks
Cargo oil	Rupture of cargo tanks	Cargo oil leaks into the sea	Oil is ignited	Fire on the surface	Fatalities, environmental damage, wrecked ship	Increase the rupture resistance of the tanks

to systematically search for deviations from the normal operation of the system that may have harmful consequences.

8.5.1 Principle

The principle (or objective) is to systematically examine the system part by part and then define the intention to each part. The intention is the way the system is expected to work. When the intentions are defined, possible deviations from the system's intentions that may lead to hazardous situations can be identified. The use of so-called guiding words may assist the analyst in the identification of such deviations. For the analysis process to be successful, a team consisting of specialists in several fields should supervise the analysis.

Table 8.4. PHA for the tanker's cargo oil vapour

Hazardous element	Trigging event 1	Hazardous condition	Trigging event 2	Potential accident	Effect	Corrective measures
Cargo oil vapour	Leakage in pump room	Explosive gas mixture	Ignition of gas	Explosion	Fatalities, environmental damage, material damage	Install pumps in the individual tanks
Cargo oil vapour	Cargo vapour in loaded cargo tanks	Explosive gas mixture	Ignition of gas	Explosion	Fatalities, environmental damage, material damage	Inert gas system
Cargo oil vapour	Cargo vapour in empty cargo tanks	Explosive gas mixture	Maintenance causes ignition of gas	Explosion	Fatalities, material damage	Inert gas system
Cargo oil vapour	Cargo vapour in empty cargo tanks	No oxygen in tanks	Maintenance personnel enters tank	Personnel asphyxiated	Fatalities, injuries	Oxygen level measurement

8.5.2 Approach

The first task is to get an overview of the system using a system description as described in Section 8.3 of this chapter. The system has to be divided into sections with independent intentions, and the intention of each part has to be carefully defined. In a real system all sections or subsystems are dependent on each other to a greater or lesser extent, and these dependencies must be identified.

When the intentions for each part of the systems have been defined, the system description is complete. Then one can start identifying deviations for each part of the system. Guidewords may assist the creativity of the analyst in order to establish as many deviations as possible, and these guidewords are applied one at a time. When the deviations are identified the causes of the deviations can be found and the reasons for the occurrence of the causes can be identified. The identification of causes results in greater/increased problem understanding and based on this safety measures can be established. These safety measures can be related to changes in processes, process parameters, design, routines, etc. The whole procedure is repeated for each part and section of the system as shown in Figure 8.7.

8.5.3 Elements

The most important resource for a HAZOP analysis is a detailed system description as well as access to complete part intention knowledge. When these resources are established,

Table 8.5. PHA for the tanker's kinetic energy

Hazardous element	Trigging event 1	Hazardous condition	Trigging event 2	Potential accident	Effect	Corrective measures
Kinetic energy	Loss of navigational control	Tanker sails on random course	Another ship is on the tanker's course	Collision, rupture of cargo tanks	Fatalities, environmental damage, damage to hull	Improving navigational standards
Kinetic energy	Loss of navigational control	Tanker sails on random course	Stationary obstacle on the tanker's course	Powered grounding, rupture of cargo tank	Fatalities, environmental damage, damage to hull	Improving navigational standards
Kinetic energy	Obstacle on the tanker's course	Retardation (i.e. reverse)	Movement of unfastened material onboard vessel	Crushed personnel, material damage	Fatalities, environmental damage	Fasten material properly
Kinetic energy	Drifting/ unfastened material	Ignition source	Combustible material present	Fire, explosion	Fatalities, environmental damage, material damage	Fasten material properly, remove combustible material

a set of guidewords may assist the analyst in identifying deviations. These guidewords are presented in Table 8.6.

The HAZOP procedure is further explained in the example below.

Example

Problem

The mobility of a vessel is highly dependent on the propeller. If the propeller fails for some reason, the whole propulsion system and navigation system is put out of operation and the ship's movement is out of control. It is therefore clear that the propeller is a critical component. As part of a HAZOP procedure the controllable pitch propeller (CPP) in Figure 8.8 is identified as a part with individual intention. Perform a single loop in the HAZOP procedure for the CPP.

Solution

Assumptions:

The analysis is to emphasize loss of propeller function. The case of degraded operation is not considered here.

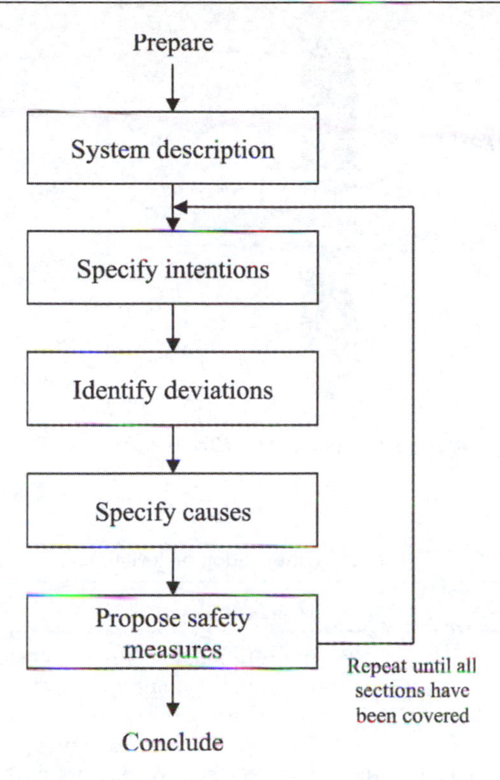

Figure 8.7. Main stages in the HAZOP procedure.

Table 8.6. Guidewords for the identification of deviation

Guideword	Description
NO or NOT	No part of the intention is achieved
MORE	Quantitative increase in flow rate or temperature, for example
LESS	Quantitative decrease
AS WELL AS	Qualitative increase. Intention is achieved plus additional activity like too much flow
PART OF	Qualitative decrease. Degraded intention achieved
REVERSE	Logical opposite of intention, e.g. reverse flow
OTHER THAN	Complete substitution. Something quite different happens

Analysis:
Definition of CPP intention: the propeller is to transform rotational energy, transmitted through the propeller shaft, into a pressure difference over the propeller blades. It is this pressure difference that accelerates and maintains the speed of the vessel. The controllable pitch's intention is to optimize this energy transformation for various operational conditions.

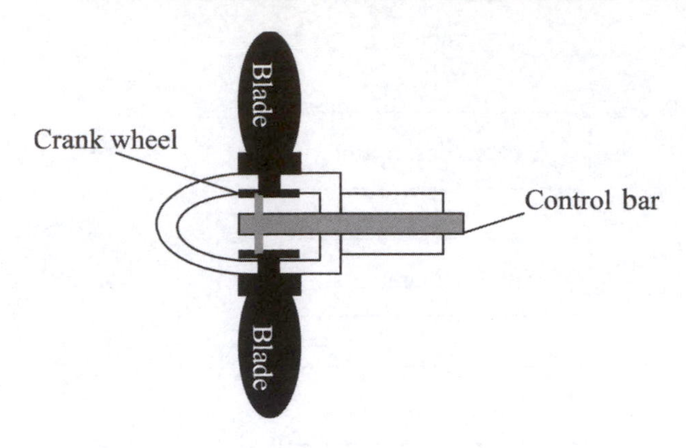

Figure 8.8. Controllable pitch propeller (CPP).

Table 8.7. Identification of deviations

No.	Guideword	Description
1	NO pitch	No rotational energy is transformed
2	NO blade	No rotational energy is transformed
3	NO control bar	All blades on random pitch, loss of operational control
4	NO crank wheel	One or all blades have independent pitch
5	NOT enough material strength	Parts of the propeller break down
6	MORE pitch than optimal	Too heavy load on propulsion system. Cavitation
7	LESS pitch than optimal	Too little load on propulsion system. Cavitation
8	LESS draft than allowed	Propeller is not sufficiently submerged. Loss of thrust
9	LESS depth than necessary	Propeller hits the ground and is damaged

Deviations are identified in Table 8.7 and their causes and safety measures listed in Table 8.8.

These tables are not exhaustive. Other deviations are possible, and to find these one must be creative and have a good understanding of the system.

Comment

The whole procedure should be repeated for all parts/subsystems of the propulsion and navigation system in a proper and comprehensive risk analysis.

8.6 FAILURE MODE, EFFECT AND CRITICALITY ANALYSIS (FMECA)

The Failure Mode, Effect and Criticality Analysis (FMECA) is a systemized inductive method of determining equipment functions, functional failure modes, assessing the causes

Table 8.8. Causes for the deviations and safety measures

No.	Causes	Safety measures
1	Operation failure or control mechanism failure, alignment mechanism defect	See 2, 3, 4 and 5
2	Object in the water breaks the blade	Implementation of propeller protection such as gratings or water jet. Sail in ice-free waters. See 7 and 8
3	Material weakness	Improve design and construction
4	Material weakness	Improve design and construction
5	Wrong design, corrosion or cavitation, alignment mechanism is defective and causes different pitch on the blades which again causes extra load on bearings and shaft line	Validate propeller design, cathodic protection, appropriate propeller material, test the propeller against cavitations, periodic alignment adjustment
6	Operation failure	Surveillance, increase operator competence
7	Operation failure	Surveillance, increase operator competence
8	Operation failure	Surveillance, increase operator competence
9	Operation failure	Technical equipment, operator competence and surveillance

of such failures and their effects (or consequences), as well as their effect on production availability and reliability, safety, cost, quality, etc., on a component level. The failure modes are normally and preferably analysed by the use of a standardized form that describes the failure, its causes and how it is detected, the various effects of the failure, as well as assessing important parameters such as failure rate, severity and criticality. FMECA is a quantitative method. However, the original version of FMECA is a qualitative version where the measured criticality is excluded, i.e. Failure Mode Effect Analysis (FMEA). Therefore, FMECA is still often described as a qualitative method in the literature.

8.6.1 Principle

The simple standardized forms used in FMECA assist the analyst to review the possible failure modes and identify their effects. The FMECA method can be used systematically to identify the most effective risk-reducing measures, which assist the process of selecting suitable design alternatives in an early design phase. As such the FMECA may be a valuable historical document for future design changes. The FMECA method is also used to form a basis for extensive quantitative reliability analyses with the objective of establishing sound maintenance strategies.

According to the Institution of Electrical Engineers (IEE), standard 352, an FMECA should give an answer to some basic questions:

- How can each part conceivably fail?
- What mechanisms might produce these modes of failures?
- What could the effects be if the failure did occur?
- Is the failure detected?
- What inherent provisions are provided in the design to compensate for the failure?

8.6.2 Approach

The first step of any risk analysis technique/method is, as described in Section 8.3 of this chapter, the system description. In general the approach to FMECA is to perform the following six stages:

1. General description of the components.
2. Description of possible failures and failure modes.
3. Description of failure effects for each failure mode.
4. Grading the failure effects in terms of frequency, and severity of consequences, as well as specifying reliability data.
5. Specifying and assessing methods for the detection of failure modes.
6. Description of how unwanted failure effects can be reduced and eliminated.

The standardized form, which is thoroughly explained later, aids the analyst's approach to the method.

8.6.3 Elements

The failure modes are important parameters in the FMECA method. A failure mode can be defined as the effect by which a failure is observed on a failed component/item. There are in principle two types of failure modes which are characterized, respectively, as unwanted change of condition and demanded change not achieved. The quantitative part of the method is given by the use of standardized terms for failure frequencies and consequences. The terms describing the failure frequencies are presented in Table 8.9. The possible failure consequences are measured using the consequence classes given in Table 8.10.

Example

Problem

The loss of propulsion power directly results in a loss of the controlled mobility of the vessel. In the HAZOP of the propeller (see earlier example) it was assumed that the controllable pitch propeller (CPP) was a critical subsystem for the propulsion system. The criticality is, however, dependent on the failure consequence and the failure

Table 8.9. Frequency classes

Frequency classes	Quantification
Very unlikely	Once per 1000 years or more rarely
Remote	Once per 100–1000 years
Occasional	Once per 10–100 years
Probable	Once per 1–10 years
Frequent	More often than once per year

Table 8.10. Consequence classes

Consequence classes	Quantification
Catastrophic	Any failure that can result in deaths or injuries or prevent performance of the intended mission
Critical	Any failure that will degrade the system beyond acceptable limits and create a safety hazard
Major	Any failure that will degrade the system beyond acceptable limits but can be adequately counteracted or controlled by alternative means
Minor	Any failure that does not degrade the overall performance beyond acceptable limits – one of the nuisance variety

likelihood. Hence an FMECA of the whole propulsion system may be appropriate. Find the elements and descriptions to be filled in the FMECA form.

Solution

A description of the propulsion system is given in Figure 8.4. The FMECA form to be used is outlined in Figure 8.9. The content of the FMECA form presented in Figure 8.9 is not exhaustive, especially in terms of failure causes.

8.7 FAULT TREE ANALYSIS (FTA)

One of the most frequently used techniques in risk analyses is fault tree modelling. In the introduction to Chapter 7 an example of a fault tree was presented. A fault tree analysis (FTA) can be used to identify the subsystems that are most critical for the operation of a given system, or to analyse how undesirable events occur. The methodology was developed in 1962 by H. S. Watson at the Bell Telephone Laboratories during the development of the 'Minuteman' rocket's combustion chamber.

8.7.1 Principle

In the context of risk analyses the FTA method is used to analyse the way an unwanted event occurs, as well as its causes. By the use of a logical diagram the relationship between

System: **PROPULSION** Performed by:
Ref. Drawing no.: Date: Page:

Description of unit			Description of failure			Effect of Failure			Failure rate	Severity Ranking	Risk Reducing Measures	Comments
Ref. No	Function	Operational mode	Failure Mode	Failure cause or Mechanism	Detection of failure	On sub-systems	On system function	Resulting state				
1	*Propulsion*	*Normal operation*	*Stop*	*No fuel feed*	*On watch (sound)*	*Gear over-load*	*Half effect*	*Reduced speed*	*Remote*	*Minor - Mayor*	*Reducing pitch, cut off*	*Engine 2 is functioning Onboard repair*
			—"—	*Crankshaft Failure*	—"—	—"—	—"—	—"—	*Occasional*	—"—	—"—	*Engine 2 is functioning Harbour repair*
			Reduced Function	*Piston running hot*	—"—	—"—	*Reduced effect*	—"—	*Very Unlikely*	—"—	—"—	*Engine 2 is functioning Onboard repair*
2	*Gear - reduce number of revolutions (i.e. RPM), transmit power*	*Normal Operation*	*No power transmittal*	*Broken cog*	—"—	*Main Engine Over-Load*	*No propulsion*	*Loss of manouver-ability*	*Remote – Occasional*	*Critical*	*Stop main engine, lock shaft line*	*Both primary and secondary gearwheels are failed Harbour repair*
3	*Shaft line-Transmit Power*	*Normal Operation*	*No power transmittal*	*Broken shaft*	—"—	—"—	—"—	—"—	*Remote*	*Critical*	—"—	*Repair in dock*
4	*Controllable pitch propeller (CPP) – transmit power*	*Normal operation*	*Reduced Function*	*Broken blade*	—"—	—"—	*Reduced effect*	*Reduced Speed*	*Occasional*	*Minor*	*Reduce main engine power, reduce propeller pitch*	*Damage to only one propeller blade Repair in dock*

Figure 8.9. FMECA form.

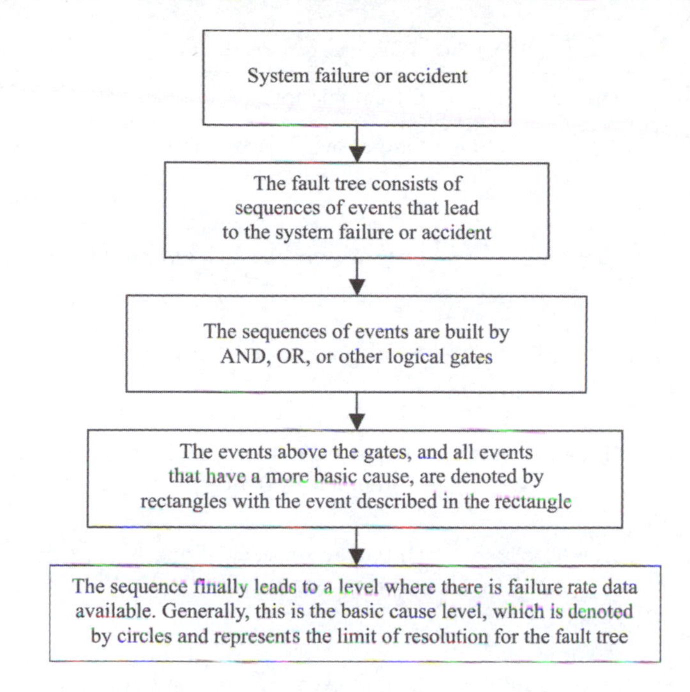

Figure 8.10. Principles of a fault tree.

the causes of the event (e.g. the failure of a certain engine component) is visualized. The method assumes binary operational modes, which means that an event either occurs or it does not (e.g. a failure alarm is given or not given). Hence, degraded operations or events are not analysed in fault trees.

The logical diagram used in an FTA consists of a set of gate symbols that describe the relationship between causes, and event symbols that characterize the causes. These gate and event symbols are described later in this section. The main principles of the fault tree analysis method are illustrated in Figure 8.10.

8.7.2 Approach

A fault tree can be analysed both qualitatively and quantitatively. These approaches are described in more detail later.

8.7.3 Elements

The fault tree is a visualization of the relationship between the failures of the analysed system (see Figure 8.10). This visualization is based on logical gates and symbols. The most common fault tree gate symbols and event symbols are presented in Tables 8.11 and

Table 8.11. Fault tree gate symbols

Gate symbol	Gate name	Casual relation
Output event / Input events	AND gate	Output event occurs if all input events occur simultaneously
	OR gate	Output event occurs if any one of the input events occurs
	Inhibit gate	Input produces output when conditional events occur
	Priority AND gate	Output event occurs if all input events occur in the order from left to right
	Exclusive OR gate	Output event occurs if one, but not more than one, input events occur
m / n inputs	*m* out of *n* gate	Output event occurs if *m* out of *n* input events occur

8.12, respectively. The use of the most frequently used gate and event symbols will be illustrated by an example later in this section.

8.7.4 Qualitative Approach: Construction

The first task of a fault tree analysis is to describe the system and its components/subsystems down to a sufficient level of detail (see Section 8.3 of this chapter). The next task is to construct the fault tree for a particular unwanted system failure using this system description. It is important that all the failures in the fault tree are given precise definitions. The unwanted event or accident target for the analysis is referred to as the top event of the fault tree. The description of the top event should give answers to *what* the event is, *where* it occurs and *when* it occurs.

The occurrence of the top event is always dependent on two or more conditions or failures on a more detailed, i.e. lower, level. The main task in the FTA approach is to systematically define and structure the conditions or causes that directly lead to the top event. These events should be defined in such a way that only a limited number of causes lead to the top event. Some literature recommends only defining two causes on the lower level at a time, but for some complex system failures this may not be realistic. The causes directly leading to the top event are at the second level in the fault tree.

Table 8.12. Fault tree event symbols

Event symbol	Meaning of symbol
	Basic event with sufficient data
	Undeveloped event
	Event represented by a gate
	Conditional event used with inhibit gate
	House event. Either occurring or not occurring
	Transfer symbol

When the events are defined and structured, the next task is to assess the logical relation between the causes. Generally, either the top event is dependent on a simultaneous occurrence of these causes on the second level, or only one of the causes may lead to the top event. In the first case an AND gate is used and in the last case an OR gate is used (see Table 8.11). This procedure is then repeated to establish the logical relations between the causes on the third level of the fault tree, and so on. When the causes are described in such a detail that failure data (i.e. failure frequency) is available, the fault tree construction is finished and ready for quantitative analysis.

8.7.5 Qualitative Approach: Minimal Cut Sets

The objective of qualitative FTA is to establish a general view and understanding of the fault tree construction. This can be achieved by establishing sets of events that have special characteristics. A set of basic events in the fault tree that triggers the top event by occurring simultaneously is called a cut set of the fault tree. For illustration purposes, a simple fault tree for the top and unwanted event of an initiation of fire can be studied (see Figure 8.11). Based on basic fire theory, a fire can occur only if three basic conditions are satisfied. These three basic conditions are the presence of a combustible material (e.g. wood, oil, etc.), oxygen, and an ignition source (e.g. flame, heat, friction, a spark, etc.). By distinguishing between combustible substances and gases, the following simplified fault tree can be constructed.

As shown in the fault tree, a fire can occur if the following set of causes are occurring: {*Combustible substance present*, *Combustible gas present*, *Oxygen present*, *Heat or ignition*

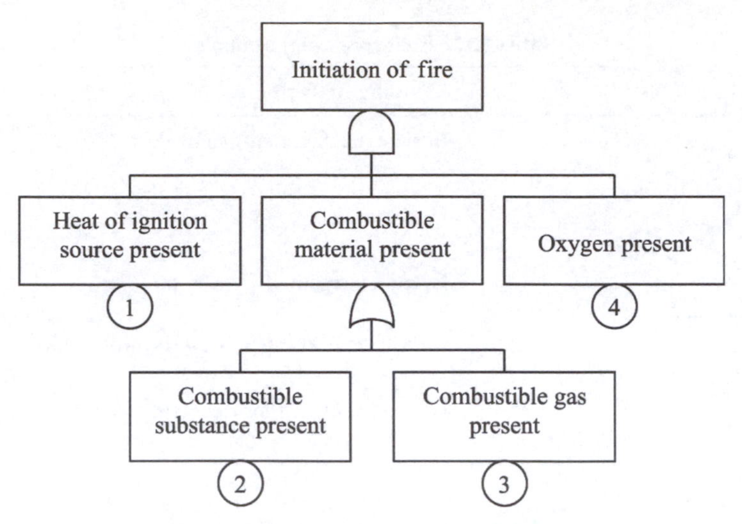

Figure 8.11. Simplified fault tree for a fire.

source present}. This is a cut set for this fault tree because the simultaneous occurrence of the four causes results in the occurrence of the top event {*Initiation of fire*}.

A Minimal cut set is a set of causes where none of the included causes can be excluded without the causes losing their status as a cut set. Hence, the following two sets of causes are minimal cut sets: {*Combustible material present*, *Oxygen present*, *Heat or ignition source present*} and {*Combustible gas present*, *Oxygen present*, *Heat or ignition source present*}.

To establish the cut sets of a fault tree a systemized algorithm called MOCUS – Method of Obtaining Cut Sets – can be applied. The MOCUS algorithm is represented by four steps:

1. Consider the top event.
2. Replace the event with the events on the second level according to the following criteria: If the events on the lower level are connected through an OR gate they are written in separate rows. If they are connected through an AND gate they are written in separate columns.
3. Perform step 2 successively for all events that are not basic events (see Table 8.12).
4. When all events are basic events the events in each row constitute a cut set.

The fault tree in Figure 8.11 can be used to illustrate the use of the MOCUS algorithm. The starting point of the algorithm is the top event according to step 1. In the fault tree in Figure 8.11 this is the following event:

{*Initiation of fire*}

This event is then replaced by the events on the lower level according to step 2. Because the events on the second level of the fault tree are connected through an AND gate, they replace the top event in three columns:

Cause 1	Combustible material present	Cause 4

The causes 1 and 4 are basic events and are not treated any further, according to step 2 in the MOCUS algorithm. However, the event of {*Combustible material present*} needs another loop of the MOCUS algorithm in order to complete the cut sets. Because the gate beyond this event is an OR gate, the causes on the third level are written in separate rows. Hence according to the MOCUS algorithm the cut sets after the second loop are:

K_1	Cause 1	Cause 2	Cause 4
K_2	Cause 1	Cause 3	Cause 4

According to step 4 of the algorithm, each row constitutes a cut set, and hence there are two cut sets, K_1 and K_2, for the fault tree in Figure 8.11. Consequently, the general conditions for a fire, i.e. the event {*Initiation of fire*}, are satisfied when, for example, Cause 1, Cause 2 and Cause 4 occur simultaneously. Because none of the causes in the two cut sets can be removed without them losing their status as cut sets, both K_1 and K_2 are minimal.

Another important term in the fault tree terminology is the so-called path set. A path set assembles a set of causes with the characteristic that non-occurrence of the causes in the path sets ensure that the top event does not occur. For the fault tree in Figure 8.11 the non-occurrence of Cause 1 {*Heat or ignition source present*} ensures that the top event does not occur. Hence Cause 1 is a path set.

Both the minimal path sets and the minimal cut sets give important information about the properties of the system. The number of elements in the minimal cut sets should be as large as possible to avoid triggering of the top event due to a few causes. Barriers may be built into the system to achieve this. The number of path sets should be large because this implies that the system is designed to have multiple ways of avoiding the top event.

8.7.6 Quantitative Approach: Calculation

The quantitative analysis of the fault tree uses the failure probability q_i of the basic events and the fault tree gates to calculate the probability of the top event Q_0. This calculation is quite straightforward. For basic events combined through an OR gate the series structure equation established in Section 8.3 of this chapter is used. For events combined through an AND gate the parallel structure equation is used (when using these equations it must be remembered that the reliability $p_i = 1 - q_i$, where q_i is the failure probability). Consequently it is easier to trigger events combined through an OR gate than events combined through an AND gate. Conditional probabilities (i.e. AND gates in fault trees) are generally very common in fault tree calculations.

8.7.7 Quantitative Approach: Assessment

In the qualitative analysis the minimal cut sets of the fault tree are established. Each of these cut sets includes one unique set of basic events, which by occurring simultaneously trigger the top event. Consequently it is important to prevent the occurrence of a basic cause (or basic event) that is present in several cut sets in order to reduce the likelihood of top event occurrence. Because the basic causes are present in several cut sets, this may be applied to calculate a measure of importance for each basic cause. A common importance measure applied on fault trees is the Vessley-Fussell measure of importance, I^{VF}. This is the probability that at least one minimal cut set that contains the basic event i is failed at time t, given that the top event is triggered at time t. This can be calculated by the following equation:

$$I^{VF}(i \mid t) = P(\textit{At least one of the cut sets containing the basic event i is failed at time t} \mid \textit{The system is failed at time t})$$

Hence:

$$I^{VF}(i \mid t) = \frac{Q_{Ki}(t)}{Q_0(t)} \quad (8.5)$$

where Q_{Ki} = the probability that one minimal cut set containing the basic cause i is failed at time t, and Q_0 = probability of occurrence for the top event.

The m minimal cut sets in which the basic cause i is present are not independent because the same basic causes may be present in more than one cut set. However, by assuming that the m cut sets are independent, the higher limit of Q_{Ki} can be estimated using the $I^{VF}(i \mid t)$ equation above and the parallel structure equation presented earlier. This assumption is implemented in the following equation:

$$I^{VF}(i \mid t) \approx \frac{1 - \prod_{j=1}^{m} (1 - Q_{Ki,j}(t))}{Q_0(t)} \quad (8.6)$$

where m = number of minimal cut sets where basic cause i is present.

Example

Problem

The failure modes of a tanker's main propulsion system have been established earlier in the chapter using a FMECA analysis. The connections and relations between the failures are unknown, and must therefore be modelled in a fault tree. Construct a fault tree where the top event is loss of propulsion power for the tanker. Then perform a qualitative and quantitative fault tree analysis using the algorithms and methods described in this chapter.

Solution

It is assumed that the information shown in Table 8.13 is commonly available and known.

Qualitative approach: fault tree construction

The top event is already defined as 'loss of propulsion for the tanker'. A simple way to break down the propulsion system is to emphasize on power transition in the main propulsion system. There are three independent events that may result in the top event. These are the 'loss of propulsion power transmission' in the shaft lines or gear, 'loss of propulsion power generation' from the engines, and 'loss of propulsion power consumption' due to propeller failure. Only one of these events has to occur in order to trigger the top event. Hence these three events have to be combined by an OR gate. The fault tree can be structured as shown in Figure 8.12.

The 'loss of propulsion power transmission' event in Figure 8.12 can be caused by gear failure and/or shaft line failure (see FMECA in Figure 8.9), and must therefore be combined through the use of an OR gate. The 'loss of propulsion power consumption' event only includes the event of controllable pitch propeller (CPP) failure. In terms of the event of 'loss of propulsion power generation', both the starboard and port engines must fail to deliver power to the gear. An AND gate must therefore be used for these two events. There are two ways each engine can fail to deliver power to the gear: by failure of the clutch and by failure of the engine itself. An OR gate must be used for these events because one is sufficient for the engine to fail to deliver power to the gear. The events of main engine failure (both starboard and port engines) in Figure 8.12 need to be treated in further detail. According to the FMECA, the causes or basic failure events 1, 2 and 3 (see Table 8.13) are all gathered in the 'main engine failure' event, and these have to be combined through the use of an OR gate since one of the causes is enough for the main engine to fail. The main engine failure modes can be arranged/modelled in a fault tree as shown in Figure 8.13.

Qualitative approach: establishing minimal cut sets

The MOCUS algorithm is applied (subscript s = Starboard, subscript p = Port):

MOCUS step 1:

> '*Loss of main propulsion power for a specified tanker under one year of normal operation.*'

Table 8.13. Failure data calculated for a sailing operation of one year (336 days)

Failure	Failure description	Reliability probability p	Failure probability q
F1	No fuel feed	0.730	0.270
F2	Crankshaft failure	0.973	0.027
F3	Piston running hot	0.984	0.016
F4	Clutch failure	0.948	0.052
F5	Gear failure	0.764	0.236
F6	Shaft line failure	0.971	0.029
F7	CCP failure	0.813	0.187

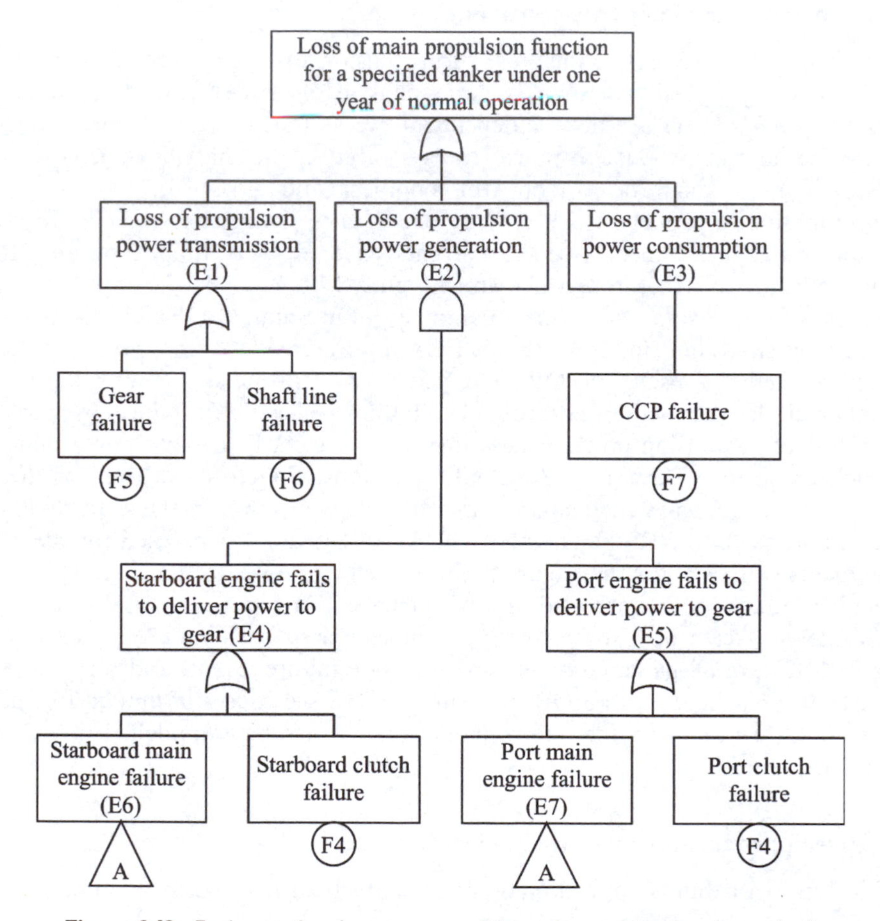

Figure 8.12. Fault tree for the top event of 'loss of propulsion for the tanker'.

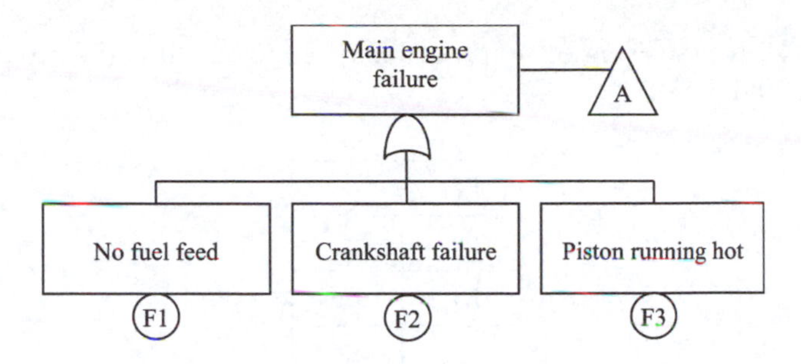

Figure 8.13. Main engine failure modes.

MOCUS step 2:

E1
E2
E3

MOCUS step 3.1 – for the 'loss of propulsion power transmission' event (i.e. E1 in the fault tree):

F5
F6
E2
E3

MOCUS step 3.2 – for the 'loss of propulsion power generation' event (i.e. E2 in the fault tree):

F5	
F6	
E4	E5
E3	

MOCUS step 3.3 – for the event that 'starboard engine fails to deliver power to gear' (i.e. E4 in the fault tree):

F5	
F6	
A_s	E5
$F4_s$	E5
E3	

MOCUS step 3.4 – for the event of 'starboard main engine failure' (i.e. E6 in the fault tree; see Figures 8.12 and 8.13):

F5	
F6	
$F1_s$	E5
$F2_s$	E5
$F3_s$	E5
$F4_s$	E5
E3	

MOCUS step 3.5 – for the event of 'port main engine failure' (i.e. E7 in the fault tree; see Figures 8.12 and 8.13):

K_1	F5	
K_2	F6	
K_3	$F1_s$	$F1_p$
K_4	$F2_s$	$F1_p$
K_5	$F3_s$	$F1_p$
K_6	$F4_s$	$F1_p$
K_7	$F1_s$	$F2_p$
K_8	$F2_s$	$F2_p$
K_9	$F3_s$	$F2_p$
K_{10}	$F4_s$	$F2_p$
K_{11}	$F1_s$	$F3_p$
K_{12}	$F2_s$	$F3_p$
K_{13}	$F3_s$	$F3_p$
K_{14}	$F4_s$	$F3_p$
K_{15}	$F1_s$	$F4_p$
K_{16}	$F2_s$	$F4_p$
K_{17}	$F3_s$	$F4_p$
K_{18}	$F4_s$	$F4_p$
K_{19}	F7	

MOCUS step 4:
There are 19 possible combinations of basic causes (or basic event failures) for the propulsion system (each row). There are mostly two basic causes in each cut set. It is advantageous to have as many basic causes in each cut set as possible, and one and two basic causes in each cut set is not much. The cut sets K_1, K_2 and K_{19} include only one basic cause. Hence the top event is triggered when one of these basic causes occurs. It would therefore be advantageous to implement redundancy or other reliability improving measures for these cut sets. For example, would the use of two independent propeller systems create redundancy and hence reduce the risk for top event occurrence? This may, however, not be practicable.

Quantitative approach: fault tree calculations

There are several interesting calculations that should be performed. The probability of the top event Q_0 is certainly of particular interest. The probabilities for each cut sets are also of interest. Normally some computerized calculation program, such as a spreadsheet, would be applied to calculate the top event probability. Here, on the other hand, the events are calculated manually using the series and parallel structure equations presented in Section 8.3 of this chapter. The series structure equation is used to calculate OR gates and the parallel structure is used to calculate AND gates (it must be remembered that the reliability $p_i = 1 - q_i$, where q_i is the failure probability). Failure data are given in Table 8.13.

As shown in Table 8.14, the probability for the top event of 'loss of main propulsion function for a specified tanker under one year of normal operation' is 0.465. This means that there is a 46.5% chance that this particular unwanted, and potentially very dangerous, event will occur.

Quantitative approach: assessment of basic cause importance

To assess the importance of the different basic causes, the cut sets' failure probability is calculated as in Table 8.15, using the given failure probability data in Table 8.13.

According to the Vessley-Fussell measure of component importance, the importance ranking of the basic causes (or failures) is established as shown in Table 8.16. The ranking of the components is the 'repairman's' ranking. If propulsion is lost, the most likely failure

Table 8.14. Calculation of top event failure probability Q_0

Q_{E7}	$= 1 - P_{E7} = 1 - [p_{F1} \cdot p_{F2} \cdot p_{F3}] = 1 - [(1 - q_{F1}) \cdot (1 - q_{F2}) \cdot (1 - q_{F3})]$	0.301
Q_{E6}	$1 - [(1 - q_{F1}) \cdot (1 - q_{F2}) \cdot (1 - q_{F3})]$	0.301
Q_{E5}	$1 - [(1 - Q_{E7}) \cdot (1 - q_{F4})]$	0.337
Q_{E4}	$1 - [(1 - Q_{E7}) \cdot (1 - q_{F4})]$	0.337
Q_{E2}	$= 1 - P_{E2} = 1 - [1 - (1 - P_{E4}) \cdot (1 - P_{E5})] = Q_{E5} \cdot Q_{E4}$	0.114
Q_{E3}	q_{F7}	0.187
Q_{E1}	$1 - [(1 - q_{F5}) \cdot (1 - q_{F6})]$	0.258
Q_0	$1 - [(1 - Q_{E1}) \cdot (1 - Q_{E2}) \cdot (1 - Q_{E3})]$	0.465

Table 8.15. Calculation of cut sets' failure probabilities

K_1	F5		$Q_{K1} = 0.236$
K_2	F6		$Q_{K2} = 0.029$
K_3	$F1_s$	$F1_p$	$Q_{K3} = 0.073$
K_4	$F2_s$	$F1_p$	$Q_{K4} = 0.0073$
K_5	$F3_s$	$F1_p$	$Q_{K5} = 0.0043$
K_6	$F4_s$	$F1_p$	$Q_{K6} = 0.014$
K_7	$F1_s$	$F2_p$	$Q_{K7} = 0.0073$
K_8	$F2_s$	$F2_p$	$Q_{K8} = 0.00073$
K_9	$F3_s$	$F2_p$	$Q_{K8} = 0.00043$
K_{10}	$F4_s$	$F2_p$	$Q_{K10} = 0.0014$
K_{11}	$F1_s$	$F3_p$	$Q_{K11} = 0.0043$
K_{12}	$F2_s$	$F3_p$	$Q_{K12} = 0.00043$
K_{13}	$F3_s$	$F3_p$	$Q_{K13} = 0.00026$
K_{14}	$F4_s$	$F3_p$	$Q_{K14} = 0.00083$
K_{15}	$F1_s$	$F4_p$	$Q_{K15} = 0.014$
K_{16}	$F2_s$	$F4_p$	$Q_{K16} = 0.0014$
K_{17}	$F3_s$	$F4_p$	$Q_{K17} = 0.00083$
K_{18}	$F4_s$	$F4_p$	$Q_{K18} = 0.056$
K_{19}	F7		$Q_{K19} = 0.187$

is related to the gear, i.e. basic cause/failure event F5, and so on. Other measures of importance should be applied at the design stage.

8.8 EVENT TREE ANALYSIS (ETA)

In the fault tree analysis (FTA) section the probability for loss of the propulsion function on a tanker was estimated. The possible consequences that may result because of the lost propulsion function, however, have not been analysed so far. If the consequences of an event or incident are to be analysed, a so-called event tree analysis (ETA) approach may be applied. The event tree splits up a given initiating event forwardly and is therefore an inductive method.

Table 8.16. Importance ranking based on the Vessley-Fussell measure of importance

	Relevant cut sets	$1-\Pi(1-Q_{Ki})$	I^{VF}	Ranking
F1	K_3, K_7, K_{11}, K_{15}	0.0966	0.208	3
F2	K_4, K_8, K_{12}, K_{16}	0.0098	0.021	6
F3	K_5, K_9, K_{13}, K_{17}	0.0061	0.013	7
F4	K_6, K_{10}, K_{14}, K_{18}	0.071	0.150	4
F5	K_1	0.236	0.507	1
F6	K_2	0.029	0.062	5
F7	K_{19}	0.187	0.402	2

8.8.1 Principle

An ETA is a logical diagram based on chains of possible events. The logical diagram used in an ETA describes the relation between an initiating event and the events that describe the possible consequences. The basic principle of the ETA approach is that each level in the chain of events leading to a consequence consists of two mutually exclusive dichotomy events. Two events are, by definition, mutually exclusive (or disjoint) if it is impossible for them to occur together at the same time. Dichotomy means that an event can only have two different outcomes. For example, the event of a tanker collision may have two possible outcomes with respect to the oil cargo tanks: non-rupture of the cargo tanks or the rupture of these tanks, the latter resulting in oil pollution and possibly also a fire. Based on this it is clear that ETA is a binary technique. An initiating event may develop into several consequences both in type and magnitude/severity. The likelihood of one event is dependent on the previous events, as well as the nature of the event. This dependency is discussed in more detail later.

8.8.2 Approach

ETA is a quantitative method for the estimation of consequence probabilities based on a given initiating event. Hence the first task of the approach is to define the initiating event, which is the first in a sequence of events leading to a hazardous situation or accident. Next, the safety systems, mechanisms and situation characteristics that function as barriers in the consequence development process are established in a chronological order. The probabilities for the outcomes of each dichotomy event (e.g. the success of a particular safety barrier/mechanism) are then estimated and an initial event tree is established. At this stage, however, the probability for each dichotomy event is independent of the previous events. Two events are, by definition, independent if the occurrence of one event does not give us any information about whether or not another event will occur, i.e. the events have no influence on each other. In reality, on the other hand, the events may to some degree be dependent on each other. As presented in Figure 8.14, these dependencies can be related to the time, their location in the event chronology, and conditional involvements of previous events. Some sort of correction for these dependencies should

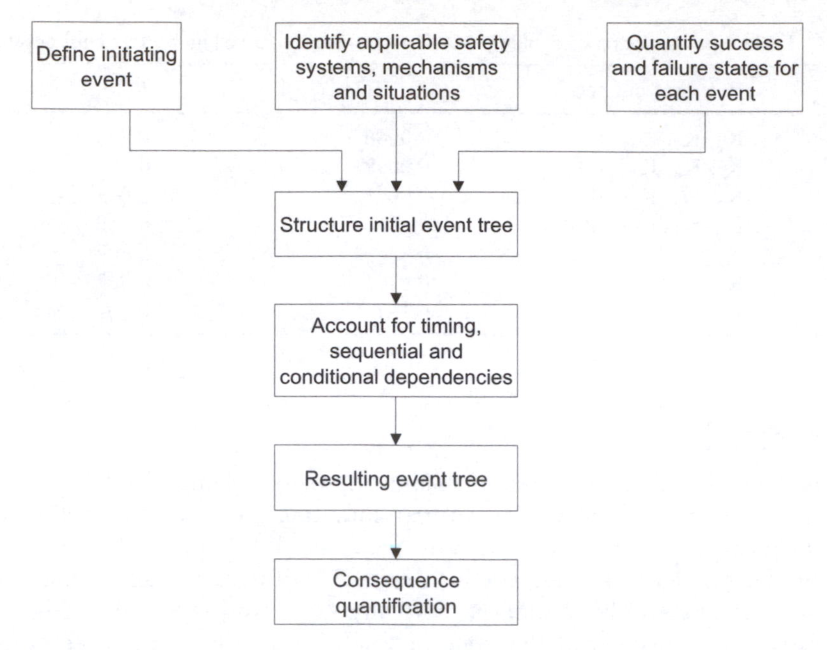

Figure 8.14. Event tree analysis approach.

be performed. It is not, however, possible to establish a general procedure for such corrections.

8.8.3 Elements

In order to perform a realistic and acceptable ETA the analyst(s) must have sufficient system knowledge and understanding. In addition, common sense, as well as logical and creative thinking, are important resources in the process of performing an event tree analysis. The design of the event tree diagram is done using the following approach.

The events are arranged in chronological order with the initiating event on the top, followed by important and relevant intermediate events and the consequence events placed at the bottom. The binary dichotomy event occurrence is visualized in the event tree by placing the unwanted event (i.e. failure) to the right and the successful/desired event to the left.

Example

Problem

The loss of propulsion power results in loss of controlled mobility for an oil tanker. The event of loss of propulsion power has been examined earlier in this chapter using the

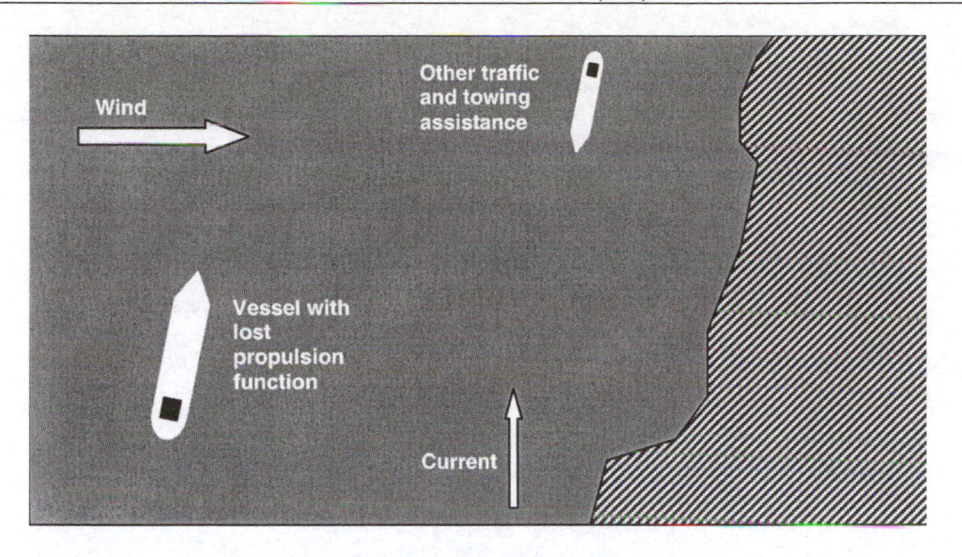

Figure 8.15. System description.

fault tree analysis (FTA) approach. The potential consequences for the loss of propulsion power have not, however, been analysed in detail. Because the oil tanker has large oil spill potential, with devastating effects on the environment, it is of great interest to estimate the likelihood of an oil spill if propulsion function is lost. Such information can, for example, be used to evaluate whether additional safety measures should be implemented to reduce the probability for such accidents. For the given system description in Figure 8.15, find the likelihood of oil spill when the propulsion function is lost.

Solution

Assumption:
The loss of propulsion power has a probability of 0.465 per year.

Analysis:
The initial event tree shown in Figure 8.16 is designed. In this initial event tree, possible dependencies between the events have not been assessed. This assessment process is, however, far from easy and is normally carried out at the discretion of the analyst. The problems involved in assessing dependencies between events exist for all quantified methodologies. The dependencies between the events in the initial event tree are assessed in the influence diagram in Figure 8.17. For example, it is found that the 'critical impact forces' event and the 'emergency anchoring failure' event are both dependent on the 'critical weather force' event. The 'critical weather force' event may, on the other hand, be dependent on the 'critical drifting direction' event.

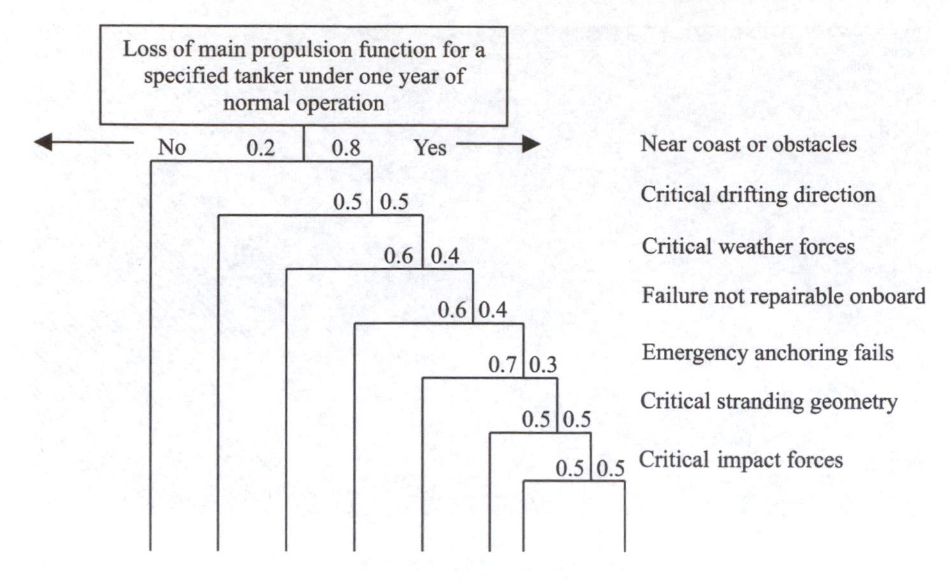

Figure 8.16. Initial event tree diagram.

Simple dependency and influence diagram Increasing likelihood of occurrence: + Decreasing likelihood of occurrence:	Critical drifting direction	Critical weather forces	Failure not repairable onboard	Emergency anchoring fails	Critical stranding geometry	Critical impact forces
Near coast or obstacles	+	+		+		+
Critical drifting direction		+				
Critical weather forces			+	+		++
Failure not repairable onboard						
Emergency anchoring fails					+	+
Critical stranding geometry						++

Figure 8.17. Influence diagram for assessing event dependencies.

Based on the influence diagram, an assessment of the probabilities is performed at the discretion of the analyst, and the event diagram shown in Figure 8.18 can then be established. The consequence probabilities are calculated by finding the product of all the events leading to the consequence, including the probability of the initiating top event in the event tree.

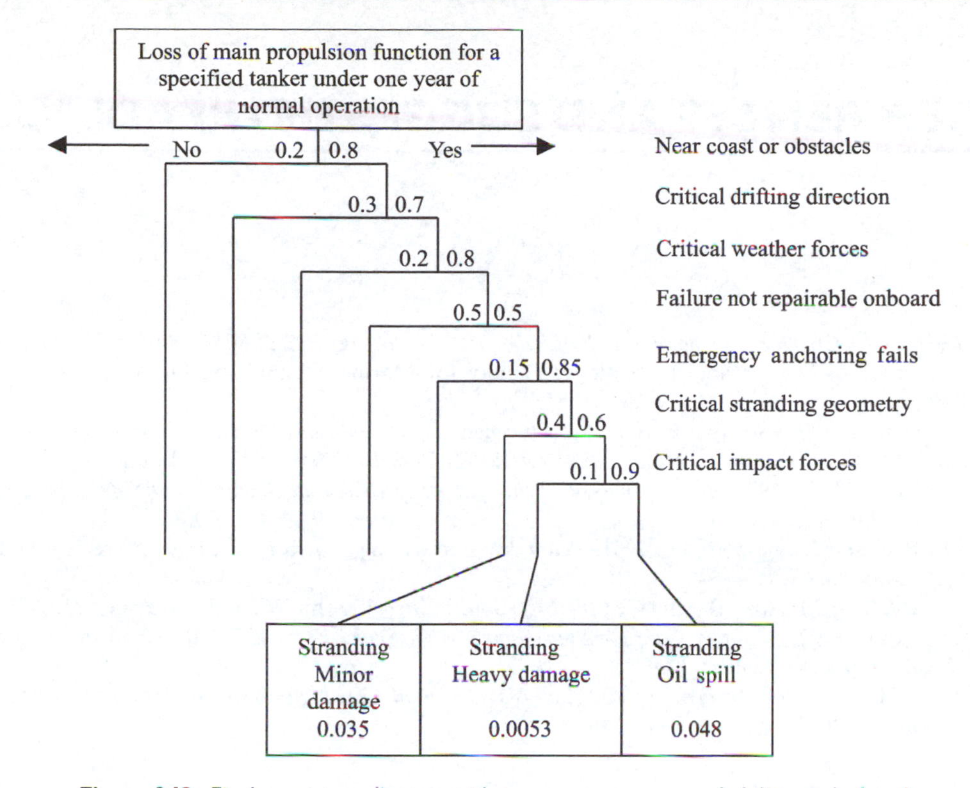

Figure 8.18. Final event tree diagram with some consequence probabilities calculated.

As can be seen from the event tree in Figure 8.18, the likelihood/probability of oil spill initiated by the loss of propulsion power, and caused by a critical stranding, is 0.048 per year of operation.

REFERENCES AND SUGGESTED READING

Dahle, E. A., 1976, *Forelesningsreferat Pålitlighetsteknikk for fag 80559, 80560, 80525*. Norwegian University of Science and Technology, Faculty for Marine Technology, Division of Marine Design.

DNV Technica, 1996, *Safety Assessment of Passenger/Ro-Ro Vessels – Methodology Report for the Joint Nordic Project on Safety of Passenger/Ro-Ro Vessels*. DNV Technica, London.

Harms-Ringdahl, L., 1993, *Safety Analysis: Principles and Practice in Occupational Safety*. Elsevier Science, Oxford, UK.

Henley, E. J. and Kumamoto, H., 1981, *Reliability Engineering and Risk Assessment*. Prentice-Hall, Englewood Cliffs, NJ.

Høyland, A. and Marwin, R., 1994, *System Reliability Theory*. John Wiley, New York.

Kuo, C. et al., 1997, *A Safety Case for Stena Line's High Speed Ferry HS1500*. Royal Institute of Naval Architects, London.

Lambert, H. E., 1973, *System Safety Analysis and Fault Tree Analysis*. Lawrence Livermore Laboratory, California.

9

COST-BENEFIT ANALYSIS

Matter will be damaged in direct proportion to its value.
("Murphy's Constant")

9.1 INTRODUCTION

Cost-benefit analysis (CBA) is basically a technique for comparing the costs and benefits of a project. The technique was originally developed to help appraise public sector projects to ensure that one achieved the greatest possible value for money, but the concept of cost-benefit analysis has now applications far beyond the public sector domain. In this chapter we are particularly concerned with CBA in the context of risk and safety assessment.

The term safety can be defined as the extent to which a system is free from danger to life, property and the environment. The concept of risk, which has the two components of probability and consequence, is used to evaluate the significance of the danger resulting from hazards, i.e. possible events and conditions that may result in severity. Reduced risk of severe accidents means increased safety. No accident is ever acceptable, and attempts should always be made to prevent them from occurring as well as to reduce the possible consequences if they occur. Applying and implementing safety measures can reduce the risks of severe accidents and hence improve safety, but these measures also, unavoidably, incur costs. At some point the cost of the implemented safety measures will make, for example, a ship or an oil rig uneconomic to operate or uncompetitive in the relevant market. As a result there must always be a trade-off between the costs of implementing safety measures and the residual risk level, because no matter what measures are taken to reduce the risks of accidents, some residual risk will remain. Risks can never be eliminated altogether, and some level of risk will always have to be accepted. It is the safety regime, the designer of the system and the system operator that must make this difficult cost-benefit trade-off, a trade-off that in reality often, but not always, consists of establishing economic criteria for acceptable risk level.

The concept of cost-benefit analysis is quite simple, and need not necessarily involve fancy mathematical tools. Actually we all perform cost-benefit analyses on a daily basis, for example when we are shopping for groceries. Most people decide on which items to buy based on a trade-off between their perceived benefits and costs, such as quality, price, personal preferences, etc. One should recognize that costs and benefits can be understood in general terms and not just in monetary terms. The costs of buying a certain product of course involve its monetary price but may also involve, for example, poor quality, health hazards, effect on the environment, etc. It is important to bear this in mind when performing all kinds of cost-benefit analyses (CBAs), as it may be difficult to examine all costs and benefits on a common scale.

This chapter will, as mentioned above, focus primarily on CBA in the context of safety and risk assessment. The first section of this chapter presents some basic theory that it is necessary to be aware of before performing CBAs in a risk assessment context. This involves a brief introduction to the ALARP (= as low as reasonably practicable) principle, the concept of risk aversion, as well as some very basic economic and cost-optimization theory that enables us to calculate monetary costs and benefits. The second section of this chapter looks closer at CBA in a risk assessment context. The main principles and a general approach to such a CBA are presented together with some useful cost-benefit analysis methodologies. The third part of this chapter looks at some alternative problem-solving approaches to CBA, predominantly methods involving ranking of different concepts. Finally, a CBA example analysing design spill prevention measures for tankers is presented.

9.2 BASIC THEORY

9.2.1 The ALARP Principle

When applying cost-benefit analysis (CBA) concepts in a risk assessment context, a major challenge is to find an appropriate balance between costs and risk. The UK Health and Safety Executive (HSE, 1992), which is the United Kingdom's governmental safety department, developed the so-called ALARP principle (or concept) to provide some guidance on finding such an appropriate balance. ALARP is an abbreviation for 'as low as reasonably practicable', and the main principle is that the risks related to a system regarding possible damage to life, property and the environment should be reduced to a level that is as low as reasonably practicable (i.e. ALARP). For example, if the risk reduction achieved by implementing particular safety measures is insignificant compared with the costs of these proposed measures, meaning that there is a gross imbalance between the risk reduction and the related costs, it would not be reasonably practicable to implement them. On the other hand, if a significant risk reduction can be achieved for an acceptable cost, meaning that 'low-priced' safety may be gained, it would be reasonably practicable to implement the risk-reducing measures. Hence, the ALARP principle states that a safety (or risk-reducing) measure should be implemented unless it can be demonstrated that the costs of implementing the safety measure are grossly disproportionate to the expected safety improvements (i.e. benefits).

The ALARP principle can be illustrated by Figure 9.1. The application of the principle is based on the evaluation of risk. Through the use of a risk acceptance criterion the risks of specific hazards, or the total risk of the system under consideration, can be regarded as intolerable, tolerable or negligible. Intolerable risks are by definition unacceptable and must be made tolerable through the implementation of safety measures. The ALARP principle states that risks in the tolerable risk region can only be accepted if they are made as low as reasonably practicable using risk reduction measures. For risks in the negligible

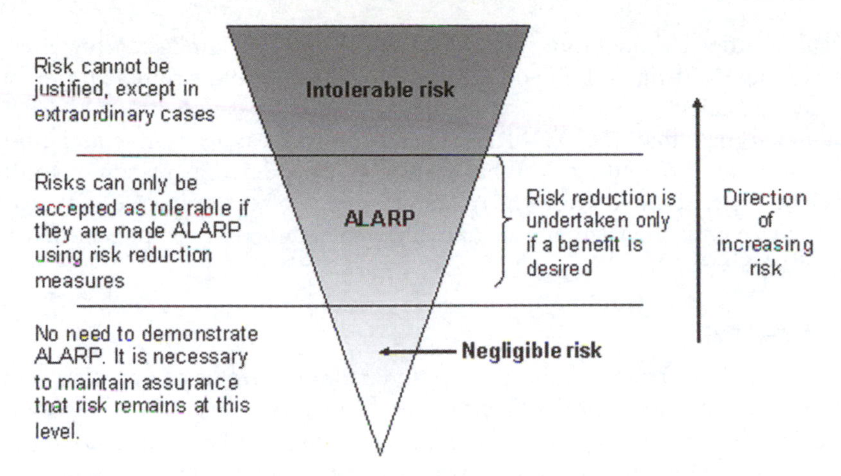

Figure 9.1. The ALARP principle.

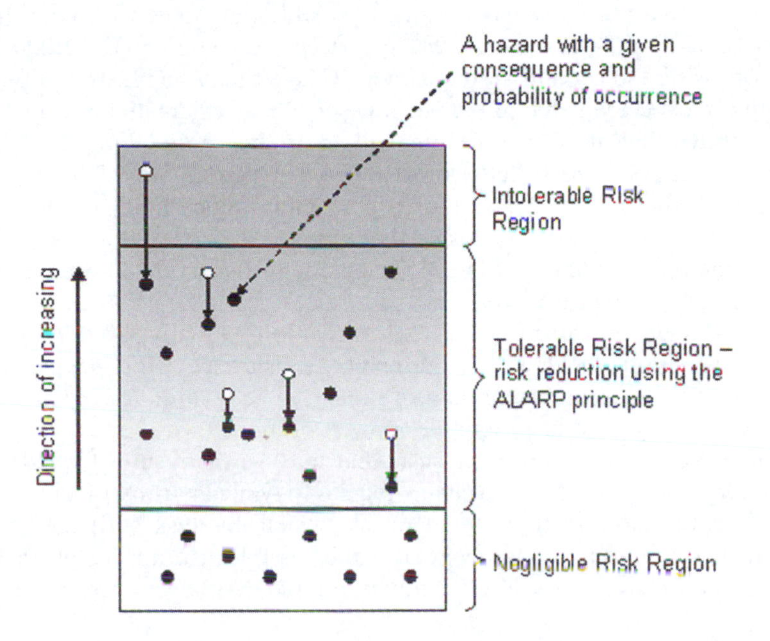

Figure 9.2. Risk reduction using the ALARP principle.

risk region, there is no need to demonstrate ALARP. Risk reduction using the ALARP principle is illustrated in Figure 9.2.

The ALARP risk region (see Figure 9.1) is dependent on what is exposed to the hazards (and their corresponding risks) under consideration. If people are exposed, the level where costs are grossly disproportionate to the achieved safety improvements is extremely low, as serious injuries and fatalities are regarded as unacceptable. Depending on the system under consideration, the ALARP regions for damage to the environment and property may be defined differently. For physical assets and equipment (i.e. property) net present value calculations may, for example, be used to establish the ALARP region.

It can be argued that the ALARP principle is too vague as it could possibly be interpreted in many different ways. What should be considered as reasonably practicable will always be a matter of discussion. A major benefit of the principle is that it can be easily adapted to individual situations and different contexts for application in a wide range of industries.

9.2.2 Risk Aversion

The concept of risk aversion is important to understand in terms of applying cost-benefit analysis (CBA) in the context of risk. Briefly explained, risk aversion is the tendency for people (e.g. managers, consumers, decision-makers, etc.) to avoid undertaking risks and to choose less risky alternatives. Risk aversion can be illustrated by an example related to gambling.

Consider a bet/gamble that has only two possible outcomes with equal probability. Given a bet of USD 100, there is a 50% probability of winning USD 200 and a 50% probability of losing the bet and all the money. This bet may be illustrated by Figure 9.3.

Although the expected value of the bet is USD 100, as calculated in Figure 9.3 below, experiments show that most people are willing to bet only about half of this, i.e. approximately USD 50. The difference between what people are willing to bet and the expected value of the bet is a measure of risk aversion. Since USD 100 is quite a lot of money for most people, the risk of losing that amount is perceived as too large to offset the possible benefits of winning USD 200. The grade of risk aversion does, however, vary with social and psychological factors.

People's risk aversion can be recognized in relation to many different activities. For example, risk aversion is the underlying principle of insurance. Most people are willing to pay a fixed small and acceptable amount of money to be secured (or insured) against the probable loss of a large and unacceptable amount of money. Insurance companies secure their customers against large losses or costs that may be the result of accidents or other undesirable events/situations by receiving regular payments from them. This payment is calculated on the basis of the estimated risk, which involves both the probability of the undesirable event and the consequence (i.e. the possible insurance payment), and a risk premium that compensates for the insurance offered to the customers. It is this risk

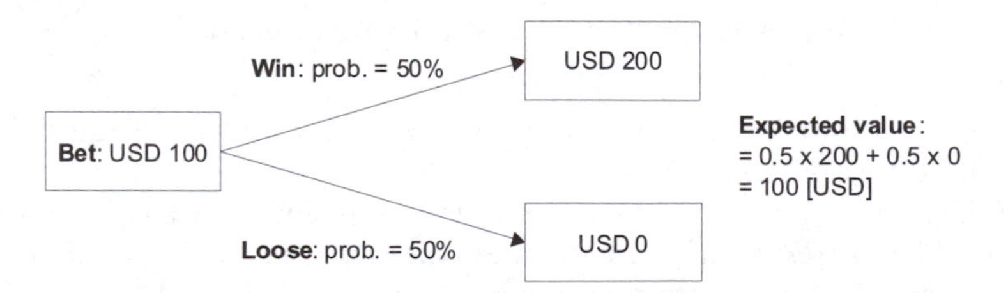

Figure 9.3. A bet of two possible outcomes with equal probability.

premium that makes it possible for the insurance companies to earn money in the long term. Risk aversion is also evident in business investment projects. Where an investment project risks making a loss that is so large as to endanger the existence of the company in question, the managers of that company would normally act with great risk aversion and thus require a large premium to induce them to take the risk.

The characteristics of risk aversion described above have an analogous relation in risk analysis. As was seen in the previous chapter, the risk (R) is normally defined as the product of the probability (P) of an event and its expected consequence (C):

$$R = C \cdot P \tag{9.1}$$

Based on this equation, the following three alternatives are of equal risk:

- 1 fatality per year ($R = C \cdot P = 1 \cdot 1 = 1$).
- 10% probability (per year) of an accident resulting in 10 fatalities ($R = C \cdot P = 10 \cdot 0.10 = 1$).
- 1% probability (per year) of an accident resulting in 100 fatalities ($R = C \cdot P = 100 \cdot 0.01 = 1$).

In reality, however, this is not the case. Based on our knowledge of people's risk aversion, it can be shown that the perceived consequence is larger than the calculated risk expresses. This situation can be illustrated by Figure 9.4, in which the perceived consequence is sketched as a function of the number of fatalities. The more severe the objective (or real) consequence is, the farther the perceived consequence moves away from the linear risk relationship, which is due to people's risk aversion.

Based on the relationship between the perceived and objective consequence shown in Figure 9.4, the basic risk equation (i.e. $R = C \cdot P$) has to be modified. The risk equation should include the perceived consequence instead of the objective consequence because

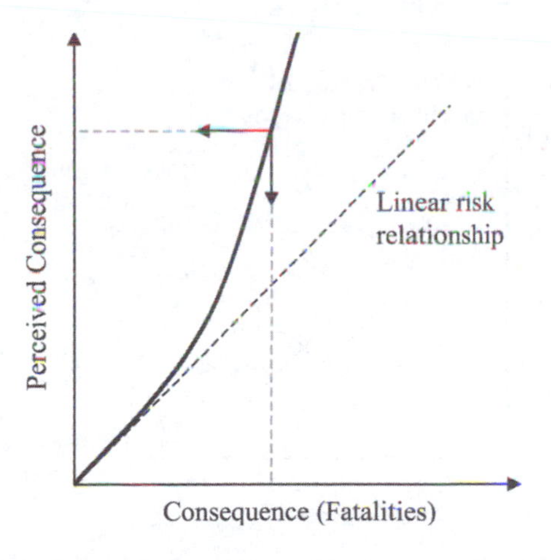

Figure 9.4. Perceived consequence as a function of the number of fatalities.

this would more correctly reflect the willingness to pay for risk reduction. The perceived consequence C_P can, for example, be expressed as a function of the objectively calculated consequence C_O:

$$C_P = f(C_O) \tag{9.2}$$

where C_P = perceived consequence and C_O = objectively calculated consequence. This modification implies that the curve for constant risk has to be modified to fit the perceived consequence, as shown in Figure 9.4. Such a modification is shown in Figure 9.5. The constant risk curve based on perceived consequences is, as expected, steeper than the constant objective risk curve.

9.2.3 Economic Theory

In relation to cost-benefit analyses it is important to study monetary costs and benefits on a common scale. There may also be non-monetary costs and benefits involved in a CBA, but in this part of the chapter we are not concerned with these. Both costs and benefits may arise at different points in time, and because the value of money changes over time due to inflation and 'time value of money' factors, one must be able to calculate the value of all monetary costs and benefits at one specific point in time. In order to calculate the value of a present amount P at some future point in time, the following simple equation can be used:

$$F = P \cdot (1 + i)^n \tag{9.3}$$

where:

P = Present monetary amount (currency)
F = The value of P after n years/periods (currency)

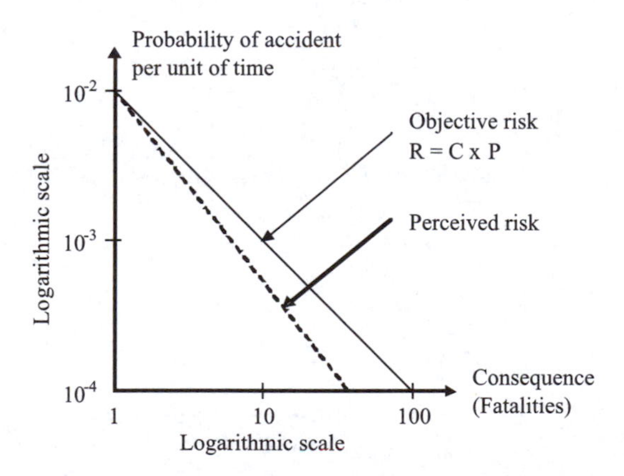

Figure 9.5. Constant risk curves for perceived and objective risk.

i = Rate of interest per year/period (corrected for inflation) given as a decimal fraction (e.g. 5% = 0.05)
n = Number of years/periods

With the use of Eq. (9.3) the value of a present amount some time in the future can be calculated. To perform the opposite calculation, i.e. to calculate the present value P of a future amount F, the following equation can be directly deducted from Eq. (9.3).

$$P = \frac{F}{(1+i)^n} = F \cdot (1+i)^{-n} \tag{9.4}$$

where:

P = Present value (currency)
F = Future amount that is incurred n years/periods into the future (currency)
i = Rate of interest per year/period (corrected for inflation) given as a decimal fraction (e.g. 5% = 0.05)
n = Number of years/periods before the future amount F is incurred

If a chain of n equal future amounts F are incurred at regular intervals (e.g. inspection costs, maintenance costs, loan payments, etc.), and the rate of interest i for those intervals can be assumed as constant, it can be shown that the present value P of this chain of future amounts F is as follows:

$$P = F \cdot \left[\frac{(1+i)^n - 1}{i \cdot (1+i)^n}\right] = \sum_{i=1}^{n} F \cdot (1+i)^{-n} \tag{9.5}$$

These equations are simple but very important tools in relation to CBA.

9.2.4 Cost Optimization

Safety measures are implemented into a system in order to improve its safety by reducing the inherent risks. Through such measures future undesirable events are made less likely, less severe, or a combination of the two. The costs involved in implementing safety measures may as such be understood as preventive costs, i.e. costs related to preventing danger to people, property and the environment. Safety can be improved through the use of a wide range of measures such as physical safety equipment/systems, organizational safety programmes, improved operating procedures, etc., and preventive costs would therefore typically include:

- Costs related to the design and development of safety measures/programmes
- Cost of equipment and installation
- Costs related to the inspection and maintenance of safety equipment in all its life
- Staff operating costs
- Training costs

- Enforcement costs
- Inspection and auditing costs
- Administration costs, etc.

On the other hand, the implementation of safety measures also results in benefits of reduced cost of losses related to the economic consequences that are more likely to be avoided because of reduced risks. In a maritime context, typical cost of losses includes:

- Total loss of ship, additional costs of getting a new vessel into operation
- Degraded operability/operation resulting in unscheduled delays
- Loss of future income (due to total loss or ineffective operation)
- Repair costs
- Fines and penalties
- Compensation to third parties
- Negative publicity (may be difficult to quantify), etc.

Based on the above discussion, a pure economic exercise can be performed to establish the optimal level of preventive safety measures that should be implemented. This can be done by studying the total safety cost, being the sum of preventive costs and cost of losses, and finding where this total cost is at its lowest. Such a cost optimization is illustrated in Figure 9.6. For this to be valid it must be possible to directly value the costs and benefits of safety measures in economic terms. In practice the economic consequences of accidents have a dominant position, but as mentioned earlier other factors may be taken into consideration as well.

The cost curves for preventive costs (C_P) and cost of losses (C_L) in Figure 9.6 are symmetrical. This does not, however, have to be the case, and the curves will generally vary greatly depending on the type of system under consideration.

In this section we have focused on establishing the cost-optimal safety level, but the type of cost optimization illustrated by Figure 9.6 has many applications, e.g. in terms of

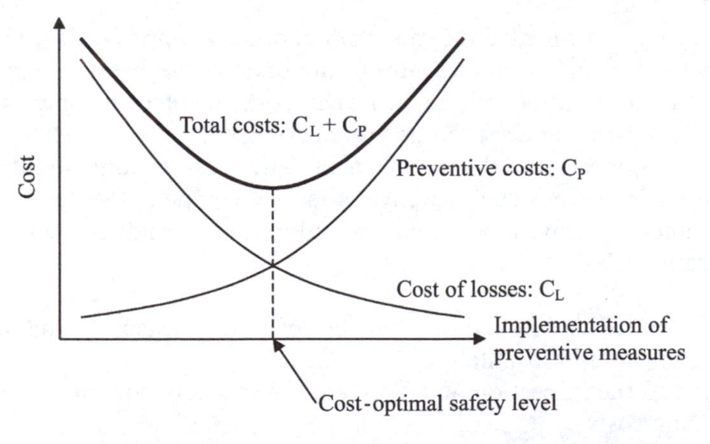

Figure 9.6. Optimal implementation of safety measures.

establishing an economically optimal level of preventive maintenance. Cost optimization can be a very useful tool in relation to cost-benefit analysis. Below some results from a risk assessment study of bridges with respect to ship collisions are presented (Sexsmith, 1983). This case shows how basic mathematical theory can be used to perform simple cost-benefit analyses.

Example

Problem

A bridge designer is to establish the optimal energy capacity of a bridge design that is exposed to ship traffic (Sexsmith, 1983). The energy capacity in 10^6 joule (MJ) denotes the energy that the bridge can absorb without collapsing. That is, if the impact energy from a ship colliding with the bridge is larger than the bridge's energy capacity, the bridge will experience catastrophic collapse. Increasing the energy capacity of the bridge requires the implementation of protective measures that come at a considerable cost. Table 9.1 gives information about the different bridge design alternatives.

There is also an assumed loss of USD $2 \cdot 10^8$ in the event of catastrophic collapse, including both the economic losses associated with the bridge structure and losses that occur as a result of loss of use of the bridge. The real rate of interest (i) is assumed to be $3\% = 0.03$ (per year).

Develop a mathematical model that enables the calculation of the most cost-optimal bridge concept of the three presented in Table 9.1.

Solution

Assuming that a loss C_f will occur at a definite time in the future, the present value of this loss can be expressed as follows:

$$C_0 = C_f \cdot e^{-it} \tag{9.6}$$

where:

C_0 = The present value of the future loss
C_f = The future loss in present monetary units (not inflated)

Table 9.1. Crucial bridge design parameters and cost of protective measures (C_C)

	Bridge energy capacity (MJ)	Return period for exceedence T (years)	Annual rate of exceedence $\lambda = 1/T$ (–)	C_C (USD)
Concept 1	800	500	0.002	$3.0 \cdot 10^6$
Concept 2	1000	1000	0.001	$6.0 \cdot 10^6$
Concept 3	5000	5000	0.0002	$12.0 \cdot 10^6$

Source: Sexsmith (1983).

i = The real interest rate (excluding inflation)
t = The time to the future loss

When the time to occurrence of the loss is a random variable (i.e. loss is stochastically distributed), the present expected value of the future loss (i.e. C_0) can be expressed as follows:

$$C_0 = E(C_f \cdot e^{-it})$$
$$C_0 = C_f \cdot \int_0^T e^{-it} \cdot f(t) \cdot dt \tag{9.7}$$

where:

$E()$ = The expected value function
$f(t)$ = The probability density function for the time t (i.e. the time to occurrence of catastrophic collapse)

Ship collisions with bridges are rare and independent random events in time. The events can therefore be considered as Poisson events, and the time to first occurrence is therefore exponentially distributed:

$$f(t) = \lambda \cdot e^{-\lambda \cdot t} \tag{9.8}$$

where:

λ = Annual rate of exceedence of bridge energy capacity
t = The time to the future loss

It then follows that the present value of the loss can be expressed as follows:

$$C_0 = C_f \cdot \int_0^T e^{-it} \cdot \lambda \cdot e^{-\lambda t} \cdot dt$$
$$C_0 = \frac{C_f \cdot \lambda}{i + \lambda} \tag{9.9}$$

Equation (9.9) can be implemented into a CBA to establish the optimum total cost consisting of the cost of loss (C_0) and the cost of protective measures (C_C), both costs that will depend upon the concept selected:

$$\text{Total cost } C_T = C_0 + C_C$$

By increasing the implementation of safety measures, the exceeding of the bridge's energy capacity will statistically happen more seldom, resulting in decreasing cost of

Table 9.2. Calculation of total cost

	Bridge energy capacity (MJ)	Return period for exceedence T (years)	Annual rate of exceedence $\lambda = 1/T$ (–)	C_C (USD)	C_0 (USD)	$C_T = C_C + C_0$ (USD)
Concept 1	800	500	0.002	$3.0 \cdot 10^6$	$12.5 \cdot 10^6$	$15.5 \cdot 10^6$
Concept 2	1000	1000	0.001	$6.0 \cdot 10^6$	$6.5 \cdot 10^6$	$12.5 \cdot 10^6$
Concept 3	5000	5000	0.0002	$12.0 \cdot 10^6$	$1.3 \cdot 10^6$	$13.3 \cdot 10^6$

losses, while the cost of the preventive measures increases. The total cost (C_T) for the three design concepts is calculated in Table 9.2. Here C_0, i.e. the present value of future loss of the bridge, is calculated using Eq. (9.9) with C_f= USD $2 \cdot 10^8$ and $i = 3\% = 0.03$.

Based on the developed model, the second concept is the most cost-optimal of the three concepts, resulting in an energy capacity of 1000 MJ for the bridge. However, it must be recognized that such a model is quite crude and very sensitive to the assumed parameters.

9.3 CBA IN A RISK ASSESSMENT CONTEXT

9.3.1 Principle

As briefly mentioned in the introduction to this chapter, the implementation of safety measures directed at reducing the risks of severe accidents related to a system unavoidably incurs costs. However, there are also benefits related to improved safety, predominantly the reduced cost of losses. In this context there are two dominant views on cost-benefit analysis (CBA), which are both described below.

The first view on CBA recognizes that for all systems there will be an economic limit to how much can be spent on safety measures before the system (e.g. an oil tanker) becomes uneconomic or uncompetitive, and there will therefore always be a trade-off between the costs of implementing safety measures and the residual risk level. In this context the challenge will be to achieve the best possible safety level within given economic limits. The second view on CBA is based on weighing the costs of implementing a safety measure against the benefits gained through this implementation with the objective of implementing cost-effective measures. Cost optimization can be a useful tool in this regard as it also enables us to estimate the optimal amount to spend on preventive (i.e. safety) measures and the economic break-even point for such measures.

Both views on CBA are important. However, given the very competitive business environment in shipping and the fact that maritime activities happen within a relatively strict safety regime of prescriptive regulations/requirements, only cost-effective safety measures will normally be implemented. In a risk assessment context this favours the second view on CBA, in which the main principle is to find safety measures that cost-effectively reduce risk. With regard to risk reduction one often has several different possible safety measures to choose from, and CBA enables us to identify for implementation those that are cost-effective.

9.3.2 Approach

The principle/objective of CBA in a risk assessment context is to identify cost-effective safety (or risk reduction) measures. This part of the chapter will present an approach directed at achieving this objective. In the previous chapter an approach to risk analysis and risk assessment was presented, and this will now be very useful in terms of understanding the necessary tasks to be performed.

A cost-benefit analysis is easy to perform as long as the costs and benefits for the suggested safety measures are known. The challenge in relation to risk assessment is how to establish these costs and benefits on the basis of the risk analysis model developed. The costs of safety measures are mainly associated with the costs of implementing, operating (including inspections, audits and maintenance) and administering the safety measure. Estimating the benefits of safety measures is, in general, more complicated and difficult. The benefits of safety measures are related to the value of averting and/or reducing the consequences of undesirable hazards becoming reality. Based on the previous chapter, the effect of safety measures on the probability of occurrence and consequences can be studied using the developed risk analysis model, i.e. by estimating their effect on the fault tree analysis (FTA) and event tree analysis (ETA). If these benefits can be quantified in monetary terms, a cost-benefit ratio (i.e. the costs divided by the benefits) can be calculated, and such ratios for the different safety measures under consideration can be used to make a decision on whether it is advantageous to implement such measures, and in that case which measures to implement. It must, however, be remembered that there are usually great uncertainties related to both fault and event tree analysis, and sensitivity analyses should be performed to test the robustness of the initial recommendation. This CBA approach can be illustrated by Figure 9.7.

CBAs regarding safety-related matters are normally based on marginal considerations, which means that the preventive measures are implemented as long as the estimated benefits of reduced risk at least equal the expected costs (i.e. costs $\leq$ benefits).

A problem related to assessing the benefits of averted and/or reduced consequences, as a result of introduced safety measures, is the often tremendously large number of consequence types that may be affected and the fact that the safety-improving effects of a particular safety measure may vary strongly between the consequences (Roland and Moriarty, 1990). One particular safety measure may, therefore, affect many types of damage extents for several accident types as illustrated by the general accident model given in Figure 9.8. The total effect of the measure can as a result be difficult to establish and quantify for CBA applications.

Example

Problem

A shore-based oil refinery receiving crude oil from shuttle tankers has recently carried out risk analyses for the parts of its operation that may result in oil spills. One area where such spills occur is during tanker offloading. The emergency response unit at the refinery is well prepared to initiate fast and effective clean-ups of smaller spills during offloading, but

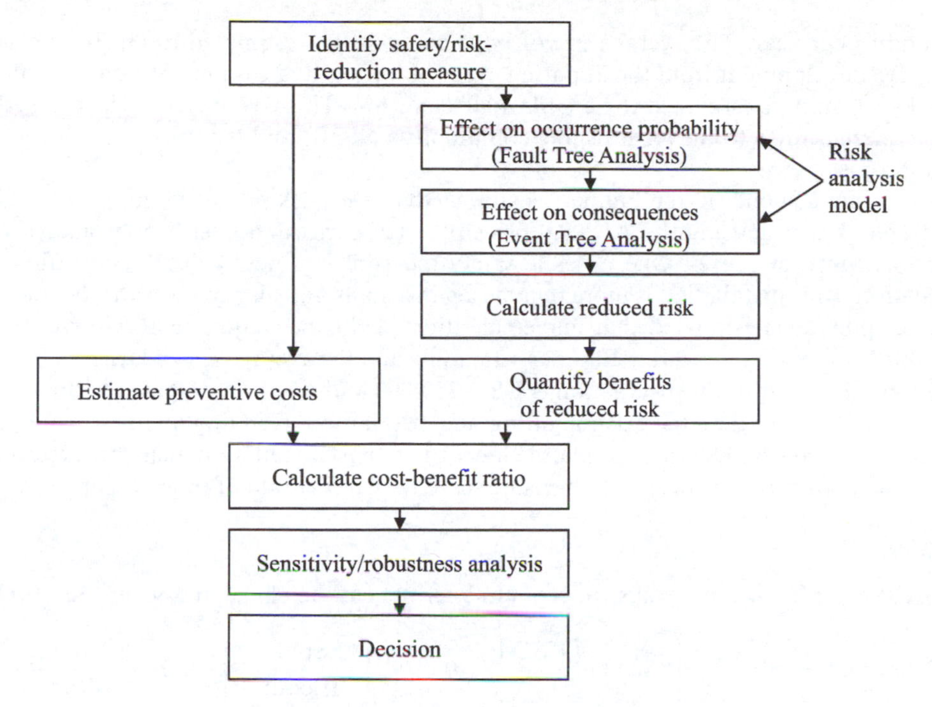

Figure 9.7. CBA in a risk assessment context.

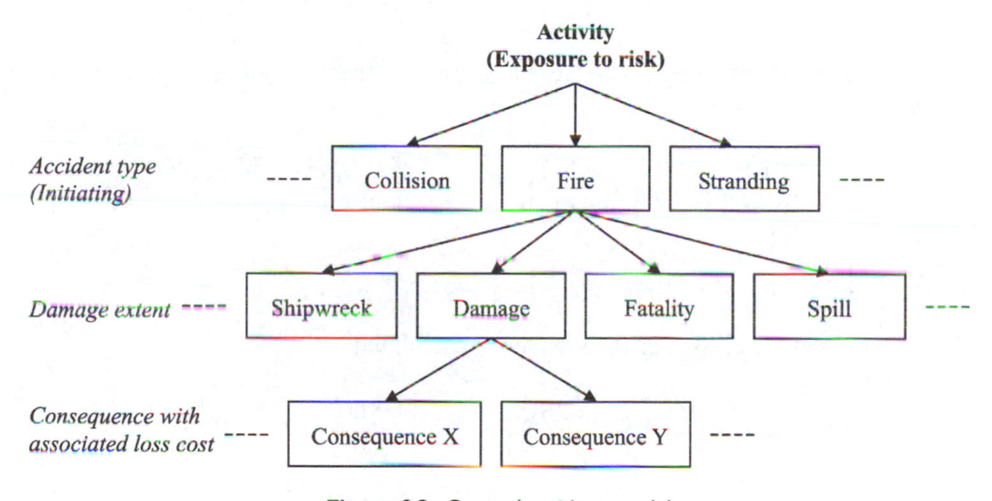

Figure 9.8. General accident model.

spills larger than 10 tons are a major concern as they will often be difficult to contain and the harm to the surrounding environment can be serious. The probability of such large spills was estimated in the risk analysis to be $2.0 \cdot 10^{-4}$ per offloading using fault tree analysis of the offloading equipment. The refinery has an average of 120 tanker offloadings per year. The average size of large oil spills was estimated to be 50 tons using event tree development from the initiating event of 'offloading equipment failure', and the cost of such spills is estimated to be USD 40,000 per ton. This cost includes clean-up costs, fines, compensation to the local fishing community and affected landowners, extra public relations costs, etc.

The management at the refinery believes that the risk involved in offloading is acceptable, but practising the ALARP principle they perform a cost-benefit analysis for available safety measures. One possible safety measure is a newly developed offloading installation. This installation is more reliable against spills and also reduces the average size of large spills. It is estimated that implementation of this new equipment will reduce the probability of large spills to $1.2 \cdot 10^{-4}$ per offloading and the average size of large spills down to 15 tons. The cost of the installation is USD 750,000 and it has an expected lifetime of 12 years. It is far better than the existing offloading installation, resulting in an estimated cost reduction of USD 20,000 per year, maintenance of the new installation included. Determine the cost-benefit ratio for the safety measure, assuming a 5% rate of interest per year.

Solution

The existing risk for large spills (in economic terms) can be calculated using Eq. (9.1):

$$R_0 = C \cdot P = 50\left[\frac{\text{Tons}}{\text{Spill}}\right] \cdot 40{,}000\left[\frac{\text{USD}}{\text{Ton}}\right] \cdot 2.0 \cdot 10^{-4}\left[\frac{\text{Spills}}{\text{Offloading}}\right] \cdot 120\left[\frac{\text{Offloadings}}{\text{Year}}\right]$$
$$= 48{,}000\left[\frac{\text{USD}}{\text{Year}}\right]$$

The risk after implementation of the safety measure, i.e. the offloading installation, will be as follows:

$$R_1 = C \cdot P = 15\left[\frac{\text{Tons}}{\text{Spill}}\right] \cdot 40{,}000\left[\frac{\text{USD}}{\text{Ton}}\right] \cdot 1.2 \cdot 10^{-4}\left[\frac{\text{Spills}}{\text{Offloading}}\right] \cdot 120\left[\frac{\text{Offloadings}}{\text{Year}}\right]$$
$$= 8640\left[\frac{\text{USD}}{\text{Year}}\right]$$

The benefits of reduced risk can then be easily calculated:

$$\Delta R = R_0 - R_1 = 48{,}000\left[\frac{\text{USD}}{\text{Year}}\right] - 8640\left[\frac{\text{USD}}{\text{Year}}\right] = 39{,}360\left[\frac{\text{USD}}{\text{Year}}\right]$$

The net present value P_b of this benefit for the 12-year lifetime of the equipment can then be calculated using Eq. (9.5):

$$P_b = F \cdot \left[\frac{(1+i)^n - 1}{i \cdot (1+i)^n}\right] = 39{,}360 \cdot \left[\frac{(1+0.05)^{12} - 1}{0.05 \cdot (1+0.05)^{12}}\right] = \text{USD } 438{,}340$$

The present value P_c for the costs of the offloading installation will be as follows:

$$P_c = 750{,}000 - 20{,}000 \cdot \left[\frac{(1+0.05)^{12}-1}{0.05 \cdot (1+0.05)^{12}}\right] = \text{USD } 527{,}266$$

The cost-benefit ratio will then be:

$$\frac{C}{B} = \frac{P_c}{P_b} = \frac{527{,}266[\text{USD}]}{438{,}340[\text{USD}]} = 1.20$$

The cost-benefit ratio is larger than 1.0, and hence the proposed new offloading installation is not found to be cost-efficient and the conclusion is that it should not be implemented.

9.3.3 Cost-Benefit Analysis Methodologies

All suggested risk control measures result in different risk reductions, different benefits and negative effects, and different implementation costs. Without some kind of method to evaluate these risk control measures against each other on a similar basis it would be very difficult to select the most cost-effective measures for implementation, i.e. the measures that result in the greatest benefits compared to costs. As long as most of the costs and benefits can be quantified in terms of monetary values, one should attempt to evaluate the measures on a similar basis.

The adaptation of the possible measures on to a common scale is one of the most important characteristics of cost-benefit assessment (CBA) methodologies. Different approaches may be used to achieve this, and the main principles of two popular approaches will be reviewed here, i.e. the Cost per Unit Risk Reduction (CURR) and the Implied Cost of Averting a Fatality (ICAF).

The Cost per Unit Risk Reduction (CURR) methodology was initially developed for use in the international context of the IMO where there might be expected to be a great disparity of views on how reduced fatalities and injuries should be valued. The approach adopted is to value all the cost and benefit items, except from the economic benefits of reduced fatalities, in monetary terms and to separately establish the number of equivalent lives lost over the lifetime of the measure assuming an equivalence between minor injuries, major injuries and death (e.g. 100 minor injuries accounts for 10 major injuries, which again accounts for one death). The net present value (NPV) of implementing a risk control measure is calculated using the following equation:

$$\text{NPV} = \sum_{t=0}^{n} \left[(B_t - C_t) \cdot (1+r)^{-t}\right] \tag{9.10}$$

where:

C_t = The sum of costs in period t

B_t = The sum of benefits in period t (excluding economic benefits of reduced fatalities)

r = The discount rate per period
t = Measure of time horizon for the assessment, starting in period (e.g. year) 0 and finishing in period n

The resulting NPV is then used to calculate a Cost per Unit Risk Reduction (CURR) by dividing NPV by the benefit of the estimated number of reduced equivalent fatalities. The CURR values for the different risk reduction measures can then be compared for cost-effectiveness in improving human safety.

All estimates of costs and benefits involve some uncertainty, and this uncertainty should be evaluated and taken into account. Uncertainty may for instance be evaluated by performing a sensitivity analysis on the parameters involved in order to study how changes in these affect the total net present value (NPV).

The Implied Cost of Averting a Fatality (ICAF) methodology is a much-used methodology for studying risk control measures on a common scale. The methodology calculates/estimates the achieved risk reduction in terms of cost using the following equation:

$$\text{ICAF} = \frac{\text{Net annual cost of measure}}{\text{Reduction in annual fatality rate}} \tag{9.11}$$

The net annual cost of a measure is calculated by distributing all the costs related to the implementation and operation of a measure over the measure's lifetime. This is achieved by calculating the yearly annuity. ICAF may also be calculated by dividing the net present value for the whole lifetime of the safety measure by the total reduction in fatalities for that particular period of time. The ICAF value can be interpreted as the economic benefits of averting a fatality. A decision criterion must be established for this value in order to evaluate whether a given risk control option/measure is cost-effective or not, and this criterion would in a way involve pricing a human life. DNV has, for example, indicated that risk control measures with an ICAF value of less than USD 3 million generally should be considered as cost-effective and therefore implemented. A method for calculating the ICAF value is presented below.

A third and less comprehensive method for adaptation on to a common scale is to only examine the net present value (NPV) of the different safety measures. Safety measures (or risk control options) with a positive NPV, which do not have other negative effects on the system under consideration, should always be implemented. However, only a few safety measures will normally have a positive economic NPV, and the method has a major weakness in not addressing the relative differences in risk reduction effect between different safety measures.

9.3.4 Calculation of ICAF

The benefits of averting a fatality are difficult to quantify. Some even assert that such quantification is impossible because it involves associating a value to human lives.

Nevertheless, an economic criterion of this kind is of great value in risk analysis, and not using such a criterion may even be counter-productive in relation to safety because the benefits of averting a fatality can be an important incentive in implementing costly risk reduction measures.

According to Skjong and Ronold (1998), the ICAF value can be calculated through analysis of a so-called Life Quality Index, which is a compound social indicator. Skjong and Ronold define the Life Quality Index as follows:

$$L = \gamma^{w} \cdot \varepsilon^{1-w} \tag{9.12}$$

where:

L = Life Quality Index
γ = Gross domestic product per person per year
ε = Life expectancy (years)
w = Proportion of life spent in economic activity (in developed counties $w \approx 1/8$)

The principle of ICAF as a criterion for risk reduction is to implement safety measures as long as the change in L is positive. By partially differentiating L and requiring that $\delta L > 0$, the following equation is established:

$$\frac{\Delta\varepsilon}{\varepsilon} > -\frac{\Delta\gamma}{\gamma} \cdot \frac{w}{1-w} \tag{9.13}$$

It can be assumed that the prevention of a fatality will on average save $\Delta\varepsilon = \varepsilon/2$, which equals half of the life expectancy. The largest change in gross domestic product, $|\Delta\gamma|_{max}$, is gained by implementing this expression for $\Delta\varepsilon$ in Eq. (9.13). This can be interpreted as the optimum acceptable cost per year of life saved. The optimum acceptable implied cost of averting a fatality, $ICAF_0$, can then be calculated by Eq. (9.14) below:

$$ICAF_0 = |\Delta\gamma|_{max} \cdot \Delta\varepsilon = \frac{\gamma \cdot \varepsilon}{4} \cdot \frac{1-w}{w} \tag{9.14}$$

where:

$ICAF_0$ = Optimum ICAF value
$|\Delta\gamma|_{max}$ = $-\gamma \cdot (1-w)/2 \cdot w$ based on Eq. (9.13)
$\Delta\varepsilon$ = Years saved by averting a fatality $= \varepsilon/2$
γ = Gross domestic product per person per year
ε = Life expectancy (years)
w = Proportion of life spent in economic activity (in developed counties $w \approx 1/8$)

Based on this criterion, proposed/suggested safety (or risk control) measures should be implemented as long as the estimated ICAF value does not exceed $ICAF_0$ (i.e. the criterion). The 1984 value of $ICAF_0$ was about GBP 2 million for developed countries. The $ICAF_0$ value does, however, vary with time. References to gross domestic products and life expectancy can be found at http://www.oecd.org/.

9.4 ALTERNATIVE PROBLEM-SOLVING APPROACHES

In this section some alternative problem-solving approaches to cost-benefit analysis will be presented, showing that there are several ways of performing such analysis. Which method to apply will largely depend upon the amount of information available regarding the activity or system under consideration.

9.4.1 Ranking of Concepts

For risk-based CBA, as illustrated by Figure 9.7, there are several tasks that must be performed. First, the effect of a specific safety measure on occurrence probability and the consequences must be assessed. This can be done by analysing the effects of the safety measure on the probabilities in the fault trees and event trees. However, utilizing these techniques require a lot of detailed information about the activity or system under consideration, as well as a comprehensive risk analysis model for that activity/system. For an existing system, such information can often be obtained and risk analysis models developed, but this may be costly and the uncertainties involved in the information and models are often relatively large. In addition, for an activity not yet carried out or for a system being designed, neither detailed information nor risk analysis models may be easily available. The implication of this is not that it is undesirable or impossible to perform a CBA comparing different safety measures, because other techniques may come to the rescue. One such technique is the ranking of different concepts (or alternatives). The term concept should here be understood in a broad manner, and could for instance include design concepts, safety measures, operational procedures, etc.

Different concepts (e.g. design concepts or safety measures) have different influences on cost and benefit factors. In a risk assessment context the different influences may be in terms of implementation or investment costs, operating costs, accident frequency, spill volume, etc., and it may be difficult to transform these influences on to a common scale (e.g. economic/monetary values) for comparison. A solution to this problem is to make no attempt to perform such a transformation and instead rank the different concepts on the basis of a set of carefully selected parameters. The parameters are the costs and benefits that are to be taken into account and compared as part of the CBA, and it is important that the number of cost and benefit parameters is balanced. The ranking is performed by giving each parameter for each concept a grade that reflects how good the concept is relative to the other concepts for the various parameters. This may sound a little vague at this point, but it will all become clearer through a simple case.

Consider an offshore development project in which there are three possible design concepts for the oil production. In the initial phase of the design process for this development a preliminary CBA is to be performed in order to single out the best concept and show how the different design concepts relate to each other in terms of costs and benefits. The project management have selected a set of parameters or criteria on which they want the concepts to be assessed, and these parameters include central economic, technical and safety-related factors that should give a reasonably good total picture of how the different concepts (or alternatives) differ. The assessment parameters, and the information estimated/gathered on each of these for the three design concepts, are shown in Table 9.3.

The different concepts must be weighted against each other for each of the assessment criteria/parameters. This can, for example, be done by implementing an ordinal grade scale for each parameter ranging from 1 to 3, where 1 denotes the best concept, 2 the second best concept and 3 the worst concept. This is done for all the assessment criteria and then the sum of all the grades is found for each concept. Consequently, the concept with the lowest total grade is considered the best one. The ranking of the three offshore development concepts is performed in Table 9.4.

By assuming that all of the assessment criteria/parameters have equal importance, the concept involving the fixed platform is considered the best concept, slightly better than the underwater production concept. This does not mean that the fixed platform is 'approved'. The project management and its associated team of engineers should now make improvements on all the concepts on the basis of the information provided by Tables 9.3 and 9.4. Later in the design process, when the concepts have been improved and more information is available about them, a more detailed ranking process should be performed, maybe involving more detailed assessment criteria/parameters.

The main advantage of the ranking technique described above is its simplicity. The technique is, however, very sensible to the assessment parameters that are included. In addition, all the parameters have been given equal weighting (i.e. importance), which may not be a correct reflection of reality. For example, the estimated risk of fatality may in reality be considered more important than the annual spill volume. Finally, the

Table 9.3. Concept alternatives for offshore oil production

Assessment criteria	Design concepts		
	Fixed platform	Floating platform	Underwater production
Prod. cost per barrel	USD 14.00	USD 12.00	USD 9.00
Investment cost	NOK 5 billion	NOK 6 billion	NOK 4 billion
Development time	3 years	4 years	5 years
Technological status	Established	Quite well known	Problematic
Availability	0.99	0.94	0.90
Est. accident frequency	10^{-3}	10^{-2}	10^{-4}
Est. risk of fatality	10^{-5}	10^{-4}	10^{-10}
Fatalities, large accident	30	50	0
Annual spill volume	10 tons	22 tons	30 tons
Accidental spill volume	5000 tons	100 tons	500 tons

Table 9.4. Concept alternatives for offshore oil production

Assessment criteria	Design concepts		
	Fixed platform	Floating platform	Underwater production
Prod. cost per barrel	3	2	1
Investment cost	2	3	1
Development time	1	2	3
Technological status	1	2	3
Availability	1	2	3
Est. accident frequency	2	3	1
Est. risk of fatality	2	3	1
Fatalities, large accident	2	3	1
Annual spill volume	1	2	3
Accidental spill volume	3	1	2
Sum	18	23	19

comparison between the concepts becomes quite rough because the absolute differences on each of the assessment parameters have little impact. For example, if the accidental spill volume for the fixed platform concept was ten times as high, this fact would not have changed the total result of the ranking. The latter drawbacks may be diminished if a grading scale that opens for an assessment of absolute differences within a certain parameter is implemented. Such a grading scale can, for instance, be defined as follows:

1. Very good performance
2. Good performance
3. Acceptable performance
4. Poor performance
5. Very poor performance
6. Unacceptable performance

The use of such a scale could possibly have resulted in a different end result of the ranking performed in Table 9.4. Below a somewhat more sophisticated ranking technique is presented.

9.4.2 Relative Importance Ranking

In previous chapters it has been shown that it is often practical to express safety by a set of consequence parameters. Such parameters may be fatalities per 10^8 working hours, economic and material loss, spill volume, etc. The relative importance of these consequence parameters is, however, not intuitive and introduces difficulties for the analyst. One possible method that may be used to estimate the relative weights of importance is presented here. This method is not ideal, but deals with the problem in a concise manner. The lack of consideration of the relative importance of different assessment parameters was one of the main drawbacks of the unsophisticated ranking technique presented above.

Let us consider a case in which the safety of oil transport from an offshore installation to a shore-based refining facility is to be analysed. It is supposed that the safety parameters relevant in the transportation of oil are as follows:

- Spill volume (i.e. oil pollution of the environment)
- Economic/material loss (i.e. damage to and loss of vessel)
- Number of fatalities aboard
- Population exposed by an explosion

Two different oil transportation concepts are to be assessed on their safety characteristics. Other parameters than safety-related ones may have been included in the analysis, such as the investment cost and development time, but these are not considered of importance in this particular analysis.

The first task of the cost-benefit assessment process is to establish the relative weights of importance that the assessment criteria/parameters are to be given. One possible approach in estimating these relative weights of importance is to gather/organize a group of experts holding excellent system knowledge about the systems and activities under consideration, as well as substantial familiarity with the preferences of governments and of society at large (i.e. the public). How these experts perceive the relative importance of the different assessment parameters, which are here exclusively related to safety, can then be measured by the use of, for example, questionnaires. Based on this information the relative weights of importance for the group of experts as a whole can be established. This final result is then assumed to reflect reality. Table 9.5 presents a questionnaire that is applied when estimating the relative weights of importance for the different safety parameters related to the case of oil transportation. So-called paired comparison is applied in this technique. A value of 0 implies equal importance, while values of 1 to 5 favour the relative importance of one of the parameter concerned. The values given in Table 9.5 are examples of answers that might have been collected from the experts, and these answers indicate, for example, that the parameters of exposed population and number of fatalities are both far more important than economic loss.

Table 9.5. Questionnaire for ranking of safety parameters

	5	4	3	2	1	0	1	2	3	4	5	
Spill volume				×								Economic loss
Spill volume									×			Number of fatalities
Spill volume								×				Exposed population
Economic loss										×		Number of fatalities
Economic loss											×	Exposed population
Number of fatalities									×			Exposed population

By allowing each member of the group of experts to answer such a questionnaire, a quantitative estimate of the resulting relative weights of importance for the safety parameters can be calculated. In practice this calculation is executed by normalizing the matrix in the questionnaire (i.e. Table 9.5). The relative weights of importance are then calculated on a common scale and may be presented as shown in Figure 9.9.

The next step is to establish a utility function for each of the four safety parameters. In this context, utility denotes the decision-maker's scale of preference, and a utility function describes graphically how the perceived utility changes with changing consequences. It is assumed here that the utility can be expressed as a continuous function with values between 0 and 1. The higher the utility (i.e. the closer to 1), the more acceptable a specific consequence is for a given accident. Similarly, as a consequence becomes more and more unacceptable and undesirable, the lower the utility will be (i.e. closer to 0). Thus, the utility functions express the perceived risks, with the risk increasing as the utility number decreases. The utility functions can also be established using a group of experts. The utility functions for the four safety parameters in the oil transportation case are shown in Figure 9.10. The graphs presented in Figure 9.10 require the following comments:

- *Spill volume:* No spill volume gives the highest utility (i.e. 1.0) as no environmental damage can be regarded as acceptable. However, small oil spills do not cause substantial damage to the environment, and the utility remains relatively high. Medium oil spills in the order of 100–1000 tons will, on the other hand, reduce the utility significantly as the consequences to the environment increase dramatically. Even larger spills (i.e. > 1000 tons) will have very detrimental effects on the environment and cannot be accepted, hence the low utility value for such spills.
- *Economic loss:* Economic losses less than NOK 10 million have relatively little effect on the utility, which remains relatively high. This shows that such material losses in this particular case are regarded as quite acceptable. One possible explanation of this is that such costs may be involved even in smaller accidents (i.e. a threshold cost). Losses larger than NOK 10 million have a significant effect on the utility function, and costs in the order of NOK 100 million are considered totally unacceptable – hence the utility value of 0.
- *Fatalities aboard:* Accident statistics reveal difficulties in avoiding 1–3 fatalities, and although very undesirable the utility therefore stays relatively high. Accidents of

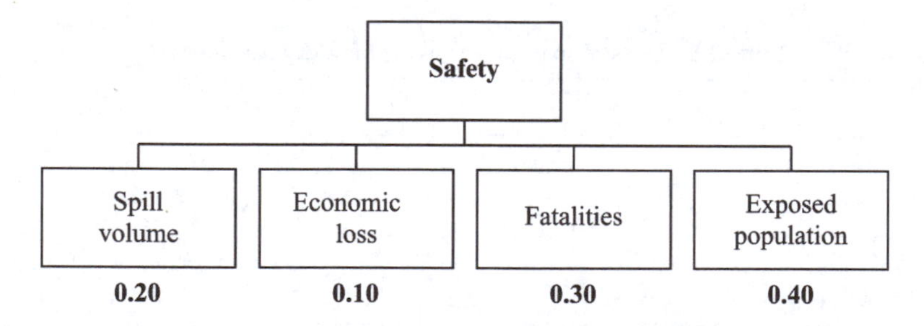

Figure 9.9. Relative weights of importance for the set of safety assessment parameters.

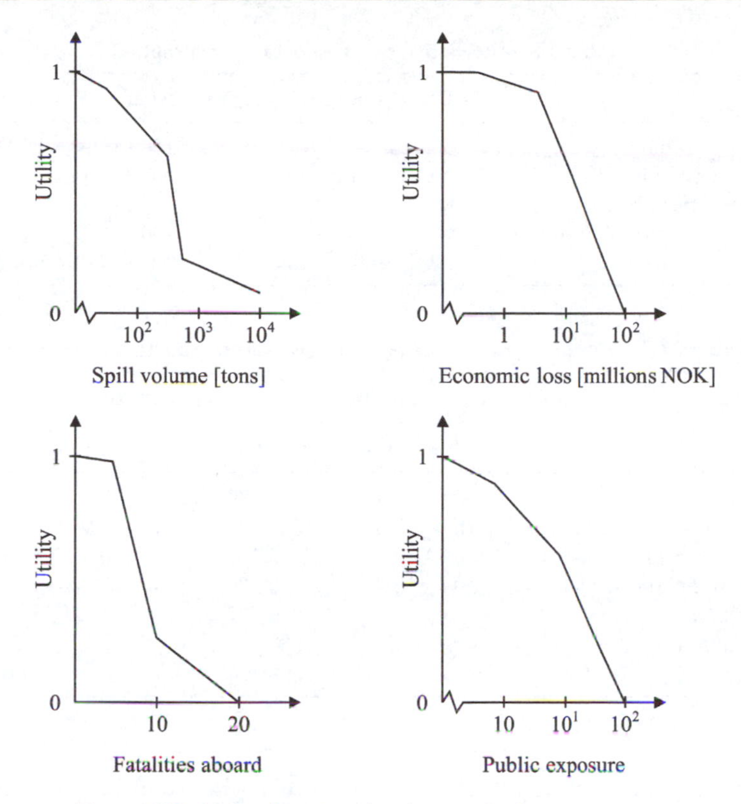

Figure 9.10. Utility functions for the set of safety parameters.

size 3–10 fatalities are considered much more serious, hence a dramatic reduction in the utility value. Accidents with more than 10 fatalities are catastrophes that must be avoided. As can be seen in Figure 9.10, the utility value is set at 0 for accidents involving 20 or more fatalities.

- *Public exposure:* With regard to public exposure, the utility function is close to being a linear function of the logarithm of the exposed population. An exposed person may in this context be defined as a person within a given distance from the centre of a fire or explosion, resulting in that person being subjected to serious danger. As would be expected, the utility decreases considerably as the number of exposed individuals increases.

The relative weights of importance and the utility functions for the selected set of safety assessment parameters gives us the necessary foundation for estimating the cost-benefit ratio for different system concepts (e.g. designs of the oil transportation vessel/system). Let us assume that two concepts are to be compared against each other on the basis of the four safety parameters and the defined utility functions. The two oil transportation concepts are described in Table 9.6. The oil is loaded from a buoy offshore and shipped with two shuttle tankers to a shore-based terminal and refining facility.

Table 9.6. Alternative oil transportation concepts

	Basis concept	Alternative
Spill volume	500 tons	1500 tons
Economic loss	NOK 16 million	NOK 4 million
Fatalities aboard	8	2
Exposed population	50	100
Total costs	NOK 800 million	NOK 770 million

Table 9.7. Cost-benefit analysis (CBA) of the two oil transportation concepts

Criterion	Priority	Basis concept			Alternative		
		Value	Utility	Weight	Value	Utility	Weight
Spill volume	0.20	500	0.95	0.190	1500	0.65	0.130
Economic loss	0.10	16.0	0.80	0.080	4.0	0.96	0.096
Fatalities aboard	0.30	8	0.55	0.165	2	0.98	0.294
Exposed population	0.40	50	0.75	0.300	100	0.60	0.240
Total utility				0.735			0.760
Costs (million)				800			770
C-B ratio				**1088**			**1013**

The basis concept is clearly the most favourable when considering environmental damage as the expected spill volume is only one-third of the spill volume for the alternative concept. In addition, the expected number of the exposed population is only half for the basic concept. The alternative, on the other hand, is more oriented in reducing ship damage in terms of expected economic loss, and the expected number of fatalities aboard is also considerably lower. In Table 9.7 a cost-benefit analysis of the two concepts is performed using the weights of importance and the utility graphs. The total utility, which is considered as a measure of the benefit/quality of a specific concept, is calculated and compared to the total costs, i.e. the net present value of the necessary investments and future operational costs. A cost-benefit ratio is calculated, enabling a comparison to be made on a common scale. In Table 9.7 the 'priority' is the relative weights of importance for the safety parameters/criteria, and the utilities are read off the graphs in Figure 9.10 depending on the characteristics of the two concepts. The weight is the product of the priority figure and the utility. Table 9.7 shows that the alternative concept has a lower cost-benefit ratio than the basic concept, and is hence considered better in this particular analysis.

9.4.3 Valuation of Consequence Parameters

In earlier sections of this chapter it is assumed that the potential consequences or losses for a particular concept can be expressed as concise values such as, for example, the number of fatalities. In reality, however, this is sometimes problematic and not always desirable, especially in relation to accidents that involve both injuries and fatalities. In the context of CBA it is often necessary to establish an economic cost of both injuries and fatalities in order to compare different costs and benefits on a common scale. This section will look more closely at the valuation of people/personnel-related consequence parameters. In analyses of the safety of people/personnel, the following consequence parameters may be considered:

- Fatality
- Permanent disability
- Temporary disability

The costs associated with these people-related consequences may be valued according to the following factors:

- Insurance payments
- Estimated remaining life-income
- Claim for compensation
- Implicit social costs

The valuation of people-related consequences can be extremely controversial, both politically and ethically. In Table 9.8 a possible valuation method for such consequences is applied on two different concepts having different potential consequences in terms of fatalities as well as permanent and temporary disability. Unitary economic costs are applied to express the cost of one fatality, one permanent disability and one temporary disability. These values can be calculated on the basis of the costs stated above. In Table 9.8 these values are only chosen randomly to illustrate the method, and it will be described in detail how these values may be established. The total costs for the two concepts are calculated as the sum of the preventive costs (i.e. costs of implementation) and the consequences (i.e. average losses in an accident). Concept 2 gives the lowest total costs in this particular study.

Table 9.8. Total people-related loss costs of two different concepts

Safety parameters	Unitary cost (1000 USD)	Concept 1		Concept 2	
		Number	Cost	Number	Cost
Fatality	1,000	2	2,000	1	1,000
Permanent disability	400	15	6,000	10	4,000
Temporary disability	20	60	1,200	100	2,000
Calculated people-related loss cost			9,200		7,000
Costs of implementation			5,500		6,100
Total costs			14,700		13,100

McCormick (1981) introduced the term social costs, which expresses the costs to society of an injury or fatality. The following equation was suggested:

$$\begin{aligned} \text{Social cost} &= N \cdot C \cdot (1+i)^t, \quad \text{when } t < 6000 \\ &= N \cdot C \cdot (1+i)^{6000}, \quad \text{when } t \geq 6000 \end{aligned} \tag{9.15}$$

where:

N = Number of injuries or fatalities
C = Cost of damage per day
i = Daily rate of interest
t = Duration of damage or sick leave in days (6000 days is equivalent with a fatality)

This particular method of calculating the social cost of injuries and fatalities is only one of many different models that may be used. A problem not solved with Eq. (9.15) is that of establishing the cost of damage per day (i.e. C), which may vary considerably depending on the type of injuries suffered, the country in question, etc.

In an investigation made by O'Rathaille and Wiedemann (1980) an attempt was made to establish the average social cost for ship collisions and groundings based on statistical consequences. It was focused on oil spills and loss of lives, and the statistical data basis is presented in Tables 9.9 and 9.10. These tables show that the likelihood of oil spills is largest in groundings, while collisions much more frequently result in fatalities.

Based on the experience of known accidents, the cost of oil spills per accident was estimated to £5100–£6100 for collisions and £50,000–£280,000 for groundings, reflecting

Table 9.9. Ship accidents and oil spills (1976)

Primary cause	Number of accidents		Pollution rate, % of accidents leading to pollution
	Total	Of which led to pollution	
Collisions	44	1	2.27
Groundings	121	14	11.57

Table 9.10. Fatality risk in ship collisions and groundings (1976)

Primary cause	Number of accidents		Number of lives lost per accident
	Total	Number of lives lost	
Collisions	44	41	0.93
Groundings	121	4	0.03

that groundings are more likely to result in oil spills and that the spills on average are larger. However, as these figures show, the costs related to groundings tend to vary greatly.

In an assessment combining both economic and non-economic factors related to fatalities, the cost of a fatality was estimated to be £85,170–£98,105 in 1977 prices. On the basis of this, the average total costs of collisions and groundings, respectively, are estimated in Table 9.11. It can be concluded from Table 9.11 that the average total social cost of a grounding accident seems to be higher than that of a collision.

Insurance payments in the aftermath of accidents to people as well as the company/organization involved must also be considered a cost related to accidents. Insurance companies give compensation to the bereaved, and tend to vary considerably from case to case depending on the circumstances and the insurance schemes. There also tend to be quite different insurance practices in different countries. All these aspects make it difficult

Table 9.11. Average total costs of collisions and groundings (1977)

Primary cause	Cost of spills per accident (£)	Cost of fatalities per accident (£)	Average total cost per accident (£)
Collisions	5,100–6,100	79,363–91,416	84,463–97,516
Groundings	50,000–280,000	2,816–3,243	52,816–283,243

Table 9.12. Cost-benefit ratios (C/B) for different safety measures, USD 10^6 per life saved

Industry/activity	Safety measure	C/B
Nuclear industry	Radwaste effluent treatment systems	10
	Containment	4
	Hydrogen recombiners	> 3000
Occupational health and safety	OSHA coke fume regulations	4.5
	OSHA benzene regulations	300
Environmental protection	EPA vinyl chloride regulations	4
	Proposed EPA drinking water regulations	2.5
Fire protection	Proposed CPSC upholstered furniture flammability standards	0.5
	Smoke detectors	0.05–0.08
Automotive and highway safety	Highway safety programs	0.14
	Auto safety improvements, 1966–70	0.13
	Airbags	0.32
	Seat belts	0.08
Medical and health care programs	Kidney dialysis treatment units	0.2
	Mobile cardiac emergency treatment units	0.03
	Cancer screening programs	0.01–0.08

to generalize about insurance payments. The same accounts for claims of compensation that often surface in the wake of accidents. Such claims are often based on the lost (entirely or partly) remaining life income by reaching nominal age. Methods used in calculating such figures are often referred to as human capital methods.

The willingness to pay for preventive safety measures geared towards reducing fatalities differs between industries and types of activities. Table 9.12 shows an American overview of estimated preventive measures and costs per saved human life for different activities. The table shows that the nuclear industry, for example, is willing to pay more to save a human life than most other activities.

When studying the cost-benefit values in Table 9.12 it must be recognized that such values usually have a limited period of validity because of factors such as regulatory changes, new technology, changed public risk perception, etc. Cost-benefit values must therefore be used or referred to with great care.

9.5 CBA OF OIL SPILL PREVENTION MEASURES FOR TANKERS

Example

Problem

Preventing pollution from maritime activities has been a major priority in recent decades. On an international basis, concerns about the environmental impact of shipping have resulted in MARPOL, i.e. the International Convention for the Prevention of Pollution from Ships. MARPOL, which is one of the more important agreements achieved within the context of the International Maritime Organization (IMO), comprises design and operational regulations and requirements geared towards reducing pollution to both air and sea from shipping. In addition to MARPOL there exist several regional agreements and regulations on the prevention of pollution from shipping. One such set of regulations is the US Oil Pollution Act of 1990 (OPA 90), which was established as a direct consequence of the *Exxon Valdez* grounding accident in 1989 that resulted in a spill of 33,000 tons of crude oil in Prince William Sound on the coast of Alaska. OPA 90 gives shipowners full economic liability for spills in US coastal waters. There are several safety measures that may reduce potential oil spills as a result of ship collisions and groundings. The US National Research Council therefore performed a cost-benefit analysis on some of these possible safety measures for tankers, and this example is a summary of this analysis.

Solution

The objective of this analysis is to calculate the cost effectiveness of alternative designs for oil spill prevention. These are compared to a standard MARPOL tanker with protectively located segregated ballast tanks (PL/SBT). Segregated ballast tanks (SBT) means that there shall be designated tanks for ballast and that ballast is not to be carried in cargo tanks (except in very severe weather conditions in which case the water must be processed and discharged in accordance with specific regulations). Protective location (PL) of SBT means that the required SBT must be arranged to cover a specified percentage of the side and the bottom shell of the cargo section in order to provide protection against oil outflow in the case of groundings and collisions. The alternative designs (i.e. safety measures) studied in this analysis are explained below:

1. Double bottom (DB)

The double bottom (DB) constitutes the void space between the cargo tank plating, often referred to as the tank top, and the bottom hull plating. MARPOL requires a DB of 2 m or $B/15$, whichever is less (B = breadth of the vessel). The double bottom space gives protection against low energy grounding. Another benefit is the very smooth inner cargo-tank surface which facilitates discharge suction and tank cleaning. Drawbacks to a double bottom include increased risks associated with poor workmanship, corrosion, and obstacles to personnel access (to the DB). Other drawbacks are related to reduced side protection relative to the PL/SBT configuration, and increased explosion risk related to cargo flow into the DB. The double bottom configuration is shown in Figure 9.11.

2. Double sides (DS)

The double sides (DS) constitute the void space between cargo side/wing tanks and hull side plating. The minimum width of the DS is equal to that of the DB (i.e. 2 m or $B/15$). However, in order to meet ballast requirements, the width is likely to be larger (normally $B/7$–$B/9$). The design offers good protection against collisions, and also some of the advantages from double bottoms as the side tanks protect the outboard region of the bottom. Drawbacks of the DS configuration are related to bottom damages that result in direct spills, as well as higher susceptibility to asymmetric flooding. The double sides configuration is shown in Figure 9.12.

3. Double hull (DH)

The double hull (DH) constitutes the void space between tank and hull plating in both the sides and bottom. Compared to DS (and PL/SBT) the side protection is reduced, as the width of the tanks may be less than in the double sides design because ballast can be divided among the side and bottom spaces. From a cargo operations point of view

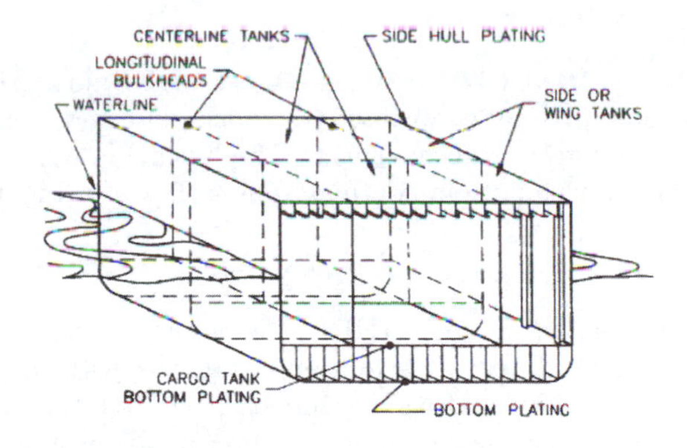

Figure 9.11. Double bottom (DB) configuration.

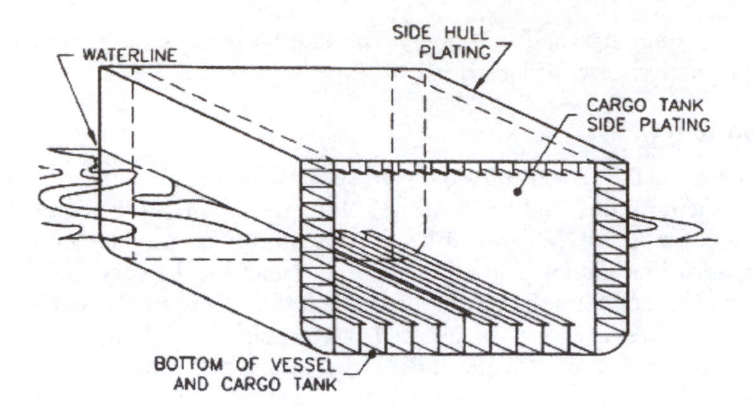

Figure 9.12. Double side (DS) configuration.

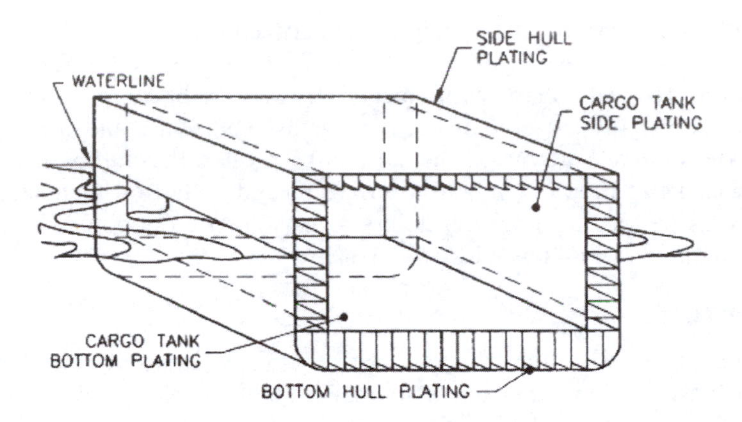

Figure 9.13. Double hull (DH) configuration.

the design is excellent. Drawbacks are similar those of DS and DB. In addition, the configuration/construction is more exposed to crack damages due to more plating. Corrosion may also be a larger problem. This puts high demands on access for inspections and maintenance. The double bottom configuration is shown in Figure 9.13.

4. Hydrostatic driven passive vacuum (HDPV)

The hydrostatic driven passive vacuum (HDPV) construction makes openings to cargo tanks airtight. This results in a progressive drop in pressure with cargo outflow, thereby reducing pollution as the oil is 'hold back'. Drawbacks are related to air tightening, and instantaneous location of damaged tanks for closure of all openings (e.g. vent pipes). Structural strengthening of the deck may be necessary in order to avoid structural collapse because of tank vacuum. The hydrostatic driven passive vacuum (HDPV) configuration is illustrated in Figure 9.14.

5. Smaller tanks (ST)

This design alternative is based on reducing the volume of the individual cargo tanks, which will reduce the potential oil outflow. The main drawback is related to an increased risk of plate cracking.

6. Interior oil-tight deck (IOTD)

An interior oil-tight deck (IOTD) greatly reduces the amount of cargo that is exposed to damage to the bottom plating and lower sides. This gives both hydrostatic favouring and reduced volume potential of oil spills if the hull integrity is broken. The need for ballast tanks makes double sides (DS) necessary, resulting in extra protection of the sides. The major drawbacks to this configuration are complex operation as well as corrosion damage. The interior oil-tight deck (IOTD) configuration is illustrated in Figure 9.15.

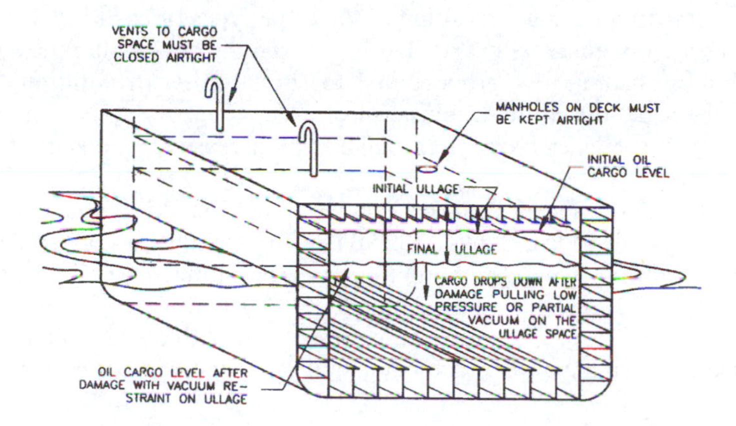

Figure 9.14. The hydrostatic driven passive vacuum (HDPV) configuration.

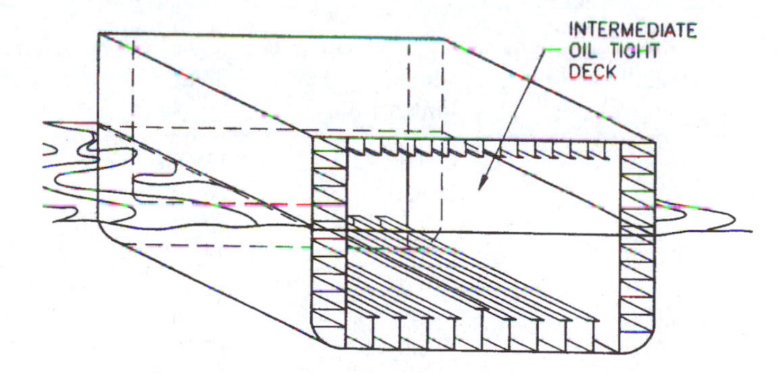

Figure 9.15. The interior oil-tight deck (IOTD) configuration.

7. Double sides and hydrostatic driven passive vacuum (DS/HDPV)

This design configuration (DS/HDPV) is a combination of both double sides (DS) and hydrostatic driven passive vacuum (HDPV). See Figures 9.12 and 9.14.

8. Double hull and hydrostatic driven passive vacuum (DH/HDPV)

This design configuration (DH/HDPV) is a combination of both double hull (DH) and hydrostatic driven passive vacuum (HDPV). See Figures 9.13 and 9.14.

The eight design alternatives presented above are different with respect to the following costs:

- *Capital cost:* The deadweight of the alternative designs is equal. However, the cost of design and construction will vary because of different complexity.
- *Maintenance and repair costs:* Some designs require more maintenance and repair because of higher exposure to salt water, resulting in more corrosion, increasing need for inspections, and higher steel replacement costs. In addition, some designs are more exposed to cracking damage, resulting in the same types of costs.
- *Insurance costs:* Insurance will vary slightly between the design alternatives in that hull and machinery insurance is proportional to capital cost. In addition, less risk for serious spill accidents may reduce insurance costs.
- *Fuel consumption:* Higher tanker lightweight will increase fuel consumption.

The alternative designs were analysed relative to a MARPOL tanker with protective location of segregated ballast tanks (PL/SBT). The volume of oil spills averted by implementing the different design configurations was estimated through an analysis of 38 large spill accidents, and the spill volume averted is considered as the benefit of the implementation. The tons of spill averted are presented in Table 9.13, and as can be seen it is distinguished between a typical and a major spill year as well as between small and large

Table 9.13. Tons of oil spill averted for the different design alternatives

Design alternatives	Typical spill year performance		Major spill year performance	
	Small tanker (tons)	Large tanker (tons)	Small tanker (tons)	Large tanker (tons)
Double bottom	2,600	4,500	13,600	24,000
Double sides	None	None	None	None
Double hull	3,300	5,300	17,600	28,400
HDPV	4,300	3,700	22,800	19,600
Smaller tanks	2,100	2,700	11,200	14,400
IOTD	4,000	5,400	21,200	28,800
DS/HDPV	3,800	5,500	17,200	29,200
DH/HDPV	3,800	5,600	20,000	29,600

tankers. The analysed accidents showed that the economic claims clustered around USD 28,000 per ton of oil spilled (1990). However, some claims could reach up to USD 90,000 per ton (i.e. *Exxon Valdez*). Because of this variation the tons of oil spill averted is not transferred into economic figures in Table 9.13.

The increased transport costs as a result of the design alternatives were calculated through a realistic weighting of three typical (and realistic) transport scenarios. All the alternative designs had higher transport costs than the base transport cost of a MARPOL tanker with PL/SBT. On the basis of 600 million tons of oil carried per year, which approximated the total US seaborne oil transport, the increased transport cost associated with each design alternative was established. The results of this analysis are presented in Table 9.14.

The cost-effectiveness of the different tanker design alternatives can be found by dividing the incremental transport costs for each design alternatives by the amount of oil each design prevents from being spilled, shown in Table 9.13. The results are presented in Table 9.15.

Table 9.14. Incremental transport costs for the design alternatives

Design alternative	Incremental cost (USD 10^6 per year)
Double bottom	462
Double sides	339
Double hull	712
HDPV	1080
Smaller tanks	430
IOTD	872
DS/HDPV	1102
DH/HDPV	2047

Table 9.15. Added transport cost per ton of oil saved

Design alternatives	Typical spill year performance		Major spill year performance	
	Small tanker (10^3 USD/ton)	Large tanker (10^3 USD/ton)	Small tanker (10^3 USD/ton)	Large tanker (10^3 USD/ton)
Double bottom	178	103	34	19
Double sides	No oil saved	No oil saved	No oil saved	No oil saved
Double hull	216	134	40	25
HDPV	251	292	55	47
Smaller tanks	205	159	38	30
IOTD	218	161	41	30
DS/HDPV	344	200	64	38
DH/HDPV	539	366	102	69

Based on Table 9.15, the most expensive ways to prevent oil spill are double sides (DS), which does not prevent any oil spill, and double hull with hydrostatic driven passive vacuum (DH/HDPV). Two of the design alternatives could be described as medium cost, namely the double sides with hydrostatic driven passive vacuum (DS/HDPV) alternative and MARPOL ships with HDPV. The most cost-effective alternatives are double bottom, double hulls, smaller tanks, and interior oil-tight deck (IODT).

Assuming that the costs of oil spills vary from USD 28,000 to 90,000 per ton of oil spilled, none of the design alternatives are cost-effective in a typical year. However, in major spill years all the design alternatives can be cost-effective. Other cost-effectiveness studies carried out on the implementation of double hull tankers have, however, shown that this measure is not cost-effective. A study performed by the Transportation Centre of Northwestern University (Brown and Savage, 1996) showed that the expected (i.e. most likely) benefits of double hulls on tankers were only about 18% of the costs expected.

REFERENCES

Brown, S. and Savage, I., 1996, The economics of double-hull tankers. *Maritime Pollution Management*, Vol. 23(2), 167–175.

HSE, 1992, *The Tolerability of Risk from Nuclear Power Stations*. Health and Safety Executive, UK.

McCormick, N. J., 1981, *Reliability and Risk analysis: Methods of Nuclear Power Applications*. Academic Press, San Diego.

O'Rathaille, M. and Wiedemann, P., 1980, The social costs of marine accidents and marine traffic systems. *Journal of Navigation*, Vol. 33(1), 30–39.

Roland, H. E. and Moriarty, B., 1990, *System Safety Engineering and Management*. John Wiley, New York.

Sexsmith, R. G., 1983, *Bridge Risk Assessment and Protective Design for Ship Collisions*. IABSE Colloquium 'Ship Collision with Bridges and Offshore Structures', Copenhagen.

Skjong, R. and Ronold, K., 1998, *Social Indicators of Risk Acceptance*. OMAE 98, Det Norske Veritas, Høvik, Norway.

10

FORMAL SAFETY ASSESSMENT

The degree to which you overreact to information will be in inverse proportion to its accuracy.
("Weatherwan's Postulate")

10.1 INTRODUCTION

The use of qualitative methods has a long history within the maritime industry. More recently, however, the use of quantitative methods has opened up an opportunity to compare different concepts and safety measures on a common scale. In the previous chapters several techniques, both quantitative and qualitative, have been described for analyses of event probabilities and consequences. In addition to these techniques, the cost-benefit analysis (CBA) described in the previous chapter makes it possible to compare and implement the results of the risk analysis in terms of costs. Hence, by applying a sequence of all these methods, decisions can be made about which concepts to choose and which safety measures to implement based on a simple assessment of the costs involved. The validity of comparing different concepts or safety measures on a common scale is, however, not only related to the scale itself but also to the analysis process. Different approaches to the problem and different assessment of details may contribute to different results. As a result of this situation a need for standardization and generalization of the analysis approach/process was brought to the surface. One proposed standard analysis approach that has gained wide recognition is the so-called Formal Safety Assessment (FSA) approach.

Formal Safety Assessment (FSA) is a designation used in a number of different contexts and industries (e.g. the nuclear industry) in order to describe a rational and systematic risk-based approach for safety assessment. In the maritime world the expression Formal Safety Assessment (FSA) is now being used by the International Maritime Organization (IMO) and its members to describe an important part of the rule-making process for international shipping. It is this maritime type of FSA that will be reviewed in this chapter. FSA is sometimes referred to as the safety case concept, which was first developed in the nuclear industry and has been used in other industries, such as in offshore.

According to IMO, FSA is a rational and systematic process for assessing the risks associated with any sphere of activity, and for evaluating the costs and benefits of different options for reducing those risks. It therefore enables an objective assessment to be made of the need for, and content of, safety regulations (see IMO's web site).

10.1.1 Historical Background

Large-scale maritime accidents, in particular the accidents with *Herald of Free Enterprise* (1987) and *Exxon Valdez* (1989), prompted a re-evaluation of the current prescriptive (i.e. rule-based) regime for marine safety. The regime was regarded unfavourably compared to the safety regimes used in other industries, which were based on more scientific methods such as risk and cost-benefit analyses. Especially within the UK, work was carried out in order to establish a more rational approach to rule development. In 1993 the UK Marine Safety Agency, renamed the Marine Coastal Agency in 1998, proposed a five-step procedure for safety analysis, named Formal Safety Assessment (FSA), to the International Maritime Organization (IMO). The main purpose of the FSA methodology was to provide a more systematic and proactive basis for the IMO rule-making process.

In 1997 IMO adopted 'Interim Guidelines for the Application of Formal Safety Assessment (FSA) to the IMO Rule-Making Process' and has since been evaluating trial applications of the technique. The use of the FSA methodology on helicopter landing areas on cruise/passenger ships in 1997 was influential in IMO's decision to abandon this risk reduction measure because the implied cost of averting a fatality (ICAF) was found to be far from cost-effective. The main principles for Formal Safety Assessment now seem to be widely accepted within the IMO.

10.1.2 The Intentions of FSA

FSA is a tool designed to assist maritime regulators in the process of improving and deriving new rules and regulation. The main intention behind the development of the FSA methodology for maritime activities was that it should be used in a generic way for shipping in general. The methodology was initially derived with two potential users in mind:

- *IMO committees:* The application of the FSA methodology can provide helpful inputs into the review process of existing regulations and into the evaluation process of proposed new regulations.
- *Individual maritime administrations:* Application of the FSA methodology can be used in the process of evaluating/assessing proposed amendments to IMO regulations. It can also be used in order to evaluate whether additional regulations, which exceed the IMO requirements, should be introduced.

It is anticipated that FSA should be relied upon where proposals may have far-reaching implications in terms of safety, cost and legislative burden. The application of FSA will enable the benefits of proposed changes to be properly established and will therefore give decision-makers a clearer perception of the scope of the proposals and an improved basis on which to take decisions.

At the present time the classification societies seem to recognize that the FSA methodology also can be used in the process of improving and developing classification rules. This should in principle not be fundamentally different from using the methodology on the IMO rules. The phrase Formal Safety Assessment has also been applied to the safety assessment of individual ships. Although the general methodology is the same in these cases, the specific aspects have been adjusted to the particulars and characteristics of the ship under consideration.

10.2 THE FSA APPROACH

The FSA approach/methodology is a standardized holistic approach to risk assessment. The approach involves several standard elements, which can be illustrated by the five-step process shown in Figure 10.1. Each step involves the use of specific methods and tasks, many of them described in detail in earlier chapters.

The interactions between the five steps of the FSA methodology are in reality not as simple as shown in Figure 10.1, which serves more as an overview of the sequential nature of the FSA methodology. The results and findings in one step are often used as feedback and input into several other steps. For example, the generation of safety measures, also known as risk control options, in step 3 of the methodology is based on both the most important hazards identified in the risk assessment step (i.e. step 2), as well as the more qualitative background and understanding of the hazards established in the step 1. These mutual interactions between the five steps of the methodology means that the FSA approach in reality looks more like the flow chart given in Figure 10.2.

Each of the steps in Figures 10.1 and 10.2 is given a more detailed description in the following text. The FSA methodology is quite complex because it involves the use of a wide range of different techniques, some which are described in earlier chapters. As a result, only the most important aspects are described and discussed here, and where appropriate references to other chapters are made.

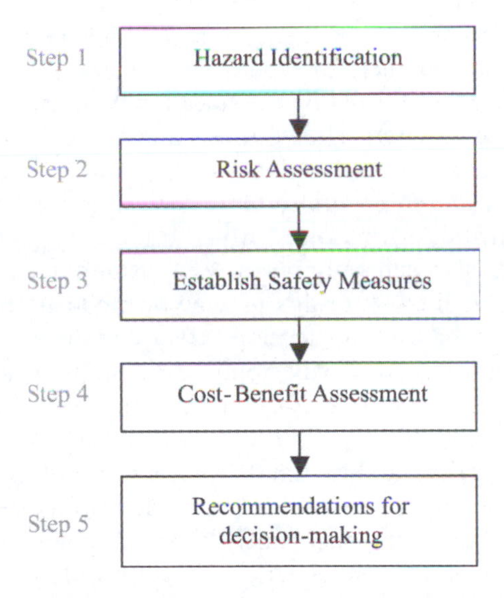

Figure 10.1. The five-step process of Formal Safety Assessment (FSA).

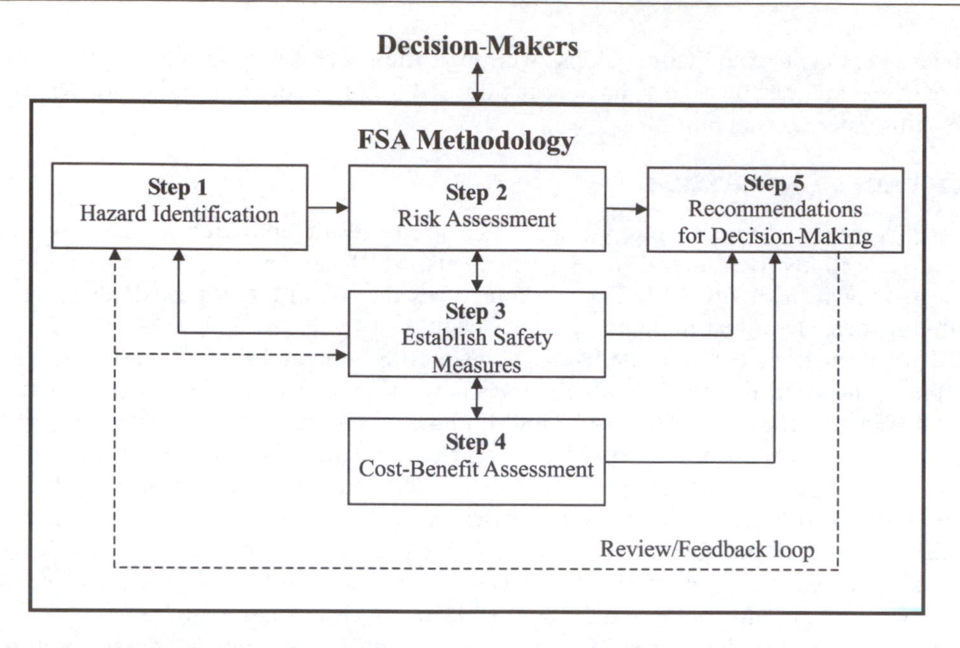

Figure 10.2. Flow chart of the FSA approach/methodology.

10.2.1 The Generic Ship

As mentioned in the introductory part of this chapter, FSA is a tool designed to assist maritime regulators in the process of developing rules and regulations. Rules and regulations must apply to ships on a general basis, and an important element of the FSA approach/methodology is therefore the use of generic ships. A generic ship should represent all the ships that are affected by the rules/regulations under consideration in the FSA process, and it should therefore have those functions, features, characteristics and attributes that are common to the vast majority of the ships relevant to the problem under concern. Hence, a generic ship model does not normally include characteristics such as cargo and detailed design features, which often vary greatly between ships. The construction of a generic ship generally means establishing a common starting point for the FSA process, and in addition it results in analytical consistency and efficiency.

FSA can, for example, be used to develop safety measures (or risk control options), such as rules and regulations, which will apply to all ships of a particular type. When performing an FSA, the generic ship will be a hypothetical vessel with characteristics that are typical or average for the ship type in question. As much effort as possible should be made in order to develop a generic ship that is representative of as many of the ships in the fleet as possible. The first efforts to establish a generic ship model aimed at creating one that would account for nearly all (merchant) ships. In later years the development has gone more in the direction of different generic models for different ship types, for example different generic ship models for oil tankers, chemical tankers, gas tankers, etc. No matter what method is used, it is of crucial importance that the generic ship model applied in the FSA methodology is applicable to the problem examined.

With reference to MSA (1995), a description of a generic ship, meant to cover most ships in international trade, is outlined below:

> The generic ship is a vessel, of monohull construction, over 500 GT, manned by competent persons, having the propelling and primary power generation machinery and associated systems located in machinery spaces within the hull. The accommodation, containing the cabins, communal areas, galley and refrigerated stores, is situated at a level above the machinery space, the upper reaches of which may be surrounded by accommodation.
>
> The primary control position, the Navigation Bridge, is situated at the forward upper area of the accommodation. A control position for machinery is located within the main machinery space. Emergency power is provided from a self-contained unit located within the emergency generator room, which is located external to the main machinery, but with a dedicated access. Emergency protection devices, such as fuel valve trips, ventilation fan stops, oil service pump stops and machinery space fire supervision systems, would also be located at a space remote from the main machinery.
>
> Mooring equipment is located at the bow and stern. Anchors are provided with the means to be deployed and recovered using windlass machinery provided at the bow.

In addition to this description, the generic ship inhabits a set of generic systems and functions, which enable it to operate and trade safely. The system categories and functions are listed alphabetically in Table 10.1.

The transport operation is characterized as a sequence of distinct phases where each phase requires the use of different functions. Hence, the need for the various functions given in Table 10.1 will vary according to a specific ship's operational cycle. The

Table 10.1. Systems and functions of a generic ship

Systems	Functions
Accommodation and hotel service	Anchoring
Communications	Carriage of payload
Control	Communications
Electrical	Emergency response and control
Human	Habitable environment
Lifting	Manoeuvrability
Machinery	Mooring
Management systems	Navigation
Navigation	Pollution prevention
Piping and pumping	Power and propulsion
Pressure plant	Bunkering as storing
Safety	Stability
	Structure

operational phases of the generic ship are defined as below:

- Entering port, berthing, unberthing, and leaving port
- Payload loading and unloading
- Passage/transit
- In dry dock

Each element of the generic ship has more detailed descriptions, and subcategories that have to be applied when utilizing the FSA methodology, but such descriptions go beyond the scope of this book. It is, however, important to notice that the defined functions and systems are closely related to all steps of the FSA.

10.2.2 Stakeholders

One problem when applying cost-benefit analysis (CBA, step 4 in the FSA methodology) in risk assessment is the number of stakeholders (or parties) involved with regard to the vessel, and their various roles in relation to the costs, benefits and risks involved. A stakeholder may be defined as a party investing risk in shipping operations. In many cases the stakeholder who imposes certain risks is not the same stakeholder who carries these risks. The extent and complexity of stakeholder groups adds complexity to the CBA analysis because each group imposes and carries different costs, benefits and risks, and has different risk perception. Each group of stakeholders therefore perceives the costs and benefits differently, making the CBA difficult and a major challenge for the analyst. Table 10.2 indicates which stakeholders incur which costs, receive which benefits, and impose and carry which risks. The so-called 'risk imposer pays' principle (or policy) implies that those stakeholders who, voluntarily or not, impose risk on others should pay for that privilege.

10.3 HAZARD IDENTIFICATION

Hazard identification is the first step of the FSA methodology. The main objective of this step is to identify relevant hazards, i.e. undesirable accidental outcomes, which could affect the ship operation under consideration. In a safety context the undesirable outcome could include injury to personnel, damage to property, and/or pollution of the environment. In addition to identifying the hazards related to an activity, this step should also include identifying the causes of these undesirable outcomes. The hazard identification step is composed of several sequential stages as may be summarized by the simple flow diagram in Figure 10.3.

The first stage in the hazard identification step is to make a precise and carefully defined problem definition. A well-defined problem definition is of great importance because it provides a desirable starting point for the FSA process. It is important that the problem definition clearly points out the objective of the particular FSA being carried out, which would include a description of the systems/activities under consideration and their relation to the rules/regulations under review or development. Identifying the boundary of the analysis is very important in this regard. When using the FSA methodology on ships,

Table 10.2. Examples of stakeholders and their risk investments

Stakeholder	Incurs costs	Receives benefits	Imposes risks	Carries risks
Owner/ charterer	Cost of vessel	Income	Choice of vessel specifications	Loss of vessel
Cargo owner	Pays for passage	Profit from trade	Dangerous cargoes	Loss of cargo
Operator	Running costs	Income	Operating practice	Loss of income
Crew	—	Employment	Lack of due diligence	Loss of life/property
Passenger	Fares	Transport, leisure	—	Loss of life/property
Flag state	Administration costs, employment	Fees	Inadequate local legislation	—
Port of call	Cost of infrastructure, operating costs	Fees	Navigational control, dredging levels	Damage to infrastructure, loss of trade
Coast state	Local navigation	—	Inadequate navigation aids	Pollution and clean-up
Insurer	—	Premiums	—	Claims
Other vessel	—	—	Impact, loss of life	Impact, loss of life
Classification societies	Operating costs	Fees	Lack of due diligence	Negligence claims
Designer/ constructor	Materials/labour	Fees	Lack of due diligence	—

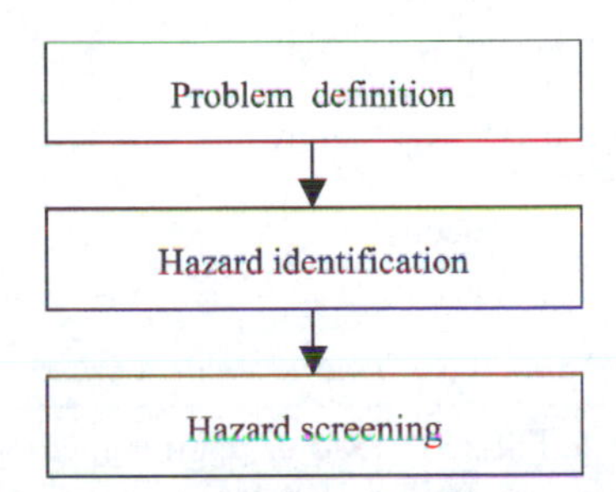

Figure 10.3. Step 1 of the FSA methodology.

the following information may be found relevant to include in the problem definition:

- Vessel type
- Relevant systems and functions
- Part of ship operation under concern
- Rules/regulations to be studied
- Geographical boundaries
- Applied measures of risk
- Type and definition of acceptance criteria

A large part of this information will be defined through the application of a generic ship model in the FSA.

When an appropriate problem definition is established, we proceed to the actual hazard identification stage. The process of hazard identification, as well as specific techniques that may be used, is described in detail in Chapter 8. Some generic accident outcomes (i.e. consequences), causes and influencing factors are outlined as part of the FSA methodology based on the characteristics of the generic ship. Some generic elements assumed to affect all ships are given in Tables 10.3, 10.4 and 10.5. These generic elements may be utilized by, for example, applying different brainstorming strategies to identify relevant hazards. One approach is to use brainstorming to identify as many hazards as possible. At the end of the brainstorming session the findings can be structured and grouped in the generic ship categories (i.e. one category covering grounding scenarios, another for collision, etc.), and a rest group categorized as ship-specific findings. An alternative and less comprehensive approach is to use the generic hazard elements (i.e. accident outcomes, causes and influencing factors) as a starting point, then find the ones that are relevant, and finally inspect the characteristics of the specific ship under consideration to identify and include ship-specific elements not covered by the generic ship characteristics.

Table 10.3. Generic accident outcomes derived from examination of historical data recordings

Collision (striking between ships)
Contact (striking between ship and other objects)
Fire and explosion
Foundering and flooding
Grounding and stranding
Hull and machinery failure
Missing
Other/miscellaneous

Table 10.4. Generic accident causes

Human causes (e.g. failure to read navigational equipment correctly)
Mechanical causes (e.g. failure of pumps)
Fire and explosion (e.g. loss of visibility due to smoke)
Structural causes (e.g. failure of bow doors)
Weather-related causes (e.g. high ambient temperature)
Other causes

Table 10.5. Generic influencing categories

Likelihood of underlying causes occurring
Likelihood of an underlying cause progressing to a major accident outcome
Magnitude of the consequence of the major accident outcome

The establishment of generic hazard elements in the development phase of the FSA becomes increasingly difficult the deeper into the accidental escalation one tries to find generic elements. Hence, the hazard influencing factors in Table 10.5 are described in categories only.

The last stage of the hazard identification step (or process) is known as the hazard screening stage and generally involves structuring the findings in the step for implementation in the later steps of the FSA methodology. It could be argued that the grouping of hazards into generic accident categories is some type of hazard screening. One more extensive approach found to be useful in some maritime applications of FSA is the use of a risk matrix where the hazards are plotted in a matrix as a function of the severity of the consequences and the probability of occurrence. This, however, means assessing the risk, i.e. both the severity and the probability of the hazards, and should in principle be included in step 2 (risk assessment) of the FSA. It is difficult to define a precise border between step 1 and step 2 of the FSA, but if we wish to be loyal to the definitions of hazard and risk (which we should in order to avoid unnecessary confusion), the construction of risk matrixes, fault trees and event trees should be included in step 2 of the FSA.

10.4 RISK ASSESSMENT

Step 2 (risk assessment) continues directly from step 1 (hazard identification). The objective of this step of the FSA methodology is to quantify the risks of loss of life, damage to property and damage to the environment. To achieve this goal, attempts must be made to identify and quantify the underlying causes and influences that affect the likelihood of initiation and progression of accident sequences (all accidents are a sequence of events). The risk assessment process may be illustrated as a sequence of stages that should be carried out in order to realize the full potential of the process. Such a sequence of stages is given in Figure 10.4.

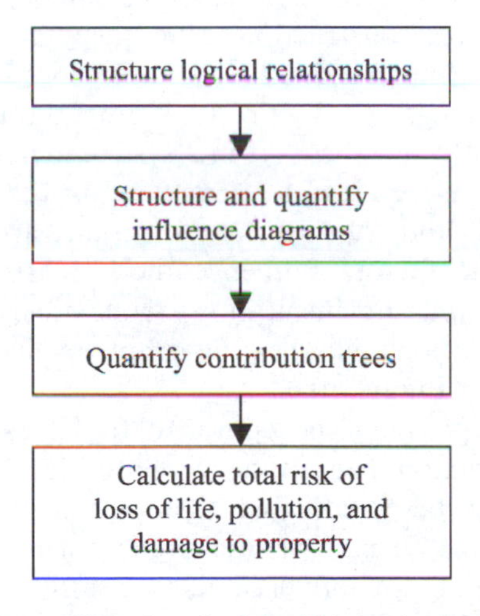

Figure 10.4. Step 2 of the FSA methodology.

Risk can generally be defined as a measure of a hazard's significance involving simultaneous examination of its consequence and probability of occurrence, and risk assessment is the process that determines where a hazard will be located on a risk scale. The risk scale can either be a continuous numerical scale or a discrete scale (risk matrix). Risk assessment can generally be divided into qualitative and quantitative risk assessment. With reference to Figure 10.4, the qualitative risk assessment involves the two first stages, while the quantitative analysis is the last two stages. It is important to obtain quantitative risk estimates in a FSA because such estimates can be used to see the effects of risk control options/measures through the application of a cost-benefit analysis.

10.4.1 Qualitative Risk Assessment

The qualitative risk assessment involves the two first stages in Figure 10.4, i.e. to structure logical relationships as well as to structure and quantify influence diagrams. This includes a structured analysis of the hazard findings from the hazard identification step of the FSA methodology.

The logical relationships underlying an accident may be constructed using fault trees (also known as logical contribution trees). Considerable knowledge of and experience with the system is needed in order to construct a meaningful fault tree. On a more preliminary basis, cause and effect diagrams may be used in order to identify the potential causes of an undesired outcome (i.e. effect) such as an accident. The International Maritime Organization (IMO) prefers so-called risk profiles, which basically are simplified fault trees, for the qualitative risk analysis. The common idea for both techniques is the deduction of the underlying causes (and their relationship) of an accidental outcome, but in comparison to a fault tree the risk profile is simpler because there are no logical gates between the underlying causes. Risk profiles are deduced mainly from historical accidental outcomes rather than from underlying causes/failures, which is the case for fault trees. Figure 10.5 shows the risk profile for the accidental outcome of a collision. It is recommended the common fault tree construction technique is applied instead of risk profiles if a detailed analysis is performed/required.

Influence diagrams, i.e. stage 2 in Figure 10.4, are diagrams constructed to illustrate the regime of factors that influence the risks in a system/activity. It is usual to distinguish between regulatory influences, corporate policy influences, organizational influences, operational influences, etc., and the influence diagrams are often interrelated in very complicated patterns. Some factors influence the performance of a system directly (e.g. organizational policies and implementation), while other factors are more underlying influences. Some IMO regulations may be an example of the latter. A rough draft of an influence diagram is given in Figure 10.6.

Influence diagrams have proved to be a powerful tool in establishing how the regulator and the managing organization can influence both the likelihood and outcome of accidents. The diagrams can be constructed as purely qualitative diagrams or they can be quantified by assessing the current significance/importance of each influence. If it is found to be possible to quantify the influence diagrams, this can provide a useful basis for judging the effectiveness of the safety measures (or risk control options) derived in step 3

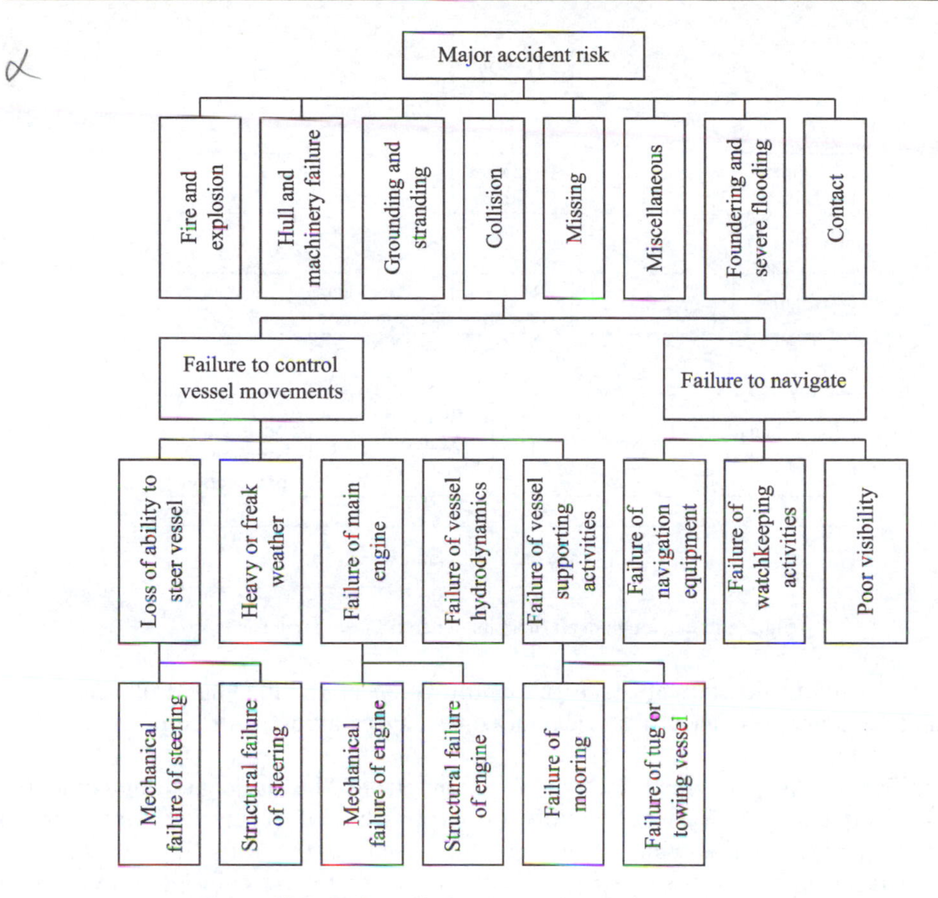

Figure 10.5. Risk profile for the accidental outcome of a collision.

of the FSA methodology. The diagrams may also reveal influencing factors that have potential for improvements. In a sense the underlying objective of the FSA methodology is to change/modify the influence diagrams.

10.4.2 Quantitative Risk Assessment

In order to be able to focus on high-risk areas, both the absolute risk level and the relative importance of different causes have to be quantified. The main objective of the quantitative risk assessment, which involves the last two stages in Figure 10.4, is to establish the relative and absolute importance of the underlying causes, as well as the influencing causes. This involves calculating risk estimates such as f–N curves/diagrams (i.e. logarithmic scales showing the probability f as a function of fatalities N), PLL (Potential Loss of Life), AIR (Average Individual Risk), and similar measures. The quantification is performed through analysis of historical data and expert judgement techniques. When analysing historical data it is common to break down a number of

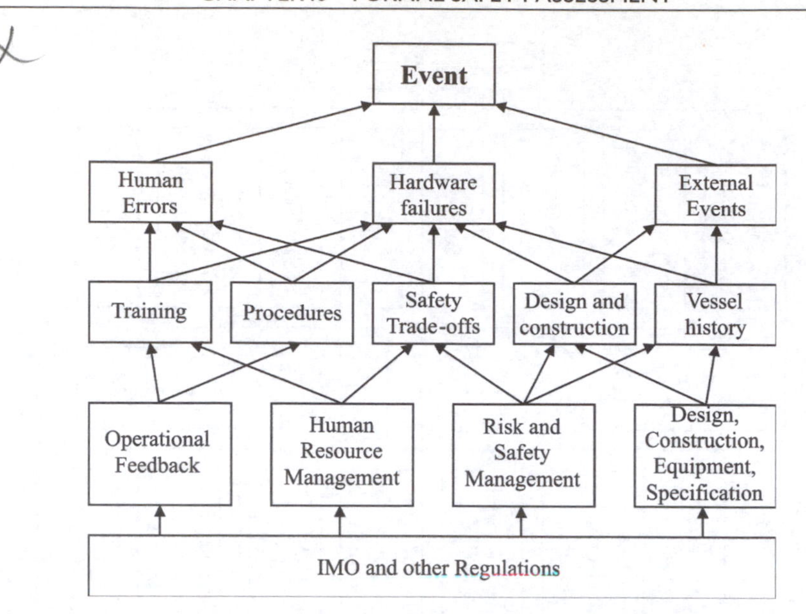

Figure 10.6. Rough draft of an influence diagram structure.

relevant accidents to develop satisfactory quantification of the likelihood of occurrence of the underlying causes. These historical measures can be adjusted or complemented by expert judgement.

Quantification is performed in two directions from the accidental outcome (or scenario): fault trees are used to quantify the probability of occurrence for a certain accident outcome/scenario, while event trees are used to quantify the frequency of different consequences given a certain accidental outcome (e.g. a collision between two ships). The potential consequences should reflect the injuries to people, as well as damage to both the environment and physical assets. Considerable knowledge of the system and situation under consideration is needed in order to create reasonable and valid event and fault trees (i.e. risk contribution trees). Influence diagrams can be used to aid the process of establishing valid risk contribution trees.

The risk is calculated by combining/multiplying the probabilities of occurrence with the severity of the consequences. If this is done for all possible outcomes of an accident scenario, the total risk picture is established. As mentioned above, the total risk can be presented in many different ways, for example as a numerical value (e.g. number of fatalities per 10^8 working hours) or as a *f–N* curve. The total risk should preferably be presented both numerically and graphically. Which risk presentation techniques are found appropriate will in general depend upon the system/activity under consideration. The process of quantitative risk assessment often used within FSA is illustrated in Figure 10.7.

In the next step of the FSA methodology, several safety measures (or risk control options) are generated and considered implemented. In order to quantify the risk-reducing effect of these options (i.e. the benefits), the quantification procedure described in this risk assessment step of the FSA has to be repeated for each risk control option considered.

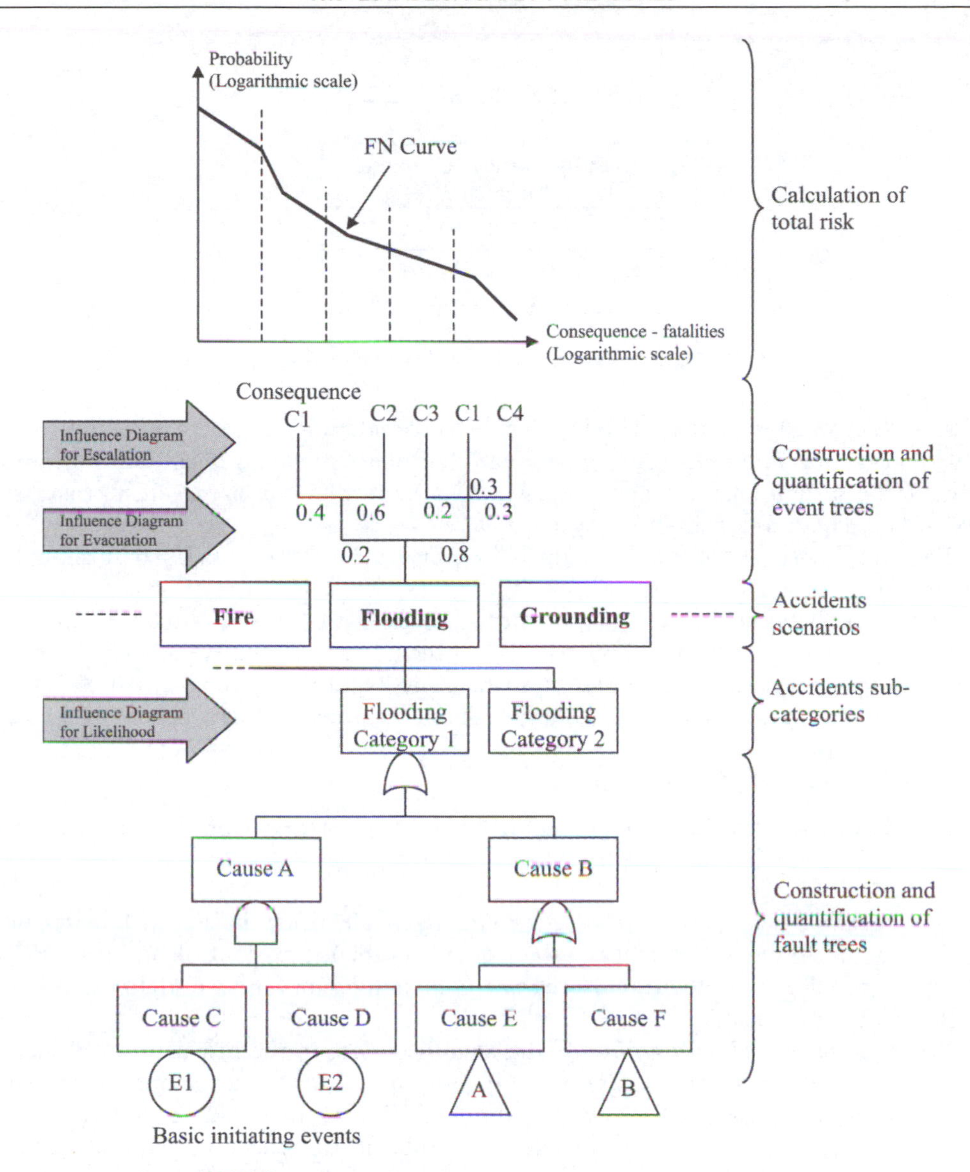

Figure 10.7. The process of quantitative risk assessment often used in FSA.

10.5 ESTABLISH SAFETY MEASURES

The output from step 2 (risk assessment) provides a risk profile showing the risks for the major hazard categories and principle subcategories. The results of the risk assessment step also include knowledge of direct contributing causes and the likelihood of alternative levels of loss, i.e. consequences. Based on this information, the objective of step 3 of the FSA methodology is to focus on the activities/systems needing control because of high

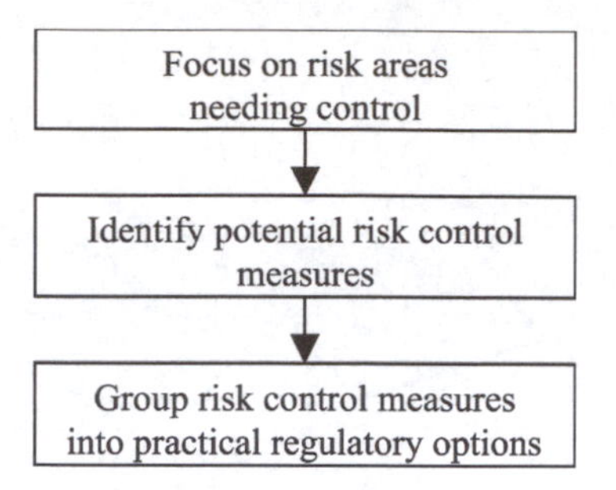

Figure 10.8. Step 3 of the FSA methodology.

risks or other weighty matters. This involves both considering new safety measures (or risk control measures) and assessing to what degree current risk management and regulations mitigate the system hazards. This process is also known as risk management because it involves managing the risks in the system.

The step 3 process is usually defined as consisting of three stages that should be performed as illustrated by Figure 10.8.

The border between the step 3 and the step 4 processes is not easily defined. Many would argue that step 3 (risk control options) should include a quantitative assessment of the effects (i.e. risk reduction and/or other possible benefits) of the options/measures, but this assessment activity will here be placed under the step 4 process of cost-benefit assessment.

10.5.1 Areas Needing Control

There are several areas needing risk control. The main aspects for risk control are as follows:

- *Unacceptable risk levels:* If some of the risks identified in the risk assessment process are found to be unacceptable (or intolerable), risk control options must be implemented in order to make the risks acceptable/tolerable and ALARP, i.e. as low as reasonably practicable.
- *Risks within the ALARP region:* If the identified risks in the risk assessment step of the FSA are within the ALARP region, cost-effective risk control options/measures should be implemented. Risks within the ALARP region should be undertaken if a benefit is desired, and are only considered tolerable if risk reduction is impracticable or if its cost is grossly disproportionate to the improvements gained. The ALARP concept is thoroughly explained in Chapter 9.
- *High probability:* If some hazard scenarios have a low severity but a high probability, they may be found to be unacceptable from an operational point of view even if they have a tolerable risk level. To be able to identify such situations, qualitative risk assessment must be made.
- *High severity:* If some hazards have a low probability but a high severity of consequences, they may be found to be unacceptable even if they have a tolerable risk level. Qualitative risk assessment is necessary in order to identify such circumstances.

- *Considerably uncertainty:* Considerably uncertainty in probability, severity, or both, could be a reason to take precautions in terms of implementing extra or redundant risk control options/measures.

10.5.2 Identify Risk Control Measures

Risk control options may take many forms, addressing the technical (engineering), human and management (organizational) aspects of an operation, and control options/measures inhabit several characteristics that are important to consider. The proposed risk control options may address both the prevention of accident/incident initiation and the mitigation of the consequence severity, i.e. the associated losses. Some risk control options may have an effect on the risks related to a single or a few hazards, while others may affect the risks related to all parts of an operation. The effect of the different options along the causal chain (Causal factor → Failure → Circumstances → Accident → Consequence) may consequently vary. Furthermore, the circumstances in which failures occur may change because of the introduction of risk control measures. For example, the likelihood of engine failure may decrease when failures due to overload vanish. The effect of risk control measures over time should also be considered. This may involve the time to full effect of the measure, as well as the duration of the effect. Because costs and effects of different safety measures may vary significantly, it is important that a wide variety of measures are considered.

According to IMO's web site, 'existing maritime safety efforts are still being primarily directed towards engineering solutions. Governments and operating companies spend perhaps 80% of available resources addressing design requirements and technical fixes. The remaining 20% are directed to the most pervasive and consistent cause of marine casualties, the human element.'

Example

Problem

The flooding of the vehicle deck on Ro-Ro passenger vessels through an open or partly open bow door may result in rapid capsizing and the loss of many human lives. This was the accident scenario in the *Herald of Free Enterprise* disaster in 1987. The risk related to this particular hazard is found to be unacceptable for a given vessel and risk control options must therefore be implemented in order to reduce the risk. Suggest some possible/potential risk-reducing control options.

Solution

Possible risk reduction options may include one (or more) of the following:

- Audible alarms on the bridge that will attract the attention of the Master when the bow door is open or not closed properly.
- Management routines onboard the vessels that control whether the bow door is closed at departure.

- Strengthen the construction of the bow door hinges in order to avoid the bow door opening in severe weather/sea conditions.
- Down-flooding of water on the vehicle deck to tanks (equipped with pumps) situated in the lower parts of the hull.
- Introduce transverse bulkheads on the vehicle decks.

This list of risk control measures/options is not exhaustive.

10.5.3 Grouping Risk Control Measures

Based on the identification of potential risk control options, a wide range of measures reflecting various areas, effects and characteristics should be forwarded to step 4 in the FSA methodology, i.e. the cost-benefit assessment. In relation to this it is often very useful to group the risk control options in different categories based on the practical type of regulatory options that can be used/implemented. It may also be useful to group the options/measures based on their effects on the system/activity under consideration. Typical effects will include preventive, mitigation, engineering, procedural, etc. Some risk control measure characteristics are given in Table 10.6.

Table 10.6. Risk control measure characteristics

Risk control	Description
Preventive	Preventive risk control is where the risk control measure reduces the probability of the undesired event under consideration.
Mitigating	Mitigating risk control is where the risk control measure reduces the severity of the undesired event outcome or subsequent events, should they occur.
Engineering	Engineering risk control involves including safety features (either built in or added on) within a design. Such safety features are safety critical when the absence of the safety feature would result in an unacceptable level of risk.
Inherent	Inherent risk control is when choices are made in the initial design process that restrict the level of potential risk.
Procedural	Procedural risk control is where the operators of the equipment/ systems are relied upon to control the risk by behaving in accordance with defined procedures.
Redundant	Redundant risk control is where the risk control is robust to failure because redundancy principles have been applied.
Diverse or concentrated	Diverse risk control is where different risk control measures are applied for different aspects of the system, whereas concentrated risk control is where similar risk control is applied across the system.

(continued)

Table 10.6. Continued

Risk control	Description
Passive or active	Passive risk control is where there is no action required to deliver the risk control measure, whereas active risk control is where the risk control is provided by the action of safety equipment or operators.
Independent or dependent	Independent risk control is where the risk control measure has no influence on other elements, whereas dependent risk control is where one risk control measure can influence another elements of the risk contribution trees (i.e. fault and event trees).
Human factors involved and critical	Human factors involved risk control is where human action is required to control the risk but where failure of the human action will not itself cause an accident or allow an accident sequence to progress. Human factors critical risk control is where human actions are vital to control the risk, and where failure of the human actions will directly cause an accident or allow an accident sequence to progress. Where human factor critical risk control exists, the human action (or critical task) should be clearly defined in the risk control measure.
Auditable or not auditable	Auditable or not auditable reflects whether the risk control measure can be audited or not.
Quantitative or qualitative	Quantitative or qualitative reflects whether a particular risk control measure has been based on a quantitative or qualitative assessment of risk
Established or novel	Established risk control measures apply currently existing technology and solutions, whereas novel risk control measures are where the measure is new. However, the measure may be novel to shipping but established in other industries.
Developed or non-developed	Developed or non-developed reflects whether the technology underlying the risk control measure is developed both in its technical effectiveness and in terms of costs. Non-developed is either where the technology is not developed but it can be reasonably expected to develop, or where the costs of the measure can be expected to decline over a given period of time.

10.6 COST-BENEFIT ASSESSMENT

The cost-benefit assessment (CBA) is important in any FSA because it decides whether or not the suggested risk control options/measures are suitable for implementation. A CBA analysis determines if the benefits of implementing a given risk control option outweigh the cost of implementation. Cost-benefit assessment was described in detail in the previous chapter, and in this chapter a more general approach is introduced by including important CBA aspects and issues relevant in an FSA context. The cost-benefit assessment used in a FSA context may be illustrated as a series of stages as shown in Figure 10.9.

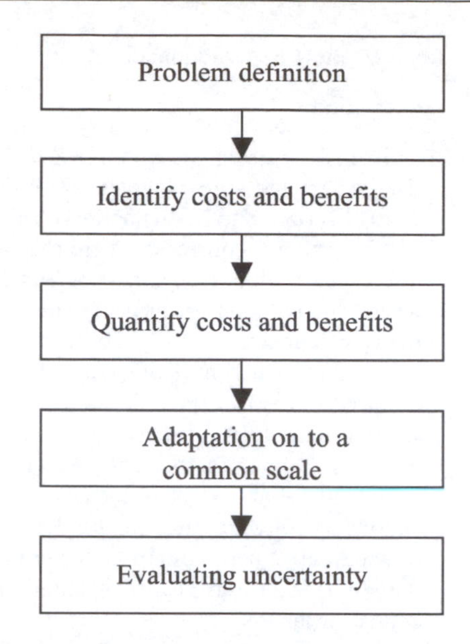

Figure 10.9. Step 4 of the FSA methodology.

10.6.1 Problem Definition

This first stage of the cost-benefit assessment (CBA) process is to make a problem definition. The boundaries for the analysis are already established in the previous steps of the FSA, and these boundaries should be implemented in the CBA in addition to some boundaries used explicitly in the CBA. One additional boundary that should be defined is the geographical boundaries of the CBA. Most safety measures may be applied world-wide and could in some cases be considered as generic, e.g. SOLAS or other IMO conventions. Another aspect of the geographical boundary is the variation in cost, and in some cases also the benefits, with geographical area. In addition some safety measures are only effective in a specific area of operation, for example in port or in transit. This may have an impact on the benefits (i.e. averted consequences) of the measure.

An important aspect to define in the problem definition stage is the base-line year to be used in the cost-benefit assessment. This base-line year must be defined in order to establish the risk improvements and the costs of implementation for a given risk control option (or measure). In most FSA applications the base year will be the time of investing in the option. If there is any limited time duration of the safety measure's effectiveness, this must be included in the problem definition because it will influence the calculations of the benefits and costs related to a risk control option. In addition, it must be taken into account that different stakeholders often incur different costs and receive different benefits (this problem is, however, not relevant for the society risk approach).

It is of crucial importance in the problem definition stage that the alternative risk control options are defined as precisely as possible, because this information will affect both the benefits and the costs related to the implementation of the option/measure. In the next two stages of the cost-benefit assessment (CBA) process the benefits and costs for the suggested risk control options are identified and quantified.

10.6.2 Identify Costs and Benefits

All the potential costs and benefits related to each of the relevant risk control options/measures should be identified in this stage of the CBA process. It is, however, equally important to identify potential negative effects that the implementation of risk control options can have on the system/activity in question. On a ship such negative effects could, for example, include longer loading/unloading times, reduced speed, more downtime due to inspections and controls, etc.

The implementation of risk control options (or safety measures) may involve many different types of costs and benefits. Typical costs involved may include one or more of the following:

- Capital/investment cost
- Installation and commissioning cost
- Operating or recurrent cost
- Labour cost
- Maintenance
- Training
- Inspection, certification and auditing
- Downtime or delay cost

There are also costs associated with not implementing safety measures, and avoiding these costs are benefits of implementing such measures. The benefits of implementing a risk control option on a ship usually include one or more of the following:

- Reduced number of injuries and fatalities
- Reduced casualties with vessel, including damage to and loss of cargo and damage to infrastructure (e.g. berths)
- Reduced environmental damage, including clean-up costs and impact on associated industries such as recreation and fisheries
- Increased availability of assets
- Reduction in costs related to search, rescue and salvage.
- Reduced cost of insurance

In the process of identifying costs and benefits related to a system, one should be careful not to double-count. Double counting may result in unbalance in the CBA assessment and will increase the uncertainties related to the assessment as a whole.

10.6.3 Quantify Costs and Benefits

When all the relevant costs and benefits related to the implementation of a risk control measure to a system/activity are identified, these must be quantified (AEA, 1995). Several approaches may be used to quantify costs and benefits. One common valuation approach is to evaluate the effect of the consequences on production factors. By applying this approach, the costs of an injury to a human being can be derived by considering the length of hospitalization and its costs, the degree of permanent disability, and the lost earnings due to this disability (see Chapter 9 for further details). Another valuation approach is based on the creation of a hypothetical market for a reduced probability of occurrence, consequence or risk, and then establishing the price individuals are willing to pay to reduce these factors. Several other valuation approaches exist, but it is outside the scope of this chapter to go into these in detail. However, most valuation approaches have in common that they result in a monetary cost of a fatality, pollution to the environment, etc. It may seem 'heartless', for example, to calculate/estimate the monetary benefits of averting a fatality because this is associated with identifying the 'value' of a human life. Such considerations are apparent, but some sort of criterion is necessary when analysing risk exposure to humans as well as property and the environment. Setting no value on the costs related to a fatality may in the worst case be counter-productive in the process of reducing the risks associated with activities and the operation of systems.

10.6.4 Adaptation onto a Common Scale

Different risk control measures result in different risk reductions, and each measure is associated with a set of distinct benefits and costs. In order to select the most cost-effective measures for implementation, it is very advantageous to evaluate these against each other on a common scale, which normally implies monetary values. In the previous chapter several approaches for cost-benefit analysis (CBA) of risk control/reduction measures were presented, and these must be applied within the framework of the FSA. In particular, the Implied Cost of Averting a Fatality (ICAF) approach/methodology is very much used in FSAs, and a detailed review of this approach can be found in Chapter 9. In essence, the ICAF methodology estimates the achieved risk reduction in terms of cost using the following equation:

$$\text{ICAF} = \frac{\text{Net annual cost of measure}}{\text{Reduction in annual fatality rate}} \tag{10.1}$$

ICAF may also be calculated by dividing the net present value of all costs for the whole lifetime of the safety measure by the total reduction in fatalities for that particular period of time. The ICAF value can be interpreted as the economic benefits of averting a fatality. A decision criterion must be established for this value in order to evaluate whether a given risk control option/measure is cost-effective or not. A method for developing such a criterion is also presented in the previous chapter. Risk control measures with an ICAF value less than the criterion should be considered as cost-effective and therefore implemented.

10.6.5 Evaluating Uncertainty

There are always uncertainties involved in a cost-benefit analysis (CBA), especially in the process of identifying and quantifying costs and benefit. The uncertainties must be taken into account in the CBA, and several approaches exist for achieving this objective. One approach is to create an interval for each specific cost (and benefit) in which it may vary. This will result in a cost-benefit ratio with a possible range of values and a most likely value. Another approach is to perform a sensitivity analysis of the parameters involved and then assess the uncertainty of the information implemented in the most sensitive parameters.

10.7 RECOMMENDATIONS FOR DECISION-MAKING

Step 5 of the FSA methodology involves proposing recommendations to the decision-makers on which risk control option(s) should be adopted in order to make the risks ALARP, i.e. as low as reasonably practicable. The recommendations will be based on the information generated in steps 1–4 of the FSA methodology. The results obtained in the cost-benefit calculations will generally form the basis for the recommendations. Of particular importance is normally the evaluation of different risk control options relative to each other using a common scale (e.g. CURR or ICAF).

The recommendations may be presented as a prioritized list of cost-effective risk control options/measures. Such a list should include a description of the options, including their cost-benefit ratios, and an evaluation of the uncertainties related to each of the options. It must be ensured that the recommendations made are fair to all the stakeholders involved in the safety management of the activity/system under consideration (e.g. a ship). The recommendations could also include suggestions for improvements to the analysis (or the FSA methodology) and advice on further work that should be carried out on the subject under consideration.

10.8 APPLICATION OF THE FSA METHODOLOGY

In the last section of this chapter, some important elements of a comprehensive FSA study on life-saving appliances for bulk carriers are presented. In addition to showing how the FSA methodology can be used in practical applications, a major objective of the example (or case) presented here is to illustrate how the methodology can be flexibly modified to suit a particular problem. This flexibility is one of the great advantages of the FSA methodology.

Example

Problem

At the 70th session of IMO's Maritime Safety Committee (MSC) the topic of life-saving appliances (LSAs) for bulk carriers was discussed, and it was decided to include LSAs as part of the formal safety assessment (FSA) process for these vessels. This case example briefly summarizes some important aspects of a comprehensive FSA project on LSAs for bulk carriers performed in Norway (Skjong, 1999; DNV et al., 2001).

This FSA study focuses solely on LSAs with the objective of identifying risk control options (RCOs) for bulk carriers that give improved life-saving capability in a cost-effective manner. The study is considered representative for all SOLAS bulk carriers over 85 metres in length.

Solution

The five steps of the FSA methodology are carried out in succession below. It must be recognized that many details are left out from the original FSA project report.

Step 1: Hazard identification

The hazard identification step of the FSA methodology begins with establishing a precise and carefully defined problem definition. The problem addressed in this study was related to the identification of effective risk control options that could bring down the fatality rates in evacuation associated with bulk carrier accidents, i.e. to improve the probability of evacuation success given different accident scenarios (e.g. a collision). Previous individual and societal risk assessments for bulk carriers have shown that the risks are high in the ALARP region, and that cost-effective risk control options therefore should be implemented. Several regulations in SOLAS 74 are affected by the recommendations of this particular FSA study. More detailed background information can be found in the reference source (DNV et al., 2001).

The process of hazard identification was carried out in multidisciplinary teams of relevant experts using the so-called 'What if...?' technique. In this method a list of potential hazards is produced for a system or subsystem by asking 'What if...?' something does not work as planned or something unexpected/undesirable happens. The hazards were identified and ranked separately for commonly used LSAs on bulk carriers, i.e. conventional lifeboats, free-fall lifeboats, davit-launched liferafts, and thrown overboard liferafts. All phases of an evacuation event were analysed for hazards, from the occurrence of the initiating event, through mustering, abandoning, survival at sea and the final rescue. Some, but not all, hazards were generic for all categories of survival crafts. Some hazards were also related to survival at sea and rescue.

The last task of the hazard identification step is to perform a hazard screening. In this particular analysis this included a ranking of the hazards based on a qualitative assessment of their importance in terms of risk. The most important hazards were given particular consideration in the risk assessment step of the FSA.

Step 2: Risk assessment

Probabilities for various accident scenarios are fairly well established for bulk carriers through incident data sources such as *Lloyd's Maritime Casualty Reports* (LMIS) and *Lloyd's Casualty Reports* (LCR). With regard to evacuations, these data sources show that a total of 115 bulk carrier evacuations were identified during the period from 1991 to 1998. The ship population exposed during this period of time was 44,732 ship years (i.e. an average fleet size of approximately 5592 ships), identified through Lloyd's Register's *World Fleet Statistics*. This gives a total evacuation frequency of $2.6 \cdot 10^{-3}$ per ship year. This is approximately the same evacuation frequency as for merchant ships in general. Distributed with respect to type of accidental event, the resulting evacuation frequencies are as listed in Table 10.7. The number of crew members on board is obtained by multiplying the number of events by the average crew size per ship of 23.7.

Table 10.7. Bulk carrier evacuation frequencies and fatality probabilities for different types of accidental events (1991–98)

Type of accidental event	Number of events	Evacuation frequency (per ship year)	Fatalities	Number on board	Probability of fatality (%)
Collision	14	$3.1 \cdot 10^{-4}$	116	332	35
Contact	5	$1.1 \cdot 10^{-4}$	54	119	45
Fire/explosion	16	$3.6 \cdot 10^{-4}$	6	379	2
Foundered	51	$1.1 \cdot 10^{-3}$	618	1209	51
Hull failure	5	$1.1 \cdot 10^{-4}$	0	119	0
Machinery failure	1	$2.2 \cdot 10^{-5}$	0	24	0
Wrecked/stranded	23	$5.1 \cdot 10^{-4}$	0	545	0
Total	115	$2.6 \cdot 10^{-3}$	794	2727	29

With regard to studying the risks involved in evacuation using life-saving appliances (LSAs), it is not of particular interest to study the causes resulting in the accidental events shown in Table 10.7. The structuring and quantification of the fault trees underlying these events is therefore not performed. In terms of studying how the use of different LSAs affects the evacuation risks, an evacuation model consisting of event trees must be constructed to model how an initiating accidental event may develop into fatalities, in particular fatalities related to evacuation. The event trees modelled in this FSA study are generic, implying that they are structurally identical for all accident scenarios. The probabilities in the event trees are, however, conditional on the initiating event defining each accident scenario. The event trees must be very detailed so that they can be used to assess (i.e. quantify) the risk-reducing effect of risk control options (RCOs). Potential loss of life (PLL) is the only decision parameter predicted by the risk assessment model that is developed, as LSAs do not have any important impact on environmental and economic risk.

As mentioned above, an evacuation model is created to analyse how the different LSAs affect PLL. The model consists of the typical sequences of events that are associated with evacuation using different LSAs. The sequence of events expected for evacuation using conventional lifeboats is shown in Figure 10.10. The model is developed by following an individual crew member. The sequence of events is slightly different for thrown overboard liferafts and davit-launched liferafts. For example, thrown overboard liferafts are launched before boarding.

The evacuation sequence gives the underlying basis for the construction of an event tree, or several event trees if the sequence is divided into several sub-sequences as illustrated in Figure 10.10. The branch probabilities in these event trees are to some degree different for the different accident scenarios. For example, there is a slight probability of fatality as a result of the initiating event in the scenarios of ship collision or fire/explosion, while this probability is negligible for the accident scenario of hull/machinery failure. Statistics show that it is the evacuation sequence rather than the initiating event that has the highest probability of contribution to fatalities. Another example is that there is a

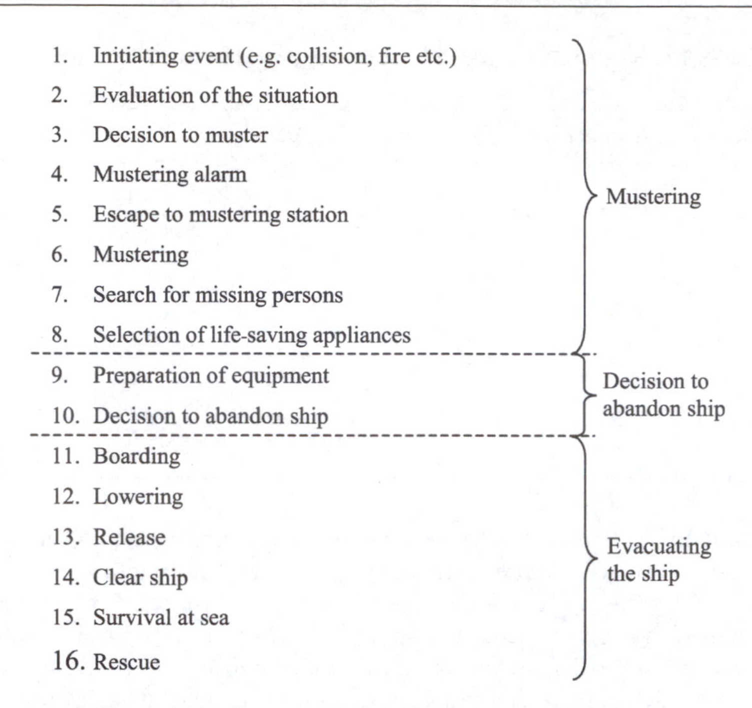

1. Initiating event (e.g. collision, fire etc.)
2. Evaluation of the situation
3. Decision to muster
4. Mustering alarm
5. Escape to mustering station
6. Mustering
7. Search for missing persons
8. Selection of life-saving appliances
Mustering
9. Preparation of equipment
10. Decision to abandon ship
Decision to abandon ship
11. Boarding
12. Lowering
13. Release
14. Clear ship
15. Survival at sea
16. Rescue
Evacuating the ship

Figure 10.10. Evacuation model for conventional lifeboats.

higher probability of not being able to escape to the mustering station in the event of foundering compared to, for example, hull/machinery failure. Using statistics and expert judgement, Table 10.8 can be constructed for the undesirable events following an initiating event until preparation of the LSA. The scenario of jumping into the sea, awaiting and being rescued may occur in the case where there is a faulty evaluation of the situation, an untimely decision to muster is taken, the crew is unable to reach mustering stations, and in the case where the search for missing personnel is not terminated in time.

Similar tables can be produced for the other parts of the event sequence (see Figure 10.10). Where different branch probabilities are expected for the different LSAs, separate tables and event trees must be constructed for each LSA. The event tree corresponding to Table 10.8 is given in Figure 10.11.

The resulting probabilities of fatality and potential loss of life (PLL) for the different types of accidental events can be summarized by Table 10.9, which shows that the established evacuation model reflects the real world data fairly well. This is mainly due to the fact that actual data are used as the basic inputs and broken down into event tree branch probabilities. The weaknesses of the model are that the statistical values are based on a very limited number of events, which result in uncertainties, and that the model does not take sufficient account of the time factor involved in evacuation (i.e. in some cases the crew have more time available than in other cases).

In Table 10.9, the PLL is calculated as the number of fatalities in the given time period (1991–98) divided by the number of ship years in that period (44,732; see Table 10.7 earlier

Table 10.8. Event tree branch probabilities for the undesirable events following a ship accident until preparation of the LSA

	Collision	Contact	Fire/ explosion	Foundered	Hull/ machinery	Wrecked/ stranded
Fatality as result of initiating event	0.0001	—	0.007	—	—	—
Faulty evaluation of situation	0.31	0.31	—	0.37	—	—
Fatality as a result of not jumping into sea – given faulty evaluation of the situation	1	1	—	1	—	—
Untimely decision to muster	0.03	0.03	0.015	0.03	0.03	0.03
Fatality as a result of not jumping into sea – given untimely decision to muster	0.90	0.90	0.90	0.95	0.95	0.95
Unable to reach mustering station	0.06	0.06	0.03	0.07	0.02	0.06
Fatality as a result of not jumping into sea – given being unable to reach mustering station	0.95	0.95	0.95	1	0.95	0.95
Not terminating search in time	0.04	0.04	0.04	0.04	0.04	0.04
Fatality as a result of not jumping into sea – given search for missing personnel not terminated in time	0.625	0.625	0.625	0.625	0.625	0.625
Fatality associated with jumping and awaiting rescue	0.358	0.323	0.420	0.970	0.420	0.323
Fatality as a result of not being successfully rescued from the sea	0.0016	0.0016	0.021	0.050	0.021	0.0016

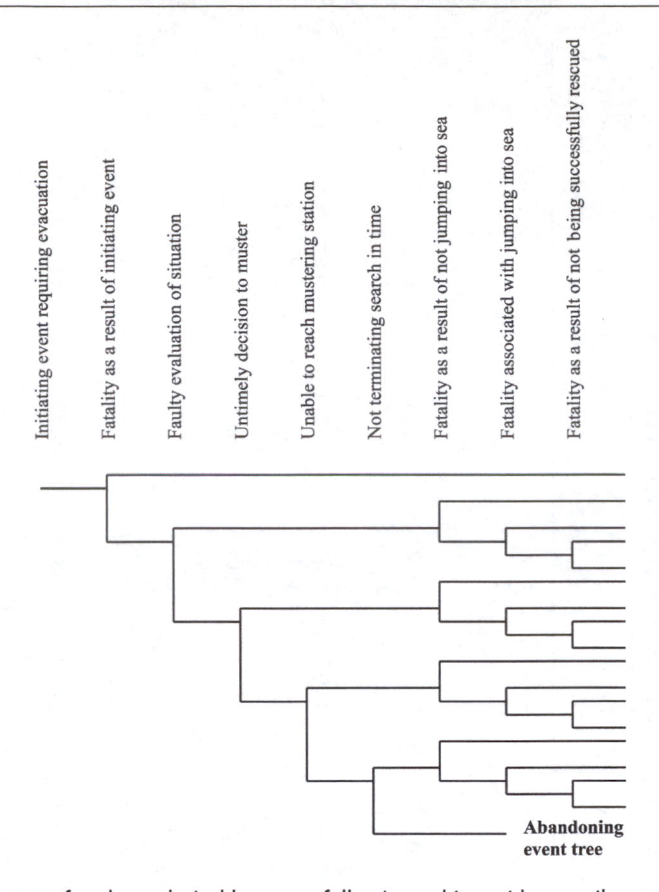

Figure 10.11. Event tree for the undesirable events following a ship accident until preparation of the LSA.

in this example). The number of fatalities is found by multiplying the statistical number of fatalities (e.g. 116 for collisions; see Table 10.7) by the probability of fatality.

Step 3: Establish safety measures (risk control options)

Risk control options (RCOs), with the objective of improving the life-saving capability of LSAs in a cost-effective manner, were identified and agreed upon by a multidisciplinary team of experts. The following RCOs were some of the measures identified for further assessment in this FSA study:

- Sheltered mustering and lifeboat area (SMA)
- Level alarms to monitor water ingress in all holds and forepeak (LA)
- Individual immersion suits for all personnel (IS)
- Free-fall lifeboat (FF)
- Marine evacuation system for thrown overboard liferafts (MES)
- Redundant trained personnel (RTP)

Table 10.9. Probability of fatality associated with evacuation

Type of accidental event	Based on evacuation model		Based on statistics	
	Probability of fatality (%)	PLL (per ship year)	Probability of fatality (%)	PLL (per ship year)
Collision	45.7	$3.392 \cdot 10^{-3}$	35	$2.6 \cdot 10^{-3}$
Contact	44.1	$1.173 \cdot 10^{-3}$	45	$1.2 \cdot 10^{-3}$
Fire/explosion	27.7	$2.347 \cdot 10^{-3}$	2	$1.7 \cdot 10^{-4}$
Foundered	55.4	$1.497 \cdot 10^{-2}$	51	$1.3 \cdot 10^{-2}$
Hull/machinery failure	16.2	$5.180 \cdot 10^{-4}$	0	0
Wrecked/stranded	20.2	$2.461 \cdot 10^{-3}$	0	0

Table 10.10. Reduction in the probability of fatality (%) with RCO implementation

	Current	SMA	LA	IS	FF	MES	RTP
Collision	45.7	−0.3	0.0	−1.2	−1.1	−0.3	−0.2
Contact	44.1	−0.3	0.0	−0.9	−1.3	−0.1	−0.2
Fire/explosion	27.7	−0.3	0.0	−4.5	+0.4	−1.0	−0.2
Foundering	55.4	−0.9	−14.8	−2.0	−4.9	−0.1	−0.4
Hull/machinery failure	16.2	−0.5	0.0	−1.6	−3.0	−0.2	−0.4
Wrecked/stranded	20.2	−0.5	0.0	−1.5	−1.8	−0.3	−0.3

The different RCOs affect the event trees modelled in step 2 of the FSA (i.e. the risk assessment step), resulting in changes to the probability of fatalities associated with evacuation. The RCOs will affect each accident scenario differently, and consequently the relative changes in the potential loss of life (PLL) value will vary both in terms of the RCOs implemented and the accidental event under consideration. Table 10.10 gives the relative change in percent for the probability of fatality in different accidental events for the RCOs listed above. This table is the underlying basis for calculating the cost-benefit relationship in the next step of the FSA methodology.

Step 4: Cost-benefit assessment

In order to establish a common and comparable cost-benefit ratio, all the potential costs and benefits related to the different risk control options (RCOs) must be identified. In this particular FSA study the cost estimation was done primarily by contacting suppliers of life-saving appliances, training centres, ship yards and ship owners. Through these sources, estimates were established on relevant costs such as investment in equipment, installation at the bulk carrier, inspection, maintenance, training of personnel to operate the equipment installed, etc. A high and a low cost estimate were established to take account of, among other things, factual variability in cost in Western Europe and the Far East, and the fact that the proposed technical solutions were not specified in detail. Depreciation of future costs was carried out at a real risk-free rate of return of 5%.

The benefits obtained by implementing the RCOs are the reduced number of fatalities, which may be calculated as a reduction in the potential loss of life (PLL) parameter using Table 10.10. For each RCO the total reduction obtained in the probability of fatality with RCO implementation was calculated as ΔPLL (in statistical terms the PLL is the expected loss of life). The ΔPLL was calculated as shown for the RCO of 'sheltered mustering and lifeboat area' (SMA) in Table 10.11.

The Implied Cost of Averting a Fatality (ICAF) was calculated using the following equation:

$$\text{ICAF} = \frac{\text{Cost of RCO}}{\text{Reduction in PLL}}$$

The cost estimates and the reduction in PLL are in this example calculated for a lifetime expectancy of 25 years for all RCOs. This is a simplification of the approach used in the reference source (DNV et al., 2001). These simplifications are made to reduce the size and complexity of this example, and result in slightly different ICAF values than those presented in the original FSA report. The final recommendations are, however, the same.

Both a high and a low ICAF value was calculated for each RCO based on the high and low cost estimates, respectively. For this particular analysis a decision criteria was based on an ICAF of £1 million. Other decision criteria may, however, have been selected, and this may have given different recommendations. Table 10.12 shows the calculation of the ICAF values. A RCO is recommended if its low (cost estimate) ICAF value is within the decision criteria of £1 million. A recommendation is considered robust if the high (cost estimate) ICAF value gives the same recommendation as the low (cost estimate) ICAF value.

Table 10.11. Calculation of ΔPLL for the RCO 'sheltered mustering and lifeboat area' (SMA)

Type of accidental event	Reduction in probability of fatality (%)	Resulting probability of fatality (%)	Resulting PLL with RCO (per ship year)	Resulting PLL without RCO (per ship year)
Collision	−0.3	45.4	$3.370 \cdot 10^{-3}$	$3.392 \cdot 10^{-3}$
Contact	−0.3	43.8	$1.165 \cdot 10^{-3}$	$1.173 \cdot 10^{-3}$
Fire/explosion	−0.3	27.4	$2.321 \cdot 10^{-3}$	$2.347 \cdot 10^{-3}$
Foundered	−0.9	54.5	$1.473 \cdot 10^{-2}$	$1.497 \cdot 10^{-2}$
Hull/machinery failure	−0.5	15.7	$5.019 \cdot 10^{-4}$	$5.180 \cdot 10^{-4}$
Wrecked/stranded	−0.5	19.7	$2.400 \cdot 10^{-3}$	$2.461 \cdot 10^{-3}$
Total PLL:			$2.449 \cdot 10^{-2}$	$2.486 \cdot 10^{-2}$
		ΔPLL = $3.7 \cdot 10^{-4}$		

Table 10.12. Calculation of ICAF values for the RCOs

RCO	ΔPLL (per ship year)	ΔPLL (per ship over 25 years)	Cost of RCO over 25 years (£ · 10^3)		ICAF (£ · 10^3)		Recommended?[a]	Robust recommendation?[b]
			Low	High	Low	High		
SMA	$3.70 \cdot 10^{-4}$	$9.25 \cdot 10^{-3}$	10	20	108.1	216.2	No	Yes
LA	$4.00 \cdot 10^{-3}$	$1.00 \cdot 10^{-1}$	14	21	140.0	210.0	Yes	Yes
IS	$1.27 \cdot 10^{-3}$	$3.18 \cdot 10^{-2}$	15	17.8	471.7	559.7	Yes	Yes
FF	$1.72 \cdot 10^{-3}$	$4.30 \cdot 10^{-2}$	−7.8[c]	18.2	−181.4	423.3	Yes	Yes
MES	$1.80 \cdot 10^{-4}$	$4.50 \cdot 10^{-3}$	4.50[d]		Criterion £ 1 m.		Not known	
RTP	$1.95 \cdot 10^{-4}$	$4.88 \cdot 10^{-3}$	8	10	1639.3	2049.2	No	No

[a]A RCO is recommended if its low (cost estimate) ICAF value is within the criterion of £1 million.
[b]A recommendation is robust if the high (cost estimate) ICAF value gives the same recommendation as the low (cost estimate) ICAF value.
[c]Free-fall lifeboats have a lower cost than more traditional LSAs (e.g. conventional lifeboats) when implemented instead of these on new ships.
[d]The maximum cost of a marine evacuation system to meet the ICAF criterion of £1 million (the costs are likely to be considerably higher).

Step 5: Recommendations for decision-making

Based on Table 10.12, the following risk control options provide considerable improved life-saving capability in a cost-effective manner and are therefore recommended:

- Level alarms to monitor water ingress in all holds and forepeak (LA)
- Individual immersion suits for all personnel (IS)
- Free-fall lifeboats (FF)

The RCO of free-fall lifeboats (FF) is only relevant for implementation on new ships as an alternative to more traditional LSAs (e.g. conventional lifeboats). The costs of fitting this RCO to an existing ship would make it unattractive in cost-benefit terms. Both the other two recommended RCOs are relevant for implementation on both new and existing ships.

If all the three recommended RCOs are implemented on a (new or existing) bulk carrier, the risk assessment procedure indicates that the evacuation success rate in the dominating foundering scenario increases from 44.6% (i.e. 1.0 = probability of fatality) to 66.2%, which is quite an improvement. Similar improvements are, however, not present for the other accident scenarios, and the general success rate in evacuation remains rather low even after implementing the recommended RCOs. This should call for additional measures, in particular measures with a focus on crew training and competence building.

REFERENCES

AEA, 1995, *Contract Draft Interim Report 1 – MSA Project 388: Step Four Methodology 'Cost-Benefit Assessment'*. AEA Technology, Deliverable 388D2, Warrington, UK.

DNV et al., 2001, Formal Safety Assessment of Life Saving Appliances for Bulk Carriers (FSA/LSA/BC). Project participants: Det Norske Veritas, Norwegian Maritime Directorate, Norwegian Union of Marine Engineers, Umoe Schat-Harding, Norwegian Shipowners' Association, International Transport Workers' Federation, Oslo.

IMO's web site, http://www.imo.org/meetings/msc/68/fsa.htm

MSA, 1995, *Project 366: Formal Safety Assessment: Draft Stage 1 Report*. Marine Safety Agences, Southampton, February.

Skjong, R., 1999, *FSA of HLA on Passenger Vessels*. Report No. 97-2053, Det Norske Veritas, Oslo.

PART IV
MANAGEMENT AND OPERATIONS

11

HUMAN FACTORS

Eternal vigilance is the price of safety.
(Fleet Admiral Chester W. Nimitz)

11.1 INTRODUCTION

In this chapter we will discuss the human factors aspects of ship operation. Although certain functions have been automated, a ship is still largely a human-controlled system. Two key concepts are sometimes used more or less as synonyms, namely ergonomics and human factors. There is, however, a certain distinction made between the concepts in some circles, although both focus on the interaction between humans and physical work environment and machines:

- *Ergonomics:* Particular emphasis is put on the design of displays, controls and the workplace. The human physical dimensions (anthropometry) and their capacity with respect to sensing and control ability are especially taken into consideration. The physical aspects of the environment such as climate, noise and vibration are also a concern.
- *Human factors:* The work situation is assessed in the light of psychological factors. A key aspect is the relation between the job or task requirements and the human capacity. Factors such as mental capacity to process information, motivation, and interaction with colleagues have to be taken into consideration.

In the rest of this chapter we will use human factors (HF) as a common term. In the course of the last five decades HF has established itself as a solid scientific discipline on the basis of theory and empirical data. For the practitioner, a number of handbooks and guides are available (Salvendy, 1987; Wilson and Corlett, 1990; McCormick and Sanders, 1983; VanCott and Kincade, 1972).

11.2 HUMAN ERROR

11.2.1 A Misused Term

It has for many years been popular to explain ship accidents by human error. There are numerous citations indicating that 75–90% of accidents are rooted in human error. Sometimes one may feel the element of condemnation or that the error is a result of negligence. Often our understanding of vague concepts is biased towards our own professional background:

- Legal: focus on negligence or criminal conduct
- Technical: the operator misused the system
- Psychological: the operator is inhibited by trauma
- Socio-technical: the operator is estranged
- Ergonomic: incapacitated

This means that for instance the explanation of a stranding by 'navigation error' may have different interpretations, e.g.:

- Negligence due to low morale
- The control system did not give any feedback
- Wrong assessment due to lack of skills
- Inadequate performance due to lack of procedure
- Electronic disturbance
- Perception error due to low arousal

The lesson from this is that the human error concept in itself is of limited value unless it is described in a broader context.

11.2.2 Earlier Studies

It often seems difficult to keep in mind that an accident seldom has only one single explanation but rather a number of them. This is disturbing given the fact that this was pointed out more than 30 years ago (Stewart, 1973). By discussing a classical stranding scenario (Figure 11.1), he demonstrated the interaction between various causal factors.

The factors that led to the stranding were:

1. Vessel was approaching a dangerous headland.
2. Wind was blowing strongly from north.
3. Attempt was made to adjust for setting to the southward.
4. No allowance was made for leeway.
5. Despite a later position fix, heading was not corrected.
6. Echo sounder was not in use.
7. Decreasing visibility near land.
8. Light picked up to starboard.

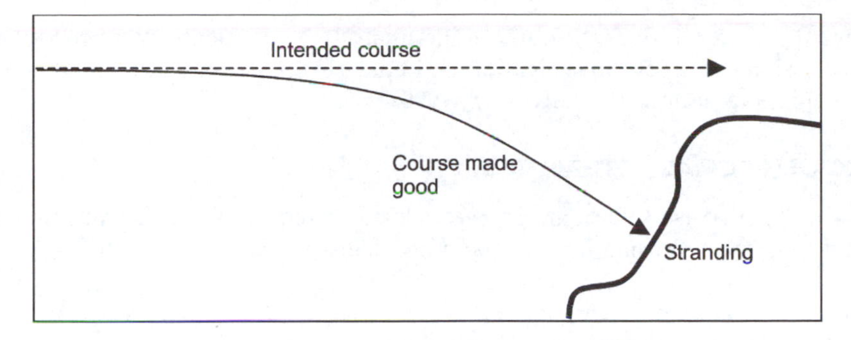

Figure II.1. A stranding scenario. (Adapted from Stewart, 1973.)

9. Assumed to be the headland light.
10. Stranded at full speed.

One of the first studies of human error in ship operation (NAS, 1976) focused on the following factors:

- Inattention
- Ambiguous pilot–master relationship
- Inefficient bridge design
- Poor operational procedures
- Poor physical fitness
- Poor eyesight
- Excessive fatigue
- Excessive alcohol use
- Excessive personnel turnover
- High level of calculated risk
- Inadequate lights and markers
- Misuse of radar
- Uncertain use of sound signals
- Inadequate rules of the road

It is interesting to notice that most of these problems are just as relevant today. We can see that the investigation had a broader view which took workplace factors, procedures, fatigue, health and management into consideration.

Fukushima (1976) took an even broader view by addressing the effect of external conditions. He saw accidents as combinations of the following complex conditions:

- Natural phenomena: weather and sea
- The route: fairway conditions, obstacles and visibility
- Ship: stability, manoeuvrability and technical standard
- Traffic congestion
- Navigator: knowledge, skills and health

It can be concluded that we have an understanding of why accidents happen that involves the operator, technology, work conditions and organization. Despite this we still seem to take a more narrow view both in design and planning of operations. This fact is witnessed by bridge design shortcomings, lack of safety training and ignorance in management. It even influences our routine analysis of accidents. We are still waiting to see professional accident investigation and analysis on a broad scale among even the advanced shipping nations (Caridis et al., 1999).

11.3 ACCIDENT CHARACTERISTICS

In the effort to increase safety and in particular to minimize the effect of human error, we should keep the following lessons about accidents in mind:

1. Routine:
   - Often related to normal operational situations
   - Not necessarily a result of abnormal events or conditions
2. Gradual escalation:
   - Seldom happens momentarily
   - Inability to cope with events as they surface
3. Multiple causes:
   - A combination of more than a single event
   - Causal factors are related to technology, humans and organization
4. Human error:
   - The operator is by definition near the accident events
   - The human behaviour should be viewed in broader scope
5. Presence of situation factors:
   - Physical environment
   - Workplace conditions
   - Task load
   - Mental and motivational state

It is important to have a realistic view of the time scale for the events related to an accident, as illustrated in Table 11.1. Decisions that may lead to hazardous conditions can be taken in the order of a year in advance. The critical events may develop within a day or an hour and the dramatic release of energy in the order of seconds. In a corresponding manner, the consequences in terms of breakdown of vessel, human suffering and environmental damage follow in a dramatic way but also have long-range effects. This perspective is important to keep in mind when we attack the human factors problem.

11.4 HUMAN INFORMATION PROCESSING

11.4.1 Accident-Prone Tasks

One may ask oneself whether specific tasks are more subject to error than others. This was investigated in the US nuclear reactor study (RSS, 1975). One of the lessons to be

Table 11.1. Characteristic time scale of a ship accident

Time scale	Typical event
1 year	Management decisions related to operations
1 day	Breakdown of safety function
1 hour	Initiating failure or error
1 min	Attempt to avoid threatening accident
1 sec	Release of energy
1 min	Collapse of hull and breakdown of systems
1 hour	Flooding and fire, loss of vessel and evacuation of people
1 day	Rescue of people and vessel
1 week	Pollution, hospitalization of people
1 year	Environmental damage, post-traumatic effects on people

Table 11.2. Omissions in nuclear power plant operations by task category

Task	Omissions (%)
Monitoring and inspection	0
Supervisory control	2.3
Manual operation and control	5.9
Inventory control	9.4
Test and calibration	32.9
Repair and modification	41.2
Administrative task	1.2
Management, staff planning	1.2
Other or not specified	5.9
All tasks	100

Source: RSS (1975).

learned was that non-routine tasks such as testing/calibration and repair/modification were especially prone to omission error, as can be seen from Table 11.2. The analysis of behavioural mechanisms showed that functionally isolated acts often lead to error (Table 11.3). This mechanism is typical for non-routine tasks.

Rasmussen (1982) makes an important distinction between the following basic performance levels: skill-, rule- and knowledge-based behaviour (see Figure 11.2). Skill-based behaviour is our learned and almost unconscious actions. For certain familiar situations we apply rule-based behaviour. The complex situations require some kind of problem solving called knowledge-based behaviour. Skill-based behaviour is most frequent in daily operations and also less subject to error and accidents. On the other hand we apply knowledge-based behaviour to deal with unfamiliar and difficult tasks. This behaviour may also have the largest consequences in case of error.

Table 11.3. Errors in nuclear power plant operations by behavioural mechanism

Behavioural mechanism	Errors (%)
Absent-mindedness	1.5
Familiar association	3.0
Alertness low	5.0
Omission of functionally isolated act	34.0
Other omissions	8.5
Mistake among alternatives	5.5
Strong expectation	5.0
Side effect not considered	7.5
Latent condition not considered	10.0
Manual variability, lack of precision	5.0
Weak spatial orientation	5.0
Other not specified	10.0
All behaviours	100

Source: RSS (1975).

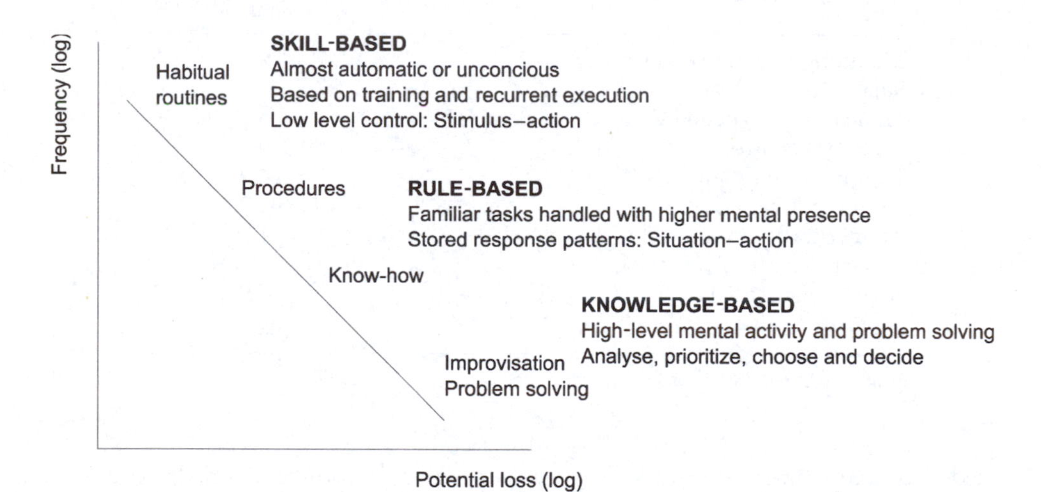

Figure 11.2. Alternative human performance levels. (Adapted from Rasmussen, 1982.)

11.4.2 A Human Control Model

Operation of vessels is still to a large degree based on human–machine control. Figure 11.3 presents a conceptual model of vessel control that outlines key topics relating to human factors that will be discussed in the rest of this chapter. The core of the control loop is the interaction between the operator function and the vessel.

The operator uses data from the visual environment and the information displays onboard. This data is processed and results in control actions and communication with

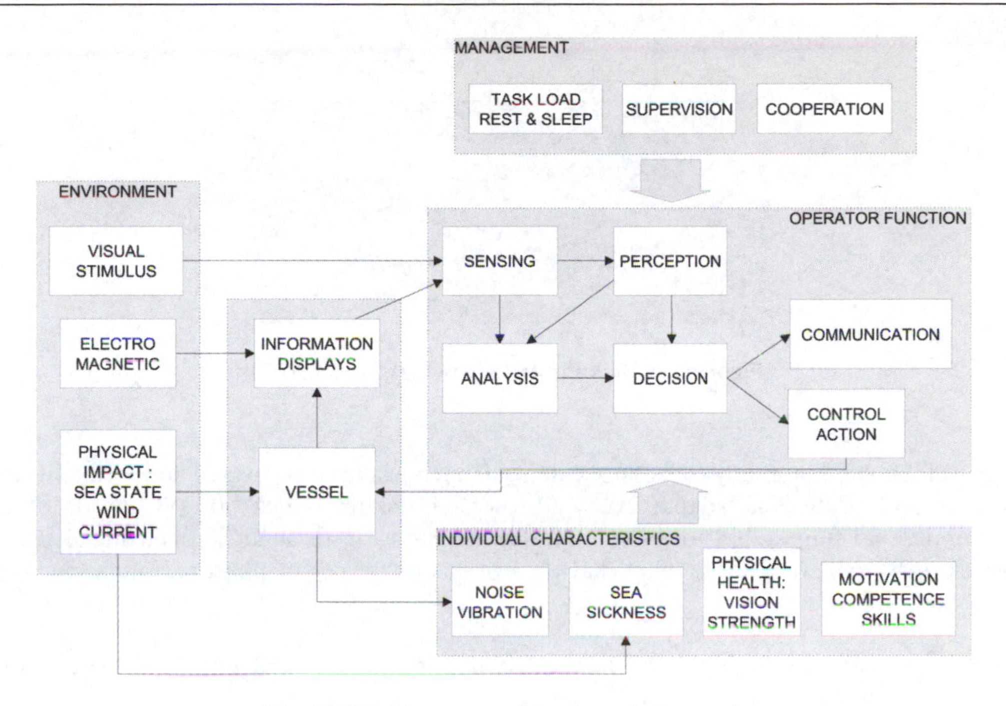

Figure II.3. Human–machine control of a vessel.

other crew members. The control of the vessel may, however, be disturbed by the physical environment. It should also be kept in mind that the performance of the operator is governed by individual characteristics and the interaction with supervisors and other crew members.

II.5 SENSING AND OBSERVATION

Navigation tasks are quite demanding for the officer on watch and the lookout. The key sources of information are:

- The seaway, landscape, sea marks and marine traffic (marine environment)
- The visual and aural displays on the bridge
- The behaviour of the vessel (movement, acceleration, etc.)

II.5.1 Night Vision

Although all vessels are equipped with radar, visual lookout is still an important method for assessing the position of the vessel in restricted seaways and detecting and monitoring the traffic. The ability to observe under night (dark) conditions is especially critical.

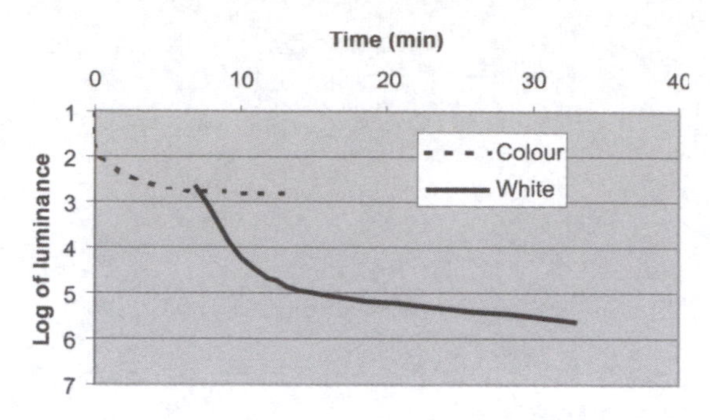

Figure 11.4. Dark adaptation curve (Jayle et al., 1959).

It is well known that many collision accidents happen just after the watch has been relieved due to inadequate dark adaptation. Madison (1974) states that it takes in the order of 25 minutes to fully adapt to darkness following exposure to sunlight or artificial light. This can be explained by the fact that the eye has two types of photoreceptors, namely *rods* and *cones*:

- Cones react to higher light levels, observation of finer details and perception of colours.
- Rods are associated with night vision and low-intensity light. They are relatively insensitive to red light, which is in the lowest frequency of the visual spectrum. Regardless of frequency of stimulus, objects are seen in shades of black and white.

Figure 11.4 shows that the dark adaption curve is discontinuous. The cones adapt in 10 minutes whereas the sensitivity of the rods is rapidly lost in high-level illumination and therefore need in the order of 25 minutes to adapt.

11.5.2 The Unique Role of Red Light

Under low-level illumination at night we are dependent on the rods. As they are relatively insensitive to red light, one may speed up the dark adaptation by wearing red goggles. The rods are not stimulated by low-level red light and start to adapt as if they were in total darkness.

In order to minimize the effect on night vision instruments on the bridge may use red illumination. This solution, however, has certain drawbacks. First, visual acuity is poorer, which means that the ability to discriminate finer details is reduced. Secondly, visual fatigue will increase. In certain instances where visual acuity is critical blue lighting is applied, as for instance on radars and CRT screens. This on the other hand will reduce the night adaptation.

Both the use of red and blue lighting may also affect the reading of charts and this should therefore also be taken into consideration. Red, orange and buff colours have a tendency to disappear under red lighting.

II.5.3 Lookout

Lookout is often a challenging task for various reasons. One factor is that at open sea there are few objects to observe and this may lead to reduced vigilance. There have been extensive discussions of what is the best lookout strategy (Madison, 1974):

- Slow rowing gaze, or
- Repeated fixations at widely separated positions

Systematic search seems to be most efficient for low contrast targets whereas free search is best for high visibility targets. It is also a well-known fact that one should not stare directly at a dim spot under dark conditions in order to avoid the interference of the blind spot in the eye. Night light sensitivity is in fact greatest 15–20° away from the centre of vision.

The lookout function may also be degraded by *night blindness*. The existence or degree of night blindness may be unknown to the person itself as it is not apparent during daytime conditions. The source may be either psychological or physiological. It has been related to neuroses, psychoses and hysteria (Jayle et al., 1959). Other pathological factors that result in night blindness are dietary deficiencies (vitamin A), diabetes, glaucoma or congenital night blindness. It has also been shown that ageing is an important factor. It will develop from roughly the age of 40 and becomes pronounced after 50.

Another phenomenon that interferes with the lookout function is so-called night myopia (near-sightedness), which is maintained as long as the eyes are night adapted. Far-sighted persons may for practical reasons therefore see better without corrective glasses, whereas a reason with normal sight will improve night vision with a corrective lens of roughly −1.5 diopters. Jayle et al. (1959) found that when night myopia was corrected the night vision threshold was improved by 50%.

II.5.4 Uniform Field Difficulty

Apart from the increasing boredom of prolonged observation of a uniform field such as calm sea, open sky or fog, there is also a physiological cost. This may lead to *blanking out* after 10–20 minutes. It results in reduced motor coordination and ability to maintain balance. Another related phenomenon is *empty field myopia*, which means that the eye accommodation is in a constant state of fluctuation due to the lack of objects to focus on. This problem may be solved by looking away at objects like masts or other objects on the vessel every 5 minutes.

II.5.5 Visual Illusions

Vision is the main source of information for the watchkeeping personnel. In light of this, it is important to keep in mind that visual illusion is extremely common. Some well-known factors are:

- Refraction
- Fog and haze
- Texture
- Autokinesis

Refraction is the phenomenon whereby the direction of the light is broken by passing through different media. During rain, water may build up on the windows of the bridge and result in refraction. This results in a misinterpretation of the relative direction (bearing) of other vessels or objects.

In fog and haze, objects appear to be smaller than they actually are and therefore seem to be farther away. This may put the vessel at risk in approaching situations by reducing the time and distance to stop or change heading. A compounding factor is also the tendency to underestimate the relative speed of other objects under marginal visual conditions. The texture of an object may be a clue about its distance from the vessel. Fine detail indicates a vessel nearby whereas a diffuse appearance indicates an object further away. For that reason unusual objects may be misjudged with respect to their distance.

The visual illusion of autokinesis can happen at night when you look continuously at a single light against an otherwise dark background. The light will apparently move, sometimes in an oscillating fashion. Both light markers and lanterns of other vessels might be the source of this phenomenon.

11.5.6 Radar Operation and Vigilance

Radar operation may be a quite demanding task for various reasons. One situation is navigation in coastal waters with heavy traffic where the operator has to monitor many targets and assess collision risk. However, owing to the invention of the ARPA device (Automatic Radar Plotting Aid), this task has been substantially eased. A quite different challenge is to maintain vigilance during radar watch in open waters with little or no traffic for longer periods. Mackworth (1950) has defined vigilance as the observer's readiness to detect infrequent, aperiodic, small changes in the environment.

It has been well established that the detection of radar targets may fail for a number of reasons:

- The signal may be weak
- The radar target is veiled by signal noise
- No warning
- Increased boredom due to the monotony of the task
- Lack of rest pauses

The fact that performance deteriorates with time is expressed in the so-called 'Mariners Law': 'Maximum vigilance can be maintained for a period of about 30 minutes – after this the performance deteriorates sharply' (Elliott, 1960). This observation has later been stated in a more precise form by Teichner (1972), who found that the probability of detecting a visual signal is a function of:

- The initial probability of detection
- The duration of the watch
- Whether the detection demands continuing adjustment of eye focus or not.

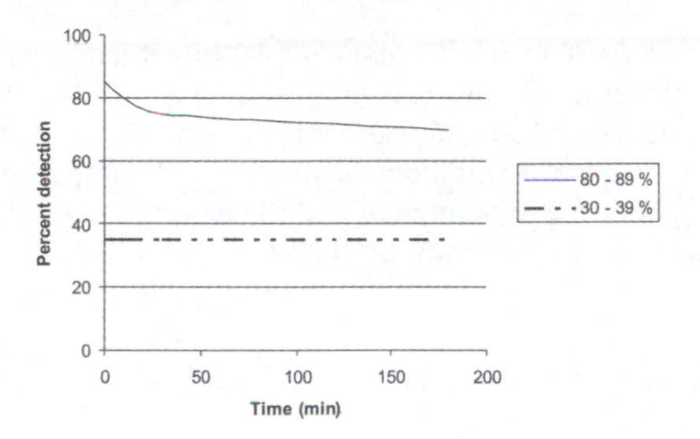

Figure 11.5. Initial detection probability versus time on watch for different initial detection probabilities. (Source: Teichner, 1972.)

Two of these factors are illustrated in Figure 11.5. This shows that, given a low initial detection frequency, there is insignificant deterioration with time, whereas a high initial detection deteriorates sharply during the first 30 minutes. This phenomenon is ascribed to the fact that the first situation reflects a demanding task requiring high concentration, whereas the other is less demanding and therefore soon results in loss of concentration.

11.5.7 Heavy Meals

Madison (1974) has summarized some of the findings of the effect of meals on human performance. The main and perhaps not surprising conclusion is that the meals should be well balanced. Experiments and questionnaire studies further indicate that:

- High carbohydrate meals have soporific qualities and give poorer performance. However, the subjective discomfort was lower.
- High fat meals seems to give a short-term gain in performance

A complicating factor is that individual preferences with respect to meals may have a major impact on performance and thereby preclude the value of the general findings cited above.

11.6 PERCEPTION AND DECISION-MAKING

By perception is meant basically the task of making sense of what you detect and observe. For observation, there are a number of sources of error. Typical questions are: do I see a vessel or a stationary object, is the light from a marker or another vessel, or is the object moving away from or towards me? Perception involves:

- Understanding the signal, applying meaningful concepts
- Relating visual input to known pictures
- Assessing movement relative to mental, dynamic models
- Giving priority to alternative information sources

Problems may arise due to the fact that humans apply sequential processing (one thing at the time), input in short-term memory is forgotten, or mental overload occurs.

The performance of human information processing rests very much also on correct decision-making: how to assess the situation on the basis of the perceived information and how to respond in light of the mission plan and the detected obstacles or constraints. Both perception and decision-making may be flawed by what we call false hypothesis and habits.

11.6.1 False Hypothesis

A rational perception of what we sense can be influenced by the so-called false hypothesis phenomenon, or what we may term anticipation or illusions. It appears in different ways or by different mechanisms:

- A tendency to take in limited information and assume the rest.
- High expectancy: long experience that things happen in a certain pattern makes you 'see' things regardless of the actual stimulus.
- Hypothesis as a defence: interprets things in way that reduce stress or anxiety (not facing reality).
- After a demanding work period, the concentration or vigilance drops and makes you more vulnerable to error.

11.6.2 Habit

Safe operation of a vessel is to a considerable degree based on sound habits or, essentially, behaviours. The objective of training is to establish habitual actions or skills. Through acquired experience the repertoire of skills is further developed. A negative aspect of time or age is that physical fitness reaches a peak in the early years. A classic question is therefore whether the physical superiority of the young is beaten by the greater experience of the mature crew member. There may, however, be a certain risk with habits under changing conditions:

- Habitual responses may be inappropriate due to change in the dynamic characteristics of the vessel, another propulsive system, and so on.
- 'Cannot make myself forget': new response patterns have to be trained, but under stress or with focus elsewhere one returns to earlier habits.
- The problem may be overcome by 'over-learning' of critical responses.
- But as new responses and improved skills are developed, the ability to stay calm is also improved.

11.6.3 End-Spurt Effect

It has been shown that for tasks demanding vigilance, knowledge about the remaining duration of the assignment has an effect on the performance of the operator.

In experiments, subjects knowing how long the time they had to stay on vigil performed superior to subjects not knowing.

11.7 PHYSICAL WORK ENVIRONMENT

Following physical work, climate factors are thought to be relevant for the performance of the crew:

- Thermal climate
- Noise
- Vibration
- Illumination

These factors were studied in depth by Ivergård et al. (1978) on Swedish vessels. The results will be summarized in this section, but it should be kept in mind that the findings are not necessarily representative of conditions today.

11.7.1 Thermal Climate

The thermal climate is a function of a number of factors: temperature, humidity, air speed (circulation) and heat radiation. The subjectively experienced climate must further be seen in relation to the work load and clothing of the individual.

The thermal climate represents a challenge on a vessel for the following reasons:

- It operates in all climate zones
- The conditions vary highly in different sections of the vessel
- The crew members are exposed to different thermal stress
- The crew can be subject to extreme thermal loads in emergency situations

In discussing thermal stress it is necessary to make a distinction between the different sections of the vessel. Table 11.4 gives a summary of the problems related to thermal control in the various sections of a vessel.

11.7.2 Temperature and Vigilance

The experience of well-being and task performance is a function of the climate. The body will be in thermal balance if the heat produced by metabolism equates the heat lost through evaporation, radiation, convection and by work accomplished. As evaporative heat loss is a key source during most conditions, humidity also plays a key role. In practice we therefore apply the *effective temperature* (ET) concept, which is a function of both dry-bulb temperature and relative humidity. The body is in balance with respect to heat loss at a dry-bulb temperature of 25°C and a relative humidity of 50% (McCormick, 1976).

Table 11.4. Thermal climate factors on vessels

Section	Situation	Solution
Living quarters	Reasonable control	Ventilation and air conditioning is standard
Navigation bridge	Thermal control up to 22–28°C Heat sources: large window panes, electronic equipment and open doors	Ventilation and air conditioning is standard
Galley	A number of heat sources: stove, etc. Long periods above thermal comfort criteria	Improved isolation of heat sources
Engine rooms	Air cooling of engines Temperature often 10–20° higher than outside temperature Heat radiation from hot surfaces Heavy repair and maintenance work High relative humidity in smaller, confined spaces	Work in engine rooms will in general take place well outside thermal comfort criteria Under winter conditions large temperature variations within the same space
Cargo tanks	Inspections and final cleaning operations may be stressing in hot outside climate. Protective clothing may contribute to the stress	Cleaning by permanently installed equipment
On deck	Both high and low temperatures depending on time of the year and geographical latitude Sun radiation Effect of wind (air speed) under winter conditions	Clothing and protection

Mackworth (1946) made one of the early studies of the temperature effect on monitoring tasks as shown in Figure 11.6. It can be seen that the number of monitoring mistakes starts to increase sharply above an effective temperature of 30°C. This corresponds to the upper limit for what is experienced as comfortable.

Obviously the question of acceptable temperature is also related to the duration of exposure. As shown in Figure 11.7, the duration of unimpaired mental performance decreases quickly as the effective temperature rises above 30°C (Wing, 1965).

Finally, it should be mentioned that there seems to be a special combined effect of warmth and sleep loss (Pepler, 1959). Observed phenomena are reduced performance in tracking tasks and gaps in serial responses. Simply stated, the following mechanisms are seen: 'Warmth reduces accuracy' whereas 'Sleep loss reduces activity'.

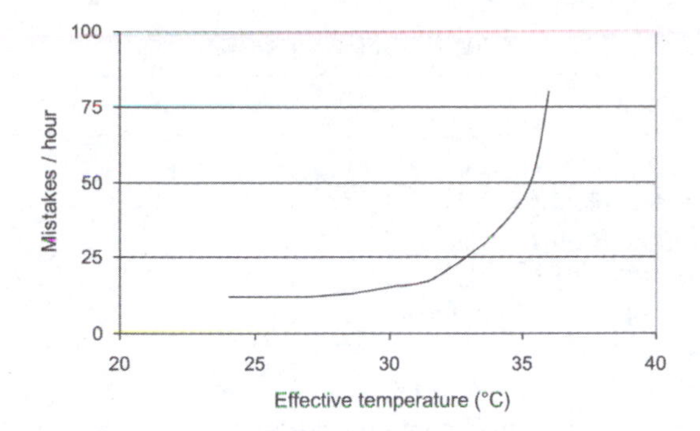

Figure 11.6. Monitoring mistakes as a function of effective temperature. (Source: Mackworth, 1946.)

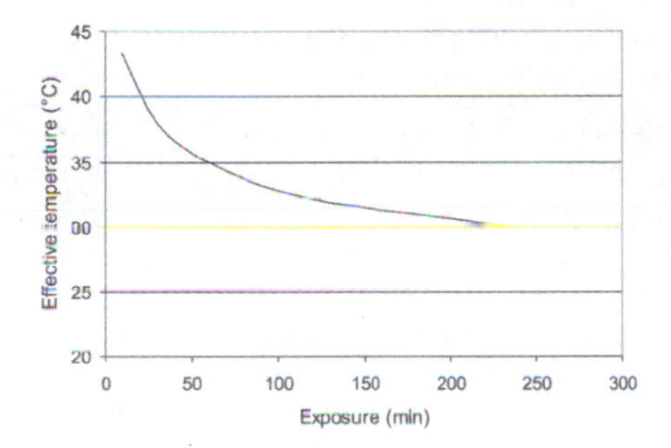

Figure 11.7. Upper limit for unimpaired mental performance. (Source: Wing, 1965.)

11.7.3 Noise

The primary noise sources onboard are: main engine, propeller, auxiliary engines and engine ventilation. The noise (or unwanted sound) is measured in terms of sound intensity and expressed as dB(A). Table 11.5 summarizes maximum recorded and recommended noise levels for the main ship sections.

11.7.4 Infrasound

Infrasound is acoustic waves below 20 Hz that are not audible by the human ear. It has, however, been found that for high intensities infrasound has negative effects on the human in terms of reduced well-being, tiredness and increased reaction time.

The relation between infrasound and possible negative effect is still only partly understood. It has been suspected that even higher intensities (above 100 dB) may have adverse effect on control of balance, disturbance of vision and choking.

Table 11.5. Noise in vessels

Section	Situation	Solution
Living quarters	Highest values: dB(A) = 58–70 Depends mainly on the distance to engine rooms Variation with ship type 28% experienced the noise as troublesome	
Navigation bridge	Highest values: dB(A) = 65–73 Some effect on direct communication and use of internal communication equipment	Lowest noise levels on vessels with bridge in the fore part (passenger, Ro-Ro)
Galley	Mean value: dB(A) = 71–77 Recommended value 65 dB(A) is exceeded due to the background noise	
Engine rooms	Highest values: dB(A) = 93–113 Recommended value: 100 dB(A) Factors: power, engine type Risk of physiological damage (reduced hearing) 50% of engine personnel report the noise as troublesome	Wear ear protection Sectioning of engine room

Typical ranges for infrasound measured on vessels are 4 Hz (55 dB) to 16 Hz (95 dB). The fact that infrasound is present both on the bridge and in the engine control room may pose a problem to safe operation.

11.7.5 Vibration

Vibration is seen by a significant part of the crew as one of the most disturbing environmental factors onboard. This is partly explained by the fact that it is experienced both at work and off-duty. This means that persons affected are given no opportunity for restitution during free hours. Vibration is expressed by acceleration (m/s^2) in selected frequency bands (Hz). Vibrations are usually measured for a range from 0.5 Hz to 125 Hz.

The main sources of vibration are the main engine and the propeller. The induced vibrations are further transmitted through the steel hull and deck houses. High superstructures are further subject to resonance phenomena. It is also experienced in the aft part as an effect of the hogging–sagging movement of the hull girder, giving very low frequency resonance (0.5 Hz).

ISO has given norm values for exposure (time) versus vibration, both with respect to 'reduced work performance' and 'reduced comfort'. Figure 11.8 illustrates the

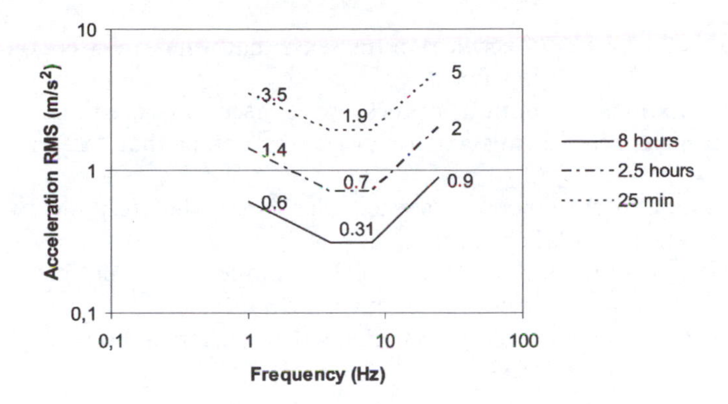

Figure II.8. Vibration tolerance limits as function of exposure time. (Adapted from ISO 2631, 1985.)

ISO norm for maintaining profiency. It can be seen that the most sensitive frequency area is roughly 4–8 Hz. For longer durations the vertical acceleration should not exceed 0.31 m/s^2.

Measurements of vibrations by Ivergård et al. (1978) can be summarized as follows:

- *Navigation bridge:* Most vessels have a vibration level below recommended values. The exception is tankers. However, crew reports some discomfort.
- *Engine room:* Most observations lie between the comfort and performance curves. Although the comfort limit is not exceeded, personnel reports problems due to the reduced opportunity for restitution during the free watch.
- *Cabin area:* Vibration level coincides with comfort line in the critical area 8–16 Hz. Confirms that restitution becomes a problem.

II.8 EFFECT OF SEA MOTION

One of the classical torments of life at sea is the so-called *seasickness* that may affect both passengers and crew. The more correct term is motion sickness as it is present in different transport modes and in certain other situations. Motion sickness may be experienced during anything from riding a camel, operating a microfiche reader to viewing an IMAX movie. How strongly one may be affected differs, but it usually involves stomach discomfort, nausea, drowsiness and vomiting, to mention a few.

The problem has recently been surveyed by Stevens and Parsons (2002) in a maritime context. As the authors point out, one of the consequences of seasickness is less motivation and concern for the safety critical tasks onboard. However, for the majority of persons there will be an adaption to sea motion over time so that the sickness fades away after a few days. There is also some comfort in the fact that susceptibility usually decreases with increasing age. The effect of motion appears sometimes in the form of the *sopite syndrome*, which is less dramatic and usually only experienced as drowsiness and mental depression.

This affection may also be safety critical in the sense that it has an effect on performance and is not always evident to the subject.

Various explanations of motion sickness have been proposed, but contemporary theory points to the conflict or mismatch between the organs that sense motion:

- *Vestibular* system in the inner ear: semicircular canals that respond to rotation and otoliths that detect translational forces.
- The *eyes* (vision system) that detect relative motion between the head and the environment as the result of motion of either or both.
- The *proprioceptive* (somatosensory) system involves sensors in body joints and muscles that detect movement or forces.

The theory assumes that under normal situations the three systems detect the movements in an unambiguous way. However, under certain conditions the senses give conflicting signals that lead to motion sickness. A typical maritime scenario is experiencing sea motion inside the vessel without visual reference to sea, horizon or land masses. In this case the vestibular detects motion in the *absence* of visual reference. Watching the waves from the ship may give rise to *conflict* between vision and vestibular system.

Extensive experiments in motion simulators found that subjects were primarily sensitive to vertical motion (heave) and that maximum sensitivity occurred at a frequency of 0.167 Hz (Griffin, 1990). Given the fact that the principal vertical frequency in the sea motion spectre is 0.2 Hz, the occurrence of seasickness is understandable. There are two methods for estimation of motion sickness: Motion Sickness Incidence (MSI) and Vomiting Incidence (VI). Both methods are outlined by Stevens and Parsons (2002). There are a number of effects of seasickness:

- Motivational: drowsiness and apathy
- Motion-induced fatigue (MIF): reduced mental capacity and performance
- Reduced physical capacity
- Added energy expenditure to counterbalance motion
- Sliding, stumbling and loss of balance
- Some interference with fine motor control tasks

The effect on cognitive tasks is more inconclusive as it has been difficult to isolate the effect of physical stress. Another problem is the fact that bridge tasks involve a number of cognitive processes and skills which may be influenced differently by the sea motion (Wertheim, 1998).

In order to minimize the risk of seasickness, it is necessary to establish operational criteria. Baitis et al. (1995) point out that it is not the roll angle in itself that limits the operation but rather the vertical and lateral accelerations associated with them. Table 11.6 shows the criteria proposed by NATO (NATO STANAG 4154, 1997), which are based on both earlier and recent principles. An alternative set of criteria make a distinction between

Table 11.6. Sea motion operability criteria

Motion sickness incidence (MSI)	20% of crew in 4 hours
Motion-induced interruption (MII)	1 tip per minute
Roll amplitude	4.0° RMS
Pitch amplitude	1.5° RMS
Vertical acceleration	0.2 g RMS
Lateral acceleration	0.1 g RMS

Source: NATO STANAG 4154 (1997). RMS, root mean square; g, acceleration of gravity.

Table 11.7. General bridge operability criteria for ships

	Merchant ships	Naval vessels	Fast small craft
Vertical acceleration (RMS)	0.15 g	0.2 g	0.275 g
Lateral acceleration (RMS)	0.12 g	0.1 g	0.1 g
Roll (RMS)	6.0°	4.0°	4.0°

Source: NORDFORSK (1987).

different ship types (NORDFORSK, 1987) as summarized in Table 11.7. It can, however, be concluded that the two sets of criteria agree fairly well.

11.9 HUMAN RELIABILITY ASSESSMENT

Risk analysis involves making quantitative estimates of the risk associated with designs and operations. Typical is estimation of the probability of having specified accident events by means of fault tree analysis (FTA) or event tree analysis (ETA). In order to make realistic estimates, human-error-related events must be incorporated. This is not a trivial problem given our limited understanding of this phenomenon and lack of systematic data. Despite this, various approaches have been developed and to some degree have also been supplemented with quantitative data. The objective of human reliability assessment (HRA) has been stated as follows by Kirwan (1992a, 1992b):

1. Human error identification: what can go wrong?
2. Human error quantification: how often will it occur?
3. Human error reduction: how can it be prevented or the impact reduced?

The key methodological steps are outlined in Figure 11.9.

Considerable experience has been accumulated with the so-called first-generation HRA methods, primarily from applications in the process and nuclear industries. However, they have come under increasing attack from cognitive scientists who point to the fundamental lack of ability to model human behaviour in a realistic manner (Hollnagel, 1998).

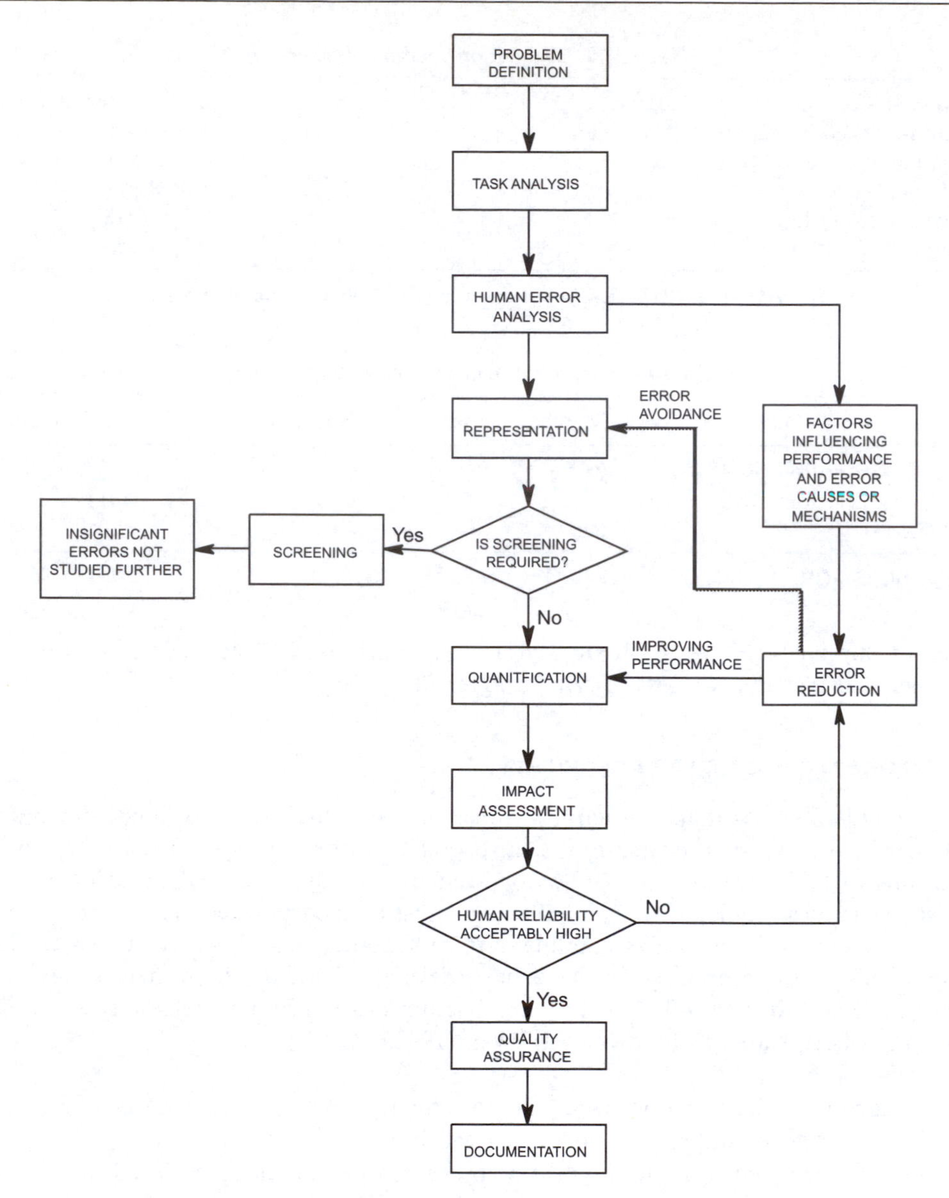

Figure 11.9. Human reliability assessment steps (Kirwan, 1992a, 1992b).

11.9.1 Therp

The Technique for Human Error Rate Prediction (THERP) is probably the best-known and most widely used technique of human reliability analysis. The main objective of THERP is to provide human reliability data for probabilistic risk and safety assessment studies. The methods and underlying principles of THERP were developed by Swain and Guttmann (1983) and are often referred to as the *THERP Handbook*.

The methodological steps are:

1. *Identify system failures of interest:* This stage involves identifying the system functions that may be influenced by human errors and for which error probabilities are to be estimated.
2. *Analyse related human operations:* This stage includes performing a detailed task analysis and identifying all significant interactions involving personnel. The main objective of this stage is to create a model that is appropriate for the quantification in stage 3.
3. *Estimate the human error probabilities:* In this stage the human error probabilities (HEPs) are estimated using a combination of expert judgements and available data.
4. *Determine the effect on system failure events:* Estimating the effect of human errors on the system failure events is the main task of this stage. This usually involves integration of the HRA with an overall risk/safety assessment (i.e. PRA/PSA).
5. *Recommend and evaluate changes:* In this stage changes to the system under consideration are recommended and the system failure probability recalculated. Possible solutions for various human factors problems include job redesign, implementation of mechanical interlocks, administrative controls, and implementation of training and certification requirements.

The probability of a specific erroneous action is given by the following expression:

$$P_{\mathrm{EA}} = \mathrm{HEP}_{\mathrm{EA}} \sum_{k=1}^{m} \mathrm{PSF}_k \cdot W_k + C$$

where:

P_{EA} = Probability of an error for a specific action
$\mathrm{HEP}_{\mathrm{EA}}$ = Basic (nominal) operator error probability of a specific action
PSF_k = Numerical value of kth performance shaping factor
W_k = Weight of PSF_k (numerical constant)
C = Numerical constant
m = Number of PSFs

The probability is a function of the error probability for a generic task modified by relevant performance shaping factors. The basic HEPs can be looked up in 27 comprehensive tables in the *THERP Handbook* (Swain and Guttmann, 1983). The PSFs are tabulated in the same fashion. The three sets of PSFs are shown in Table 11.8.

The modelling of relevant human actions in event trees is based heavily on the task analysis performed in stage 1. In the present stage a more detailed analysis is carried out

Table 11.8. PSFs in THERP

Situational characteristics	Architectural features Temperature Humidity Air quality Lighting Noise and vibration Degree of general cleanliness Work hours and work breaks	Availability of special equipment Adequacy of special equipment Shift rotation Organizational structure Adequacy of communication Distribution of responsibility Actions made by co-workers Rewards, recognition and benefits
Job and task instructions	Procedures required (written or not) Cautions and warning Written or oral communication	Work methods Plant policies
Task and equipment characteristics	Perceptual requirements Motor requirements (speed, strength, etc.) Control–display relationships Anticipatory relationships Interpretation Decision-making Complexity (information load) Narrowness of task	Frequency of repetitiveness Task criticality Long- and short-term memory Calculation requirements Feedback (knowledge of results) Dynamics vs. step-by-step activities Team structure and communication Man–machine interface
Psychological stressors	Suddenness of onset Duration of stress Task speed High jeopardy risk Threats (of failure, of losing job, etc.) Monotonous and/or meaningless work	Long, uneventful vigilance periods Conflicts about job performance Reinforcement absent or negative Sensory deprivation Distractions (noise, flicker, glare, etc.) Inconsistent cueing
Physiological stressors	Duration of stress Fatigue Pain or discomfort Hunger or thirst Temperature extremes Radiation G-force extremes	Atmospheric pressure extremes Oxygen insufficiency Vibration Movement constriction Lack of physical exercise Disruption of circadian rhythm
Organism factors	Previous training/experience State of current practice or skill Personality and intelligence variables Motivation and attitudes Knowledge required Stress (mental or bodily tension)	Emotional state Sex differences Physical condition Attitudes based on external influences Group identification

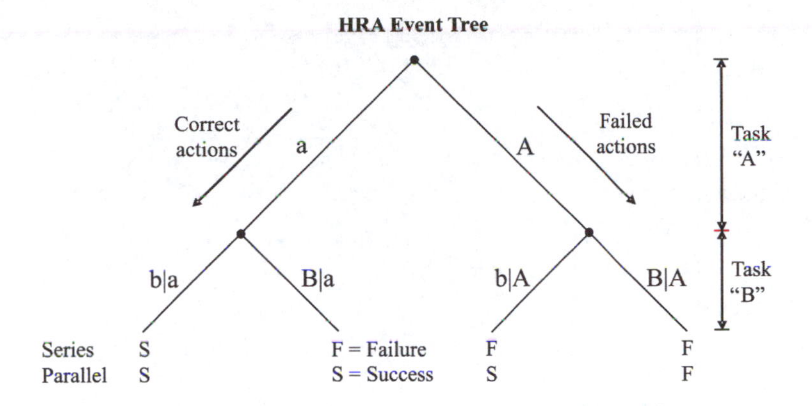

Task "A" = The first task
Task "B" = The second task

a = Probability of successful performance of task "A"
A = Probability of unsuccessful performance of task "A"
b|a = Probability of successful performance of task "B" given a
B|a = Probability of unsuccessful performance of task "B" given a
b|A = Probability of successful performance of task "B" given A
B|A = Probability of unsuccessful performance of task "B" given A

Figure 11.10. HRA event tree for two successive subtasks.

on each of the relevant human actions identified and a comprehensive description of the performance characteristics is made. Each human/operator action is divided into tasks and subtasks, and these are then represented graphically in a so-called HRA event tree (Figure 11.10).

HRA event trees model performance in a binary fashion, i.e. as being either a success or a failure. Branches in a HRA event tree show different human activities, and the values assigned to all human activities represented by the branches (except those in the first branching) are conditional probabilities. Figure 11.11 illustrates a simple HRA event tree (Swain and Guttmann, 1983). As can be seen from this figure, it is common to present the correct actions on the left-hand side of the tree and failures on the right-hand side.

11.9.2 Criticism of the HRA Approaches

It was commented above that the binary categorization of human erroneous actions used in HRA event trees may be too simple to make any claim on psychological realism (Hollnagel, 1998). First of all there is an important difference between failing to perform an action and failing to perform it correctly. Furthermore, the HRA event tree approach fails to recognize that an action may happen in many different ways and for different reasons.

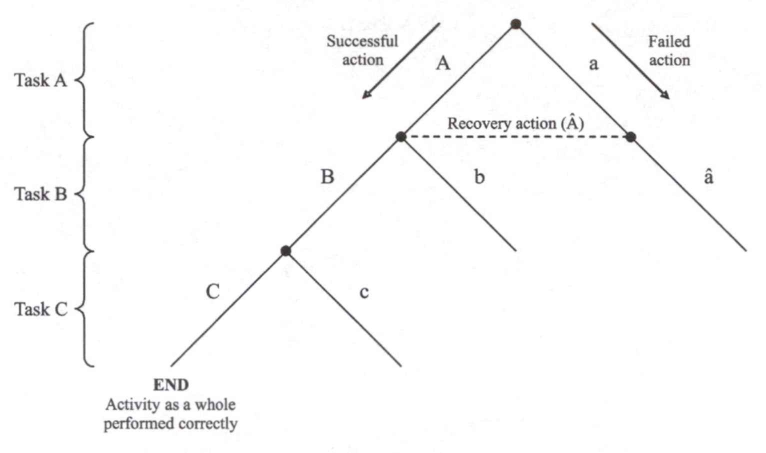

Figure 11.11. Modelling principles in THERP.

Hollnagel (1998) argues that in cases where human behaviour is a cognitive function or a mental act, rather than a manual action, the use of the event tree description does not makes sense. This is because cognitive acts are not so easily separated into distinct and sequential subtasks, as is the case with manual actions. Human cognitive functions, such as diagnosis and decision-making, and in particular the notion of a 'cognitive error', play an increasingly important role in HRA. The standard approach used to model cognitive functions in HRA event trees has been to decompose these functions into their assumed components or subtasks, e.g. problem identification, decision-making, execution and recovery. This may at first seem sensible, but there are a number of problems related to it:

- The number of subtasks that need to be modelled in the HRA event trees in order to incorporate the cognitive functions would easily become excessive, making the event trees complex and difficult to model.
- It is difficult to obtain appropriate and reliable Human Error Probability (HEP) data on each of the (assumed) components of cognition.
- There are doubts as to whether such modelling is psychologically realistic.

Below is given a comprehensive list of the shortcomings of first-generation HRA approaches. These shortcomings should be kept in mind in the overall assessment of quantitative risk estimates:

- *Scarcity of data:* Within HRA there is a general lack of appropriate data that can be used for quantitative predictions of human erroneous behaviour. Much of the data used comes from simulator studies, and there is no basic agreement on how these data reflect the real world, or how they can be modified in order to achieve this.

- *Lack of consistency in treating errors of commission:* Fragola (2000) argues that this is one of the most glaring weaknesses of conventional HRA approaches. Well-intended actions with unintended and undesirable consequences largely remain an 'unknown' area in HRA.
- *Inadequate proof of accuracy:* Demonstrations of the accuracy of HRAs for real-world predictions are almost non-existent. This goes particularly for non-routine tasks.
- *Inadequate psychological realism*: Many first-generation HRA approaches are based on highly questionable assumptions about human behaviour.
- *Insufficient treatment of performance-shaping factors:* In most first-generation HRA approaches the influence of factors such as management attitudes, safety culture, cultural differences, etc. (i.e. organizational factors) are not treated sufficiently. There is also little knowledge on how these factors actually affect performance.
- *Inadequate treatment of dynamic situations:* Conventional HRA approaches are relatively static, meaning that they do not consider the dynamic situations under which tasks are performed.
- *A mechanical view of humans:* Conventional HRAs use a decomposition approach adapted from reliability analysis of mechanical systems (i.e. hardware) on human actions. This decomposition is binary, i.e. either success or failure. In addition, this view results in a focus on the observable (or overt) aspects of human behaviour, instead of the more 'internal structure' of the problem space.
- *Quantitative rather than qualitative focus:* This is argued by some to be a weakness of many conventional HRA approaches, mainly because the quantitative estimates produced are so uncertain.
- *High level of uncertainties:* Different HRA methods may produce widely different values for the Human Error Probabilities (HEPs) when used on the same tasks.
- *Inadequate identification and understanding of error causes:* This weakness is in accordance with the weakness of 'inadequate psychological realism' described above. Most of the first-generation HRA approaches run into problems when trying to explain why humans make errors.
- *Lack of a systematic analysis of task structure:* Analysts who use HRA methods/techniques make judgements based on the information obtained from task analysis. Therefore, a systematic task analysis is essential to enhance the validity and consistency of the HRA results. Most first-generation HRA approaches lack such a systematic task analysis.
- *Inadequate provision of error reduction strategies:* Few HRA methods provide clearly defined strategies for how the estimated HEPs may be reduced in order to enhance safety.

It is only in the past decade or so that the criticism directed at the techniques of conventional HRA have resulted in various efforts to resolve the problems stated above. It should also be recognized that second-generation HRA approaches should be developed from the point of view of solving the limitations of the first-generation HRAs.

REFERENCES

Baitis, A. E. et al., 1995, *Motion Induced Interruptions (MII) and Motion Induced Fatigue (MIF) Experiments at the Naval Biodynamics Laboratory*. Technical Report CRDKNSWC-HD-1423-01, Naval Surface Warfare Center, Carderock Division, Bethesda, MD.

Caridis, P. A. et al., 1999, *State-of-the-Art in Marine Casualty Reporting, Data Processing and Analysis in EU Member States, the IMO and the United States*. CASMET Project Report C01.D.001.1.1, National Technical University of Athens, Department of Naval Architecture and Marine Engineering, Athens.

Elliott, E., 1960, Perception and alertness. *Ergonomics*, Vol. 3, October, pp. 357–364.

Fragola, J., 2000, Focus, strengths, and weaknesses of HRA. In Swedung, I. (ed.), *Risk Management and Human Reliability in Social Context*. 18th ESReDA Seminar, Karlstad, Sweden.

Fukushima, H., 1976, Factors contributing to marine casualties. *Journal of Navigation*, Vol. 29(2), 135–140.

Griffin, M. J., 1990, Motion sickness. In *Handbook of Human Vibration*. Academic Press, New York.

Hollnagel, E., 1998, *Cognitive Reliability and Error Analysis Method (CREAM)*. Elsevier Science, Oxford, UK.

ISO 2631, 1985, *Guide to the Evaluation of Human Exposure to Whole Body Vibration*. International Standardization Organization.

Ivergård, T. et al., 1978, *Work conditions in shipping – A survey. In Swedish: Arbetsmiljö innom Sjöfarten – En kartläggning*. Sjöfartens Arbetarskyddsnämnd, Stockholm.

Jayle, G. E. et al., 1959, *Night Vision*. Charles C. Thomas, Spingfield, FL.

Kirwan, B., 1992a, Human error identification in human reliability assessment. Part 1: Overview of approaches. *Applied Ergonomics*, Vol. 23(6), 299–318.

Kirwan, B., 1992b, Human error identification in human reliability assessment. Part 2: Detailed comparison of techniques. *Applied Ergonomics*, Vol. 23(6), 371–381.

Mackworth, N. H., 1946, Effects of heat on wireless telegraphy operators hearing and recording Morse messages. *British Journal of Industrial Medicine*, Vol. 3, 143–158.

Mackworth, N. H., 1950, *Researches on the Measurement of Human Performance*. Medical Research Council Special Report 268. HMSO, London.

Madison, R. L., 1974, Human factors in destroyer operations. M.Sc. Thesis, Naval Postgraduate School, Monterey, CA (NTIS: AD 775 011).

McCormick, E. J., 1976, *Human Factors in Engineering and Design*. McGraw-Hill, New York.

McCormick, E. J. and Sanders, M. S., 1983, *Human Factors in Engineering and Design*. McGraw-Hill, New York.

NAS, 1976, *Human Error in Merchant Marine Safety*. Maritime Transportation Research Board – Commission on Sociotechnical Systems, National Academy of Sciences, Washington, DC.

NATO STANAG 4154, 1997, *Common Procedures for Seakeeping in the Ship Design Process*. Chapter 7: Seakeeping Criteria for General Application.

NORDFORSK, 1987, *The Nordic Cooperative Project. Seakeeping Performance of Ships. Assessment of a Ship Performance in a Seaway*. Marintek, Trondheim.

Pepler, R. D., 1959, Warmth and lack of sleep: accuracy or activity reduced. *Journal of Comparative and Physiological Psychology*, Vol. 52, 446–450.

Rasmussen, J., 1982, Human reliability in risk analysis. In Green, A. E. (ed.), *High Risk Safety Technology*. John Wiley, Chichester.

RSS, 1975, *An Assessment of Accident Risk in US Commercial Nuclear Power Plants*. WASH 1400 (NUREG 74/014), US Nuclear Regulatory Commission.

Salvendy, G. (ed.), 1987, *Handbook of Human Factors*. John Wiley, New York.

Stevens, S. C. and Parsons, M. G., 2002, Effects of motion at sea on crew performance: a survey. *Marine Technology*, Vol. 39(1), 29–47.

Stewart, J. P., 1973, Basic causes of marine casualties. *Tanker & Bulk Carrier*, June, pp. 18–27.

Swain, A. D. and Guttmann, H., 1983, *Handbook of Human Reliability Analysis with Emphasis on Nuclear Power Plant Applications*. NUREG/CR-1278, US Nuclear Regulatory Commission, Washington, DC.

Teichner, W. H. (1972). *Predicting Human Performance III. The Detection of Simple Visual Signal as a Function of Time on Watch*. New Mexico State University, Department of Psychology Report NMSU-ONR-TR-72-1.

VanCott, H. P. and Kincade, R. G., 1972, *Human Engineering Guide to Equipment Design*. US Superintendent of Documents, Washington, DC.

Wertheim, A. H., 1998, *Mental Load is Not Affected in a Moving Environment*. Report TNO-TM-98-A068. TNO Human Factors Research Institute, Soesterberg.

Wilson, J. R. and Corlett, E. N., 1990, *Evaluation of Human Work: A Practical Ergonomics Methodology*. Taylor & Francis, London.

Wing, J. F., 1965, *A Review of the Effective Ambient Temperature on Mental Performance*. USAF, AMRL, TR 65-102.

12

OCCUPATIONAL SAFETY

'There has been an accident', they said,
'Your servant's cut in half; he's dead!'
'Indeed!' said Mr Jones, 'and please,
Send me the half that's got my keys.'
(Harry Graham: "Ruthless Rhymes for Heartless Homes")

12.1 INTRODUCTION

One of the first systematic studies of the risk picture for seafarers was done by Arner (1970). The breakdown by fatality type shown in Table 12.1 is from this study and highlighted one important aspect, namely that the loss of seafarers could be attributed to at least three different situations: loss of vessel, work-related and non-work-related.

It is interesting to observe that 11% of the fatalities are related to loss of vessel, 28% are directly work-related (explosion/fire, fall, poisoning) and the remaining 61% are in different ways more related to social and psychological factors. This means that safety work must address three main areas:

- The safety of personnel in situations where the survivability of the vessel is threatened: fire, explosion, sinking, capsizing, etc.
- The safety related to the fact that the vessel is a dangerous workplace (fall, hit by object, poisoning).
- Thirdly, the fact that seafaring is to a certain degree associated with negative social and economical factors.

These problem areas still exist, although their relative importance has changed since Arner's study (1970) was undertaken.

12.2 LIVES LOST AT SEA

Nielsen (2002) points out that although the fatality rate in shipping is relatively high compared with other occupations, there has been a certain improvement over time as shown in Figure 12.1. He has, however, pointed out that the figures are quite uncertain due to inadequate statistics on a world-wide basis. Assuming that there is under-reporting,

Table 12.1. Death by type of Norwegian seafarers, 1957–64

Main fatality type	No.	%
As a result of ship loss	110	11
Explosion, fire	73	7
Fall	134	13
Poisoning	87	8
Drowning	288	28
Traffic accident	42	4
Found in the water	47	5
Suicide	159	15
Homicide	28	3
Other	59	6
Total	1027	100

Source: Arner (1970).

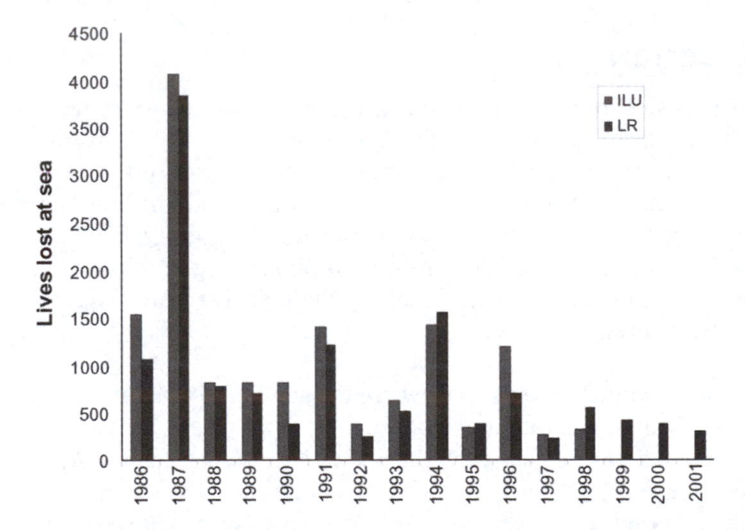

Figure 12.1. Lives lost at sea on world-wide basis. (Source: Nielsen, 2002.)

one may assess the risk level to be in the order of 10 times as high as in a comparable land-based industry.

Figure 12.2 shows how the accident frequency has developed during a 10-year period in Norwegian shipping. Although we know that the safety level has improved in a longer perspective, it is interesting to observe that there are still cyclical variations. It is difficult to explain these shorter-term variations, but is in general it is attributed to business conditions, competition, the labour market, and so on.

The fatality rate on a world-wide basis was investigated recently by Nielsen (1999). The author points out that the available data are incomplete and affected by under-reporting.

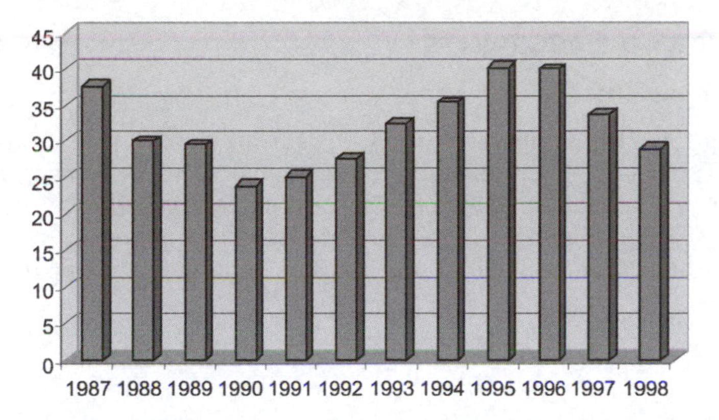

Figure 12.2. Reported work accidents in the Norwegian merchant fleet: accidents per 1000 work-years. (Source: NMD, 1999.)

Table 12.2. A world-wide estimate for the number of lives lost at sea

Death category	Average annual fatalities, 1990–94, responding states	Population at risk, responding states	Fatality rate per 1000 work-years	Annual fatalities world-wide	Relative distribution world-wide
Ship accident	84.4	1,148,822	36.7	1229	47.4
Occupational	71.4	911,140	39.2	404	15.6
illness	62.2	582,728	53.3	550	21.2
Missing at sea	11.2	569,728	9.8	101	3.9
Homicide, suicide, unexplained	7.4	526,539	7.0	72	2.8
Off-duty	4.4		11.8	122	4.7
Unclear cause	4.4		11.4	117	4.5
Total	245.4	—	—	2595	100.0

Source: Nielsen (1999).

However, he has given some interesting estimates which are summarized in Table 12.2. The estimate is based on incomplete data in different death categories, referred to as figures from 'responding states'. These figures have then been corrected for the known seafaring population world-wide. The method gives an annual world-wide loss figure of nearly 2600. In this estimate the losses related to ship accidents represent 47%, occupational accidents (work-related) 16%, illnesses 21% and off-duty accidents 6%.

In other words, it can be concluded that roughly 50% of fatalities are related to phenomena other than ship accidents. Occupational accidents represent 15% whereas the rest is related to illness, suicide, missing and events off-duty. This indicates that there are many negative aspects of the sea-going occupation.

12.3 OCCUPATIONAL ACCIDENTS

12.3.1 United Kingdom

Li and Shiping (2002) have analysed accident data for the UK for the years 1989–97. As shown in Table 12.3, the dominating occupational accident form is *slips and falls*, constituting 46% of the total number. The next largest groups are *manual handling* and *machine operation*, both accounting for 19%. It is also important to notice that the relative number of accidents related to ship casualties is only 15% compared to 47% on a worldwide basis.

12.3.2 Norway

Eriksen (2000) has compared the risk level in Norwegian shipping with similar occupations. As indicated in Table 12.4, the fatality rate is more than 10 times as high as in land-based industry. A breakdown by occupational accident category is given in Table 12.5. Data for both fatal and non-fatal accidents are shown. It confirms the picture from the UK that falls and manual handling are primary sources.

12.3.3 Accidents by Ship Type

There is limited information on how the accident rate varies among ship types. A notable exception is the recent study of the Danish foreign trading fleet (Hansen and Vinter, 2004). The key results are shown in Table 12.6. Looking at fatal accidents, it is clear that the rate for coasters is much higher than for the rest of the fleet. With respect to accidents leading to permanent injury, container ships, passenger ships and gas tankers perform

Table 12.3. Fatal and non-fatal accidents on UK-registered ships

Year	Slip/ fall	Manual handling	Machine operation	Rope/ hawse	Hit by object	Noxious substance	Electric shock	Ship casualty	Total	Population	Rate (%)
1989	212	105	104	30	32	13	2	82	500	20,958	2.39
1990	239	106	93	41	20	8	4	79	516	18,289	2.82
1991	239	86	90	31	21	4	6	87	479	16,005	2.99
1992	181	89	72	24	13	4	8	57	392	14,118	2.78
1993	154	54	75	18	23	12	3	67	341	13,351	2.55
1994	158	77	70	16	18	3	4	58	346	12,837	2.70
1995	142	54	53	12	17	12	9	40	301	12,175	2.47
1996	166	69	59	25	30	4	9	33	364	11,043	3.30
1997	153	64	57	16	19	6	3	23	320	11,044	2.90
Total	1644	704	673	213	193	66	48	526	3559	129,820	24.89
Mean	183	78	75	24	21	7	5	58	395	14,424	2.77
%	46.19	19.78	18.91	5.98	5.42	1.85	1.35	14.78	100		

Source: based on Li and Shiping (2002) and data from MAIB.

Table 12.4. Fatality rate in Norwegian industry, 1999

Sector	Shipping and fishing	Offshore	Land-based industry
Number of fatalities	31	1	57
At risk (workforce)	58,000	13,600	2,000,000
Fatality rate = fatalities per 1000 employees	0.53	0.07	0.03

Source: Eriksen (2000).

Table 12.5. Work accident types in Norwegian merchant shipping

Work accident type	Non-fatal in 1998		Fatal 1994–98	
	No.	%	No.	%
Fall, jump	289	30.3	35	44.9
Blow, squeeze, stumble	256	26.9	10	12.8
Lift, handling	107	11.2	—	
Hurt by tool or machine	86	9	2	2.6
Contact with hot/cold substance	72	7.6	—	
Hit by object	72	7.6	4	5.1
Poisonous substance	36	3.8	—	
Violence, fight	6	0.6	1	1.3
Explosion, fire	5	0.5	3	3.8
Sport, play	5	0.5	—	
Suffocating media	4	0.4	3	3.8
Traffic accident ashore	2	0.2	1	1.3
Contact with electricity	1	0.1	1	1.3
Assault	1	0.1	1	1.3
Terrorism, piracy	1	0.1	—	
Other	10	1	5	6.4
Missing	—		12	15.4
Total	953	100	78	100

Source: NMD (1999).

best. Gas tankers have an exceptionally low rate taking all reported accidents into consideration.

It is necessary to stress that the Danish data may not be representative for the world fleet as a whole.

Table 12.6. Occupational accident rate (LTI) versus ship type for the Danish International Registry (DIS), 1993–99

Ship type	Accident rate per 10^6 work hours		
	All accidents	Permanent injury	Fatal accidents
Container ships	5.59	0.37	0.04
Dry cargo	6.58	0.86	0.09
Coasters	6.19	1.10	0.28
Ro-Ro	11.51	1.08	0.13
Passenger ships	11.00	0.40	0.05
Tankers	9.45	0.90	0.06
Gas tankers	1.70	0.26	0.04
Other ship types	7.44	1.10	0.14
All ships	7.35	0.75	0.11

Source: Hansen and Vinter (2004).

Table 12.7. Cost of accidents in Norwegian shipping

Degree of seriousness	Cost in 1995 (NOK)
Fatality	16,600,000
Very serious accident	11,370,000
Serious accident	3,750,000
Less serious	500,000
Average accident	Approx. 1,500,000

12.3.4 The Cost of Accidents

Eriksen (2000) has also given an indication of the typical cost of a work accident in shipping (see Table 12.7). These cost figures consist of:

- Medical treatment
- Lost production
- Material loss
- Administration costs
- Cost of loss of welfare or quality of life

12.4 ACCIDENT SITUATION

A Swedish study in the late 1970s gave the following characterization of hazardous work conditions (Ivergård et al., 1978):

1. Deck department:
   - Mooring
   - Handling of hatch covers
   - Work on scaffolding or above floor (height work)

2. Engine department:
   - Handling tools in narrow/confined spaces
   - Handling of heavy machine objects or large tools
   - Carrying of objects on staircases and ladders
3. Hotel department:
   - Slippery floor
   - Hot surfaces and sharp objects in sea motion
4. General problems:
   - Ladders
   - Slippery and uneven floor or deck

The study also pinpointed factors that contributed to or initiated accidents. The ranking of factors was done for work and off-duty respectively, as shown in Table 12.8. The highest importance is indicated by the value 1 and the lowest by 7. The highest weight was put on sea motion and vibration, which are both typical characteristics of the workplace onboard.

A Danish study by Hansen and Vinter (2004) shows that almost 50% of occupational accidents are related to work on deck (Table 12.9). Other less critical activities are walking from one place to another on the ship and work in the engine room. The authors report relative accident rates for specific activities within these areas.

On the basis of their findings, Hansen and Vinter have pinpointed the following improvement areas:

- Conditions for walking and climbing on decks, floors, ladders and stairs
- Access between ship and shore
- Risk assessment of mooring equipment
- Lashing and unlashing of cargo
- Access and arrangement in store rooms
- Wheelhouses: sufficient handles and avoidance of sharp edges

Table 12.8. Accident-inducing factors

Factor	At work	Off duty
Noise	2.5	3
Vibration	4	1
Inadequate illumination	6	7
Inadequate climate	2.5	4
Air pollution	5	6
Sea motion	1	2
Other	7	5

Source: Ivergård et al. (1978).

Table 12.9. Working situations at time of accident: ships in the Danish International Registry (DIS), 1993–97

Working situation	Accidents without permanent disability (%)	Accidents with permanent disability (%)	Fatal accident (%)
Work on deck: cargo handling, maintenance	44.9	46.9	46.0
Work in engine room: routine operations, stores, maintenance	16.7	17.7	0
Service function: catering, work in galley	15.9	3.8	0
Walking from one place to another	10.4	21.1	14
Other functions: drills, accidents, violence	12.1	10.5	30
Total	100	100	100

Source: adapted from Hansen and Vinter (2004).

- Easy access to frequently used controls on deck and in engine room
- Operation of lifeboats and rescue boats

12.5 JOB-RELATED PROBLEMS

12.5.1 The Ship as a 24-Hour Community

Both in studies of maritime accidents and work-related problems onboard, reference is often made to the fact that the ship and its total crew function as a '24-hour community'. Both work and free hours are spent on the ship. The characteristics of the ship community can be summarized as follows:

- Closeness
- Together both on duty and on free watch
- May lead to depression and paranoia in extreme cases
- Destroys informal communication patterns
- Watch hours
- Night work
- Irregular work periods (disrupted in port)
- Sleep deprivation
- Effect of changing time zones
- Fatigue
- Inadequate organization of work
- Periods of idleness

- Lack of physical exercise
- Irregular eating hours/unhealthy food

12.5.2 Stress

It is often pointed out that there are a number of stressing factors related to work onboard. One should, however, be careful to make a clear distinction between the different forms of stress:

1. Physical:
   - Temperature (hot/cold)
   - Climate, level of humidity
   - Noise, vibration
   - Inadequate illumination
   - Weather, sea motion
2. Physiological:
   - Result of fatigue
   - Inadequate rest between watches
   - Sleep loss, being woken unexpectedly
   - Irregular and unpredictable working hours
   - Irregular eating hours
   - Drugs
3. Psychological:
   - Milder forms of mental disorders
   - Subjective experience of inability to cope with job
   - Other job-related factors such as separation from family
   - Bad social and interpersonal relations
   - Unpredictable change in the industry (uncertainty)

It has also been pointed out that the strength and duration of stressors are vital factors when assessing their seriousness. The different stressors should also be seen in relation to each other. The cumulative effect might be greater than the sum of the individual effects indicate. Finally, it is important to keep in mind that individuals might show considerable variation in ability to cope with stress.

Table 12.10 gives a summary of the findings from an Australian study (Parker et al., 1998) with respect to the specific stress of the industry. Environmental hardships such as heat, humidity and noise shows a greater reporting among engineers and crew. The relative reporting frequency is 15.4%. This kind of stress is suspected to contribute to fatigue, neurotic syndromes, arterial hypertension, and gastric and duodenal ulcers.

Extreme weather is given less concern compared to the environmental factors above. Only 7.1% reported this factor to be a problem. There were no significant variations between the occupational groups. Prolonged exposure to extreme weather is suspected to contribute to physical and mental fatigue, low-quality sleep and sore joints.

Table 12.10. Industry-specific sources of occupational stress for whole group and by category: percentage reported by mean value

Factor	Seafarers	Pilots	Masters/Mates	Engineers	Crew
Hardships at sea	15.4	10.5	12.5	17.8	15.7
Weather	7.1	6.2	7.3	7.1	7.1
Missing home	13.4	11.8	13.4	13.3	13.4
Broken rest	12.6	10.2	12.7	14.2	11.8
Long hours	12.1	10.2	12.4	12.3	11.8
Industry change	76.5	77.8	71	75.1	82.2

Source: Parker et al. (1998).

Separation from home is seen as an important stress factor by all crew groups. Missing the family and lack of contact during periods of illness at home are both typical situations which are also known from the offshore industry. The problem has been somewhat reduced during recent years through the introduction of longer off-duty periods ashore. The most constructive way to counter this problem is to let family members visit the ship for shorter or longer periods. This has become quite common in the United Kingdom and New Zealand.

12.5.3 Organizational Aspects

Keltner (1995) relates the stress of seafarers to various factors of the work organization. Parker et al. (1998) have studied the effect of organizational factors in a health perspective. We shall here give only a brief summary of the factors that may influence the performance of crew members:

1. The job itself:
   - A heavy workload with frequent deadlines
   - Excessive responsibilities without powers to meet such responsibilities
   - Limited recourse to correct such problems without endangering their career
2. Management role:
   - Conflicting job requirements
   - Insufficient feedback about their work efficiency
   - Ignorance of their personal needs
   - At times uncaring and exploitative
   - Adopting negative roles
   - Uninformed about the work conditions
3. Social factors:
   - Conflicts
   - Lack of social support
   - Coping with politics

4. Career and achievement:
   - Opportunities for personal development
   - Promotion prospects
   - Threat of redundancy and early retirement
5. Organization:
   - Lack of information
   - Lack of loyalty both ways
   - Perceived distance to the land organization
   - Little opportunity for input into administrative decisions which affect their work situation
   - Uncertainties about their future status in the organization
6. Home–work interface
   - Worry about family
   - Lack of support from home
   - Absence of stability in home life

Parker et al. (1998) studied these sources of pressure at work. The results of the investigation are summarized in Table 12.11. It appears that seafarers have a higher score on all sources compared to the normative group. The most pronounced difference was found on:

- Relationships with others, and
- Home–work interface

The relationship problem is mainly attributed to the fact that the seafarers work and live together with the same people for long periods. A land job on the other hand means that the individual has a daily shift between the professional and private sphere. The home–work interface problem has already been commented upon and involves being away from family and friends.

Table 12.11. Sources of pressure at work for normative groups, the entire maritime sample and four occupational groups

	Normative group	Seafarers	Pilots	Masters/Mates	Engineers	Crew
The job itself	30.2	32.8	27.3	33.0	33.2	31.3
Management	35.5	36.7	29.9	38.2	37.4	35.6
Social factors	30.3	34.5	25.8	36.3	34.6	33.7
Career	28.4	29.8	25.9	30.4	30.0	29.5
Organization	38.9	40.0	28.8	40.6	40.8	39.5
Home–work	30.9	33.8	31.1	33.8	32.8	34.3

Source: Parker et al. (1998).

An interesting fact was that there were no significant differences among the occupational groups on 'Career and achievement'. This was taken as an indication of the general lack of job security due to the ongoing deregulation of the Australian maritime industry.

On the other job-related factors, the Masters/Mates group reported the highest scores among the occupational groups. The pilots generally reported lower scores on all aspects of pressure from sources at work.

12.5.4 Sources of Job Satisfaction

Parker et al. (1998) identified elements under sources of job satisfaction:

1. Achievement value and growth:
   - Perception of their current scope for advancement – career opportunities
   - Value of your efforts
   - Development and growth in the job, utilization of skill level of employees
2. The job itself:
   - Satisfaction experienced with the type and scope of job
   - Level of security, kind and amount of work
3. Organization design and structure:
   - Satisfaction with communication of information around the organization, implementation of change
   - Conflict resolution
4. Organizational process:
   - Satisfaction with internal processes within an organization
   - Processes for promotion, motivation in the job
   - Style of supervision, involvement in decision-making
   - Amount of freedom and flexibility in the job
5. Personal relationships:
   - Interpersonal contact within an organization – relationships with others at work
   - Identification with the public image of the organization
   - Organizational climate
6. Overall job satisfaction – 'measures the satisfaction with the job as a whole – personal and organizational issues.'

The reported scores on job satisfaction are shown in Table 12.12. On average, seafarers reported lower overall job satisfaction than the reference group (normative group). The seafarers reported the same levels of satisfaction with 'Achievement, value and growth' as the normative group. For seafarers as a group the most satisfying factor was the job itself.

It is also interesting to observe that all occupational groups onboard reported lower levels of satisfaction with the 'Personal relationships' and that the group values showed no significant differences.

Table 12.12. Sources of job satisfaction for normative groups, the entire maritime sample and four occupational groups

Factor	Normative group	Seafarers	Pilots	Masters/ Mates	Engineers	Crew
Achievement, growth	21.3	21.6	22.9	22.6	22.8	19.9
Job itself	16.3	15.1	16.8	15.7	15.7	14.4
Organization design	16.4	16.1	19.6	16.6	16.4	15.7
Organization process	15.3	14.8	18.8	15.5	15.9	13.8
Personal relations	11.6	10.6	12.6	10.7	10.7	10.5
Overall satisfaction	82.1	77.8	22.8	19.7	19.7	18.5

Source: Parker et al. (1998).

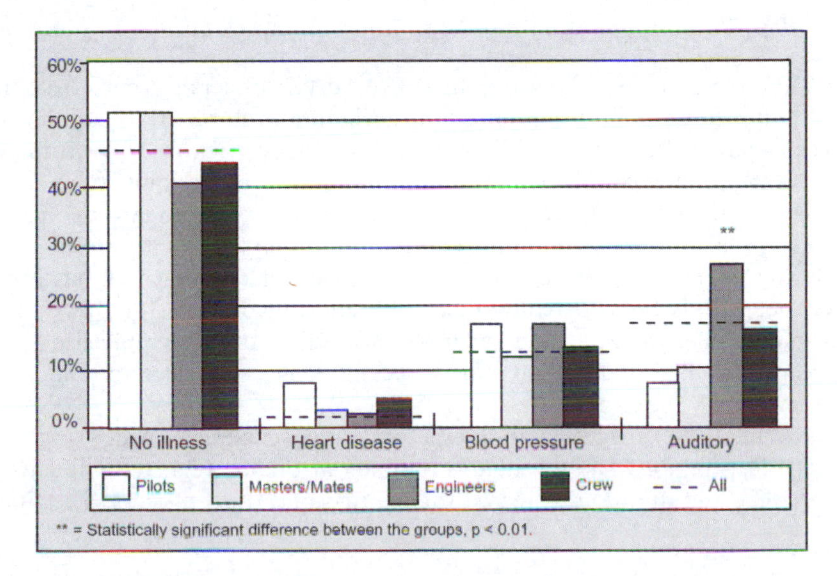

Figure 12.3. Self-reported medical conditions for occupational groups onboard. (Source: Parker et al., 1998.)

12.5.5 Health Survey

The investigation of job-related factors of Australian seafarers (Parker et al., 1998) referenced in the previous section was ultimately focused on their health situation. The self-reported health data are shown in Figure 12.3. The investigation showed that there were a number of physical and mental health problems among seafarers consistent with Australian population data. Only one exception was found, namely that of blood pressure which was slightly higher.

Some differences were identified between the occupational groups, namely:

- Engineers, and to a lesser degree crew, reported a higher incidence of auditory problems.
- Pilots displayed more cardiovascular risk factors.

12.5.6 Work Schedule and Sleep Patterns

Apart from the factors already discussed, the fatigue of the crew is suspected to be related to the watch system practised onboard. The typical system for bridge personnel is:

4 hours on watch
8 hours off
4 hours on
8 hours off

This is possible to practise with three mates (or deck officers). Apart from this, the schedule is disrupted when the vessel enters the harbour and starts cargo handling or other tasks. These negative effects are somewhat limited by the restriction on overtime working given by working regulations. However, it has been shown by Sanquist et al. (1995) that watchstanders on the 04–08 and 00–04 watch had to split their sleep into two periods (see Table 12.13). It is also interesting to note that the total sleep for watchstanders is less than for dayworking personnel. The author further pointed out that research has shown that fragmented sleep leads both to reduced alertness and impaired performance on tests of perceptual monitoring (lookout and radar watch). Other studies emphasize the quality of sleep: sleep onset before midnight is best; second comes sleep after midnight; and the lowest quality is daytime sleep.

Fletcher et al. (1988) suggested an alternative watch schedule in order to get a single, uninterupted sleep period. The schedule is outlined in Figure 12.4. Experiments showed that sleep quality and duration increased for the first and third officers. The first officer

Table 12.13. Average sleep durations for watchstanders and dayworkers aboard two US flag ships

	Watch period	Sleep 1 (hours)	Sleep 2 (hours)	Total sleep (hours)
Ship A	00–04	3:32	2:37	6:11
	04–08	5:13	1:48	5:24
	08–12	5:37	—	5:37
	Dayworkers	6:19	—	6:19
Ship B	00–04	3:32	4:21	7:41
	04–08	4:30	1:25	5:55
	08–12	5:30	0:50	6:18
	Dayworkers	7:04	1:51	7:35

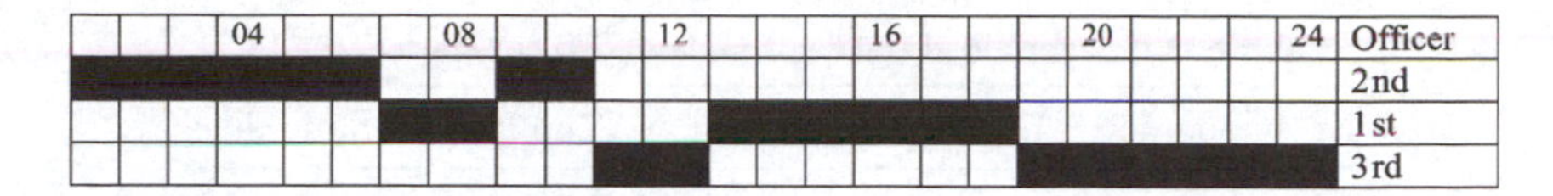

04
08
12
16
20
24
Officer
2nd
1st
3rd

Figure 12.4. Alternative watch schedule for bridge officers. (Source: Fletcher et al., 1988.)

Table 12.14. Ranking of flag standard

Flag standard	Characterization
High	Traditional maritime nations
	Centrally operated second registers
Medium high	Semi-autonomous second registers
Medium	Established open registers (seeking EU membership)
	National registers
Medium low	New open registers
Low	New entrants to the open register markets

Source: Alderton and Winchester (2001).

would start his long rest at 19 hours whereas the third officer would start at 24 hours. Obviously the improvement for the second officer is less due to the fact that his period started at 10 hours (in the middle of the day). Incidentally, the first officer still did not like the system anyway, primarily due to the fact that he had sailed for many years and was not motivated by changing his routine.

12.6 WORK CONDITIONS AND FLAG STANDARD

Despite the fact that there is a general acceptance that flags of convenience (FOC) are associated with substandard shipping and degradation of work conditions, there has been a steady growth in flagging out. Alderton and Winchester (2001) have studied the effect of flag standard on working conditions. We will here give a brief summary of their investigation. The quality or standard of the flags was assessed on the basis of a broad set of criteria and was grouped as shown in Table 12.14.

Table 12.15 shows a clear correlation between flag standard and work load. The assessment of whether working time is too long has twice as high a response percentage (31%) for low compared to high standard (14%). The same pattern can be found for the evaluation of rest and leave (Table 12.16). For the two lowest standards roughly 25% of the respondents find the conditions inadequate.

Stress seems to be a widespread problem onboard but is especially common on the most substandard ships as 37% report too high a level (Table 12.17). The general working conditions measured by four different criteria as shown in Table 12.18 confirm the same message. Personnel on low standard vessels report inadequate conditions twice as often as on high standard vessels.

Table 12.15. Work load and flag standard: percentage reported

Flag standard	Too large work load	Too long working time
High	19	14
Medium high	22	11
Medium	18	14
Medium low	20	21
Low	29	31

Source: Alderton and Winchester (2001).

Table 12.16. Rest and leave versus flag standard: percentage reported

Flag standard	Too little rest and sleep	Too short shore leave	Too short holiday
High	15	25	14
Medium high	13	24	17
Medium	15	23	19
Medium low	16	29	30
Low	25	43	35

Source: Alderton and Winchester (2001).

Table 12.17. Stress and flag standard: percentage reported

Flag standard	Too high stress level
High	22
Medium high	24
Medium	25
Medium low	29
Low	37

Source: Alderton and Winchester (2001).

It is also alarming to see that racism is quite widespread at sea and even on high standard vessels (see Table 12.19).

12.7 THE GLOBAL LABOUR MARKET

There are some concerns about the general situation in the maritime labour market, not least in view of the fact that the almost two-thirds of the world fleet is flagged under offshore registries (not national flags). As ICONS (2000) states, it is critical to maintain 'a competent, rested and well-motivated crew'. The employment of seafarers takes place in an international labour market which offers considerable flexibility in the hiring and firing

Table 12.18. General employment conditions and flag standard: percentage dissatisfaction reported

Flag standard	Salary	Job safety	Moral, culture	Social support
High	25	18	16	14
Medium high	28	25	18	22
Medium	35	27	17	24
Medium low	52	49	26	34
Low	52	39	29	43

Source: Alderton and Winchester (2001).

Table 12.19. Racism versus flag standard: percentage reported

Flag standard	Unfair treatment	Physical abuse by officer	Psychological/harassment
High	22	9	17
Low	33	19	34

Source: Alderton and Winchester (2001).

of crew. It is fair to emphasize that not everything is substandard, but there are many examples of abuse, exploitation and denial of legal rights. This negative development was started as a result of the oil crisis in the 1970s and the slump in world trade that forced shipowners to reduce their operating costs. The present labour market is characterized to a greater or lesser degree by the following (ICONS, 2000):

- Relative unattractiveness in developed economies
- Outsourcing of labour
- Questionable training standards
- Trainees failed to complete training
- Too many officers leave too early in their careers
- Reduced autonomy and standing for masters and senior officers
- Paying of unreasonable hiring fees to manning agents
- Practices of some manning agencies are criticized
- 'Passport holders': recruitment outside the authorized agencies or channels
- Fraudulent certificates
- A surplus of 224,000 ratings on a world-wide basis (BIMCO, 2000)
- Voyage-by-voyage employment
- Blacklisting for 'offences' like contacting union officials (10,000 in the Philipines alone)
- 'Quit claims': P & I agents pressure seafarers to accept compensation below what they are entitled to
- Abuse and ill-treatment
- Non-payment of wages
- Abandoning of seafarers in a foreign port

REFERENCES

Alderton, T. and Winchester, N., 2001, The Flag State Audit. *Proceedings of SIRC's Second Symposium*, Cardiff University, 29 June. Seafarers International Research Centre, Cardiff, UK. http://www.sirc.cf.ac.uk/pubs.html

Arner, O., 1970, *Fatal Accidents among Seafarers* [In Norwegian: *Dødsulykker blant sjømenn*]. Universitetsforlaget, Oslo.

BIMCO, 2000, *BIMCO/ISF Manpower Update Summary Report*, April.

Eriksen, S. I., 2000, *Do We Take the Work Conditions in the Merchant and Fishing Fleet Seriously?* [In Norwegian: *Tar vi arbeidsmiljøet i handels – og fiskeflåten på alvor?*]. Haugesundskonferansen, 15 Febuary, Haugesund.

Fletcher, N. et al., 1988, Work at sea: a study of sleep, and of circadian rhythms in physiological and psychological functions, in watchkeepers on merchant vessels. VI. A sea trial of an alternative watchkeeping system for the merchant marine. *International Archives of Occupational and Environmental Health*, Vol. 61, 51–57.

Hansen, H. L. and Vinter, M., 2004, *Occupational Accidents and Ship Design: Implications for Prevention*. Danish Society of Naval Architecture and Marine Engineering (Skibsteknisk Selskab), http://www.skibstekniskselskab.dk/download/WMTC/A18(R99).pdf

ICONS, 2000, *Inquiry into Ship Safety: Ships, Slaves and Competition*. International Commission on Shipping, Report ISBN 0-646-41192-6, Charlestown, NSW, Australia. http://www.icons.org.au/images/icons_reportmedia.pdf

Ivergård, T. (ed.) et al., 1978, *Work Conditions in Shipping – A Survey*. [In Swedish: *Arbetsmiljö innom Sjöfarten – En kartläggning*]. Sjöfartens Arbetarskyddsnämnd, Stockholm.

Keltner, A. A., 1995, Stress management on board. In Wittig, W. (ed.), *The Influence of the Man-Machine Interface on Safety of Navigation. ISHFOB '95: Proceeding of the International Symposium on Human Factors on Board*, Bremen, November, Verlag TÜV Rheinland.

Li, K. X. and Shiping, Z., 2002, Maritime professional safety: prevention and legislation on personal injuries on board ships. *IAME Panama 2002 Conference Proceedings*, 13–15 November, International Association of Maritime Economists.

Nielsen, D., 1999, Occupational accidents among seafarers. RINA Conference: Learning from Marine Incidents, 20–21 October, Royal Institution of Naval Architects, London.

Nielsen, D., 2002, Safety and working conditions in international merchant shipping. *IAME Panama 2002 Conference Proceedings*, 13–15 November, International Association of Maritime Economists.

NMD, 1999, In Norwegian: Shipping Statistics – *Work Accidents 1994–98* (*Skipsfartsstatistikk* – 4. *Personulykker*). Norwegian Maritime Directorate, Oslo.

Parker, A. W. et al., 1998, *A Survey of the Health, Stress and Fatigue of Australian Seafarers.* Australian Maritime Safety Authority, ISBN 0-642-32007-1, http://www.amsa.gov.au/SP/Fastoh/

Sanquist, T. F. et al., 1995, United States Coast Guard alertness and fatigue research program In Wittig, W. (ed.), *The Influence of the Man-Machine Interface on Safety of Navigation. ISHFOB '95: Proceeding of the International Symposium on Human Factors on Board*, Bremen, November, Verlag TÜV Rheinland.

13

ACCIDENT ANALYSIS

Life can only be understood backwards, but has to be lived forwards.
(Søren Kierkegaard, Danish author and philosopher)

13.1 INTRODUCTION

It should be more or less obvious that the basis for any kind of safety work should be firm knowledge about why accidents happen – or, stated in the opposite way, what is safe operation? A key source of such knowledge is lessons learned from accidents, incidents or near-accidents.

Both maritime authorities and operators should base their activities on such knowledge. Authorities are supposed to enforce and develop safety regulation, whereas operators should strive to improve their safety performance.

The basis for sound accident analysis is that the fieldwork is done in a reliable and thorough way:

- Who were involved: persons, vessels, agencies
- Under what conditions: weather conditions and sea state
- What happened: events in terms of errors and failures
- What triggered the events: immediate causes
- Why did it happen: basic causal factors related to humans and organizations

In this chapter we will give some background for accident analysis, mainly related to the characteristics of a maritime operation. In addition, a few different approaches to accident analysis will be outlined.

13.2 SAFETY AND LEARNING

Systematic and efficient safety work has to be based on a firm understanding of how the maritime system functions and what kind of generic behaviour will minimize the risk of the operation.

Table 13.1. Key factors in safety work

Factor	Examples
Motivation	Top management concern about safety matters View of safety as a competitive business factor Priority between safety, time and economics Individual attitudes at all levels of the organization
Knowledge	Risk level of operation Frequency of incidents of various degrees of seriousness Why do accidents happen? Human attitudes and behaviour
Methods	Risk analysis methods (FSA – Formal Safety Assessment) Accident investigation techniques Auditing and inspection methods Reliability modelling

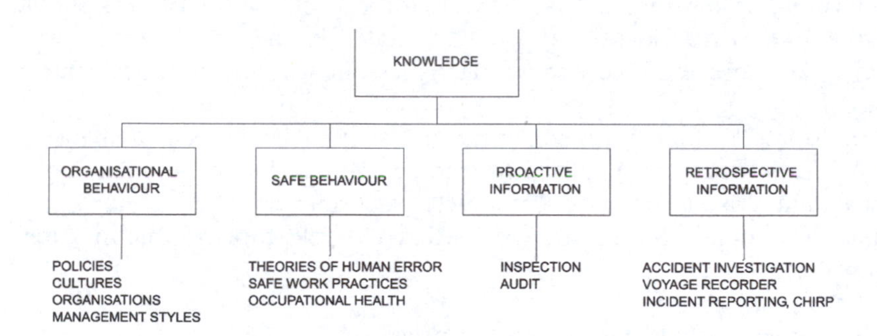

Figure 13.1. Sources of information about safe operation.

Safety work rests on three fundamental factors, namely motivation, knowledge and methods, as elaborated upon in Table 13.1.

Safety work has to be based on a broad set of information sources. They are of both a specific character and of more a general character, as indicated in Figure 13.1.

Incident-based information is also called retrospective information and is a vital source for organizational learning within the company or regulatory institution. We may make a distinction between the following events:

- *Accidents:* events that lead to damage and environmental consequences, injury/fatalities and economic loss.
- *Incidents:* events that are controlled before they lead to accidents.
- *Non-conformities:* deviations from accepted technical or operational tolerances.

The proactive approach will be discussed in Chapter 15 in relation to formal safety management.

13.3 WHY ANALYSE ACCIDENTS?

13.3.1 Regulatory Requirement

All Flag States have regulations that specify the situations in which maritime accidents will be investigated. The objectives are threefold:

- Identify causes and potential measures that will reduce the risk of similar accidents in the future
- Identify weaknesses in the current regulations
- Establish liability and criminal conduct

Caridis et al. (1999) have discussed the organization of official investigations in European countries and the USA. We will look further into this later in this chapter.

Official investigation of accidents should also be seen as an element in the continuous development and enforcement of safety regulation. It should be evident that lessons learned from accidents are an important input in this context. The key functions of Maritime Administrations are:

- Flag State inspection of sailing vessels
- Port State control
- Survey and acceptance of new buildings
- Approval of manning
- Audit of safety management system under ISM
- Monitor maritime traffic and dangerous cargoes
- Investigate and analyse maritime accidents

As discussed earlier in this book, Formal Safety Assessment (FSA) will see wider use in the future revision and development of international regulation. FSA is based on the principles and methods of risk analysis and requires data on accident patterns and frequencies. Table 13.2 illustrates the data needs for typical FSA methods.

Table 13.2. Data needs in Formal Safety Assessment

Technique	Relevant accident database information
Preliminary hazard identification (PHA)	Accident type and consequence frequency Event distribution Correlation of accident type, events and consequences
Failure mode and effects analysis (FMECA)	Failure modes for systems and components Criticality based on correlation with consequence data
Fault tree analysis (FTA)	For given accident type, distributions for events and causes Correlation of events and causes Frequency for basic events
Cost-benefit analysis	Distributions for various consequence parameters: economic and human loss

13.3.2 Management Needs

Safety management will be outlined and discussed in Chapter 15. It is sufficient here to stress that a systematic approach must be based on information and data from all kinds of non-conformities.

Typical questions that may be addressed in company accident investigations are:

- Accident frequency and severity rates by ship type
- Accident frequency trends (change over time)
- Accidents frequency by experience level or crew category
- Incidents by time of day or time into work shift
- Injuries for different jobs or tasks
- Severity of different accident types

Table 13.3 gives an idea of the various types of information that are needed for the different management functions.

13.4 THE MARITIME SYSTEM

13.4.1 Hierarchical Structure of Maritime Systems

Analysis and understanding of why casualties take place must be founded on an understanding of how the maritime system functions. Description of the maritime system may involve a number of dimensions. A crude conceptual model is outlined in Figure 13.2 (see page 366).

The system itself has two important dimensions, namely:

- *Horizontal:* Different systems elements operate in parallel and interaction. The resultant performance is a joint function of these elements. Examples: human–machine interaction; system acquisition and personnel training.
- *Vertical:* The system as a number of decision levels. Seen from the top down it goes from broad strategic aspects to operative oriented actions with a limited scope. Examples: top-level decisions on maintenance policy to low-level actions to control safe speed in harsh weather.

It should also be pointed out that there are important relations between the decision levels (Kjellén and Tinmannsvik, 1989). In general, a higher decision level puts restrictions on the functionality of the lower level. As illustrated in Figure 13.2, the way the departments are set up will influence or govern the quality of the key operative resources: systems, crew and procedures.

Finally, it should be pointed out that the performance, or rather lack of performance, will be reflected in different types of deviations or non-conformities. A high-level deviation might be management oversights, as seen in the upper right part of Figure 13.2.

A more critical kind is the fully developed casualty which in the first instance is related to the operative level. It should, however, be pointed out that different forms of deviation

Table 13.3. Data needs in safety management

Function	Relevant accident database information
Safety policy and culture	Accident/incident frequency, accident-related costs Social conditions, LTA supervision onboard
Resource allocation and priority to safety activities	Accident/incident frequency, accident-related costs LTA maintenance policy and operations Inadequate tools and equipment
Internal audits	Reported incidents and non-conformities
Safety coordination	Accident/incident frequency, accident-related costs Reported injuries and fatalities Reported incidents and non-conformities
Organization, management, business re-engineering	Less than adequate (LTA) management LTA personnel management, watch system Social relations onboard
Non-conformances and incidents	Reported incidents and non-conformities
Design, system acquisition, system modification	Reported accidents and incidents related to technical failure, dangerous material and environmental impact Inadequate tools and equipment Maintenance-related problems
Personnel management, hiring and training	Reported injuries and fatalities Human error related accidents and incidents Social relations onboard LTA personnel management, watch system LTA emergency preparedness Physical stress, inadequate workplace conditions LTA occupational health management
Maintenance management	Accident/incident frequency, accident-related costs LTA maintenance policy and operations Inadequate tools and equipment Maintenance-related problems
Emergency preparedness	Human error related accidents and incidents LTA emergency preparedness

may reflect the same inherent problem in the maritime system in question. This is also reflected in the fact that modern thinking in safety management works along a different axis:

- Experience feedback from accidents and incidents
- Inspection of systems and operation to detect hazardous states and behaviour
- Auditing of management system to detect oversights

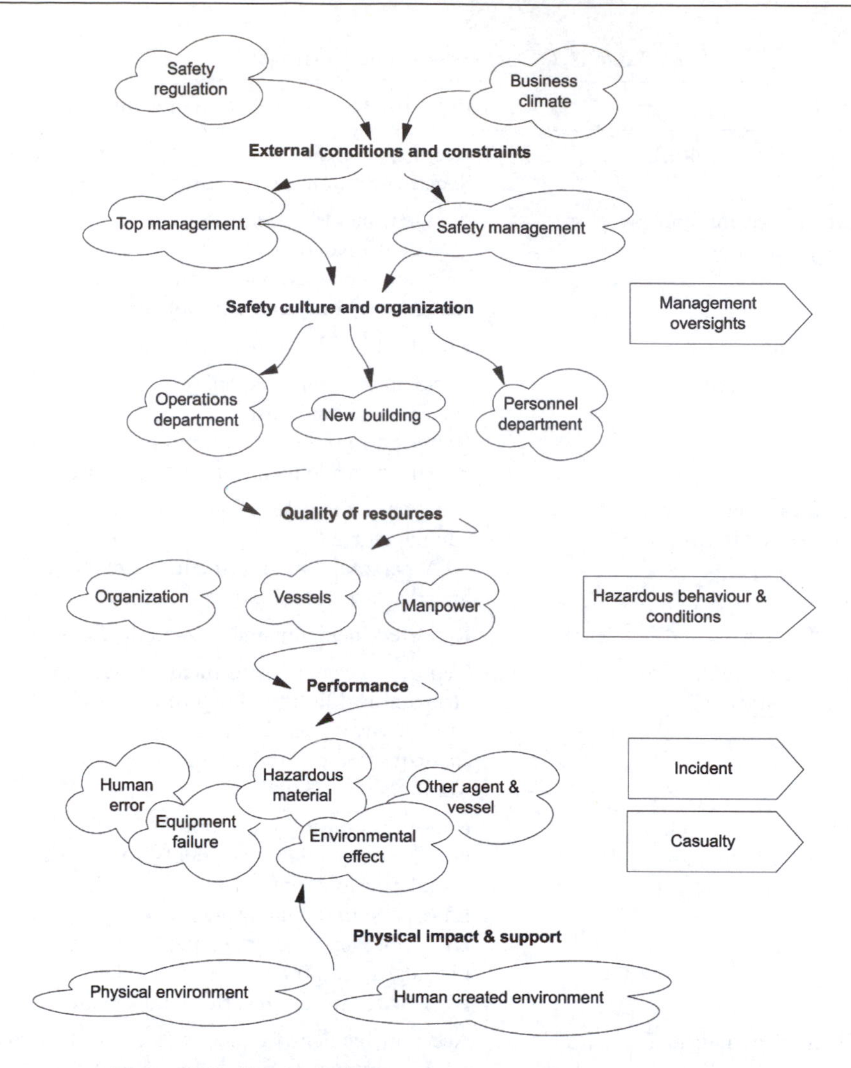

Figure 13.2. The maritime system: actors, effects and deviations.

Seen in a historical perspective, the analysis of casualties and the search for preventive measures has shifted gradually from technical problems to viewing the interaction between humans, machines and the environment. More recently there has also been an increasing focus on higher-level decisions in contrast to the last critical actions before the casualty is a fact. However, accepting that the perspective and understanding of accidents are broader today, there is still much to be done in the day-to-day practice of experience feedback. As a conclusion, it should be a requirement for a casualty database that it reflects the structure and functional characteristics of the maritime system as outlined.

13.5 ACCIDENT THEORIES

An objective understanding of an accident is difficult to achieve for a number of reasons. Accidents come without warning, the individuals involved cannot take a neutral observer's role, and too much happens in too short a time. Our feelings and beliefs will often disturb our perceptions and analysis of what actually happened. For these reasons, a number of theories regarding accidents have been proposed and argued for over time. Brown (1990) gives a comprehensive summary of the theories. We shall summarize a few of them here:

- *Pure chance:* Everybody is subject to the same risk and has the same liability to accidents.
- *Biased liability:* Having been involved in an accident will change an individual's behaviour and prudence. By contrast, a person who has never been involved in an accident may become more risk-prone.
- *Accident proneness:* Certain individuals are more often involved in accidents due to their innate personal characteristics.
- *Stress*: Accidents happen when the mental task-load on the operator increases beyond their capacity. Too much is happening and the capacity is reduced by stress factors.
- *Epidemiological:* The accident is a conjunction between operator (victim), tool (agent) and working environment (situation).
- '*Domino*'*:* Focuses on the sequential and multi-causal nature of accidents.

The epidemiological theory is promising in the sense that it addresses the factors that can be subject to improvement, namely the individuals, technology and work environment. The 'Domino' theory has also received considerable acceptance and is attributed to Heinrich (1950), who postulated that a work accident undergoes the following generic phases:

1. Ancestry and social environment
2. Individual fault
3. Unsafe act and/or mechanical hazard
4. Accident
5. Injury

This theory also put some emphasis on situational factors in the understanding of the causation mechanism. We will return to the description of the two last theories later in this chapter. But first we will discuss some general characteristics of the accident phenomenon.

13.6 WHAT IS AN ACCIDENT?

An accident can be seen from different perspectives. One may focus on what identifies the accident, the factors that describe the things that take place, or the explanation of the phenomenon. All these aspects, however, contribute to our understanding and should be incorporated in a model:

- Identification of the accident:
  - Accident type
  - Outcome or consequences
- The background:
  - Situation in physical terms
  - The vessel's characteristics
  - Manning
- The explanation:
  - Sequence of events
  - Causal factors on individual level
  - Causal factors on organizational level

The structural accident model shown in Figure 13.3 is an attempt to unite the different aspects. Hollnagel (1998) points to a common pitfall, namely mixing up these concepts. He takes the often used term *error* as an example and shows how it can be used to denote the cause, the event and the outcome:

- 'The oil spill was caused by human error' (cause)
- 'I forgot to check the water level' (event)
- 'I made the error of putting salt in the coffee' (outcome)

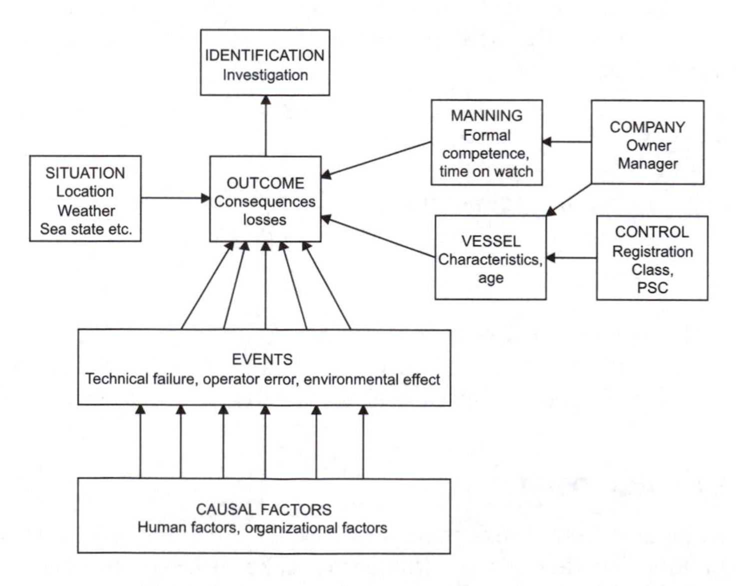

Figure 13.3. The accident description structure.

The confusion in the use of accident descriptors is seen in the DAMA database operated by the Norwegian Maritime Directorate (NMD, 1996). The DAMA causal factors taxonomy is fairly detailed but has been criticized for having serious shortcomings. Typical is the unsystematic use of the cause concept. Inspection of the taxonomy shows a mix of events, functional limitations, operator errors or basic weakness of the organization. This can be illustrated by a few examples of causal factors taxonomy:

- *Self-ignition in cargo/bunker (event)*
- *Fault with charts and publications (functional limitation)*
- *Did not use protective equipment (operator error)*
- *Fell asleep on watch (individual failure)*
- *Unfortunate design of the bridge (inadequate work condition)*
- *Technical fault with navigation equipment (technical failure)*
- *Too small a crew (basic weakness of the organization)*

13.6.1 The Capsize of *Herald of Free Enterprise*

The rapid flooding, capsize and sinking of the Ro/Ro passenger ferry *Herald of Free Enterprise*, with its catastrophic number of fatalities, had a heavy impact on public opinion (Dand, 1988; DOT, 1987). It also led to serious legal actions and subsequent changes in the regulatory regime. We will use this tragic casualty to illustrate some basic concepts in accident analysis.

Example: Summary of *Herald of Free Enterprise* Casualty

What happened

The Ro-Ro passenger and freight ferry MV *Herald of Free Enterprise* (HFE), operated by Townsend Car Ferries Ltd, left the port of Zeebrugge bound for Dover at 18:00 hours on 6 March 1987. The vessel had a crew of 80 plus 81 cars, 47 freight vehicles and approximately 460 passengers. Only a few minutes later when the vessel had turned and started to pick up forward speed, water started to enter through the bow door on to the G deck which resulted in progressive heel and capsize. The vessel did not sink completely due to limited water depth. Because of the complicated evacuation and lack of rescue resources, at least 150 passengers and 38 of the crew were lost.

Circumstances and contributing factors

As a typical Ro-Ro vessel, HFE had an enclosed superstructure above the main car deck (G). The doors were operated by hydraulic systems which were controlled manually on the car deck. The car deck was largely unrestricted as it had no sectioning or bulkheads. The deck was kept watertight with closed bow and stern doors. Smaller leaks of water on to the deck were removed by pumps (scuppers).

Immediate causal factors

The immediate causal factors for this accident were as follows. Due to high tide and mismatch between vessel and ramp design, the vessel had been trimmed by the bow in order to access the E deck. The vessel had not been trimmed to even keel before it left the port. Secondly, the bow doors were still open at the departure due to a series of failures by the crew. It was the job of the Assistant Bosun to close the doors but he had left the watch and gone to bed. The Bosun observed the situation but did not see it as his task to intervene or notify the bridge. The Chief Officer was stationed at the bridge and could not inspect the closing of the doors himself, and even more seriously did not seek to get a positive confirmation of the closing. The Master was also passive in this respect.

As the vessel had backed out and turned and started to pick up speed, the water started to flow on to the G deck which is accessible through the bow doors. The combination of trim nose down, possible overloading, increasing bow wave and squat was sufficient to overcome the remaining freeboard or clearance to the deck at the bow. The fact that the vessel was in a turn may also have contributed to the sudden heel. As the car deck had no sectioning, the water quickly started to accumulate along the deck side and thereby to build up a heeling moment, also known as 'free surface effect'.

Basic causal factors

During the investigation of the casualty it was established that the management of the company had a critical role. The Master was under considerable pressure to keep the sailing schedule although the vessel had taken over the service at short notice. It was further clear that the Master of this vessel and a sister vessel requested installation of door indicators which would allow checking of the status of the doors from the bridge. This was denied by management on two occasions. It was also established that the vessels of this company regularly sailed overloaded. The Master, however, had no practical means of monitoring cargo intake and number of passengers. The policy onboard to accept 'negative reporting' was fatal in this instance as nobody sought to positively confirm the closing of the doors. Apart from the inflowing water, the fact that the vessel was top-heavy may also have contributed to the sudden capsize.

Let us now identify some of the main factors that constitute this accident:

1. Causes:
   - Vessel replaced another vessel at short notice
   - Vessel trimmed by the bow to match ramp
   - Pressure on Master to keep schedule
   - Policy onboard to accept 'negative' reporting
   - No monitor or indicator on bow door
   - Watch system in conflict with sailing schedule
2. Events:
   - Assistant Bosun leaves watch and goes to bed
   - Bosun takes no action with respect to open door

- Bridge officers do not check closing of bow door
- Vessel backs out from Zeebrugge and turns to sea
- Inflow of water through bow door

3. Consequences:
   - Progressive heel
   - Capsize
   - Sinking in shallow water
   - Loss of 188 persons

The accident illustrated the vulnerability of Ro-Ro vessels with respect to flooding. Unlike traditional ship types they have large doors near the waterline and few bulkheads to restrict water on deck. Experience has shown that even minor operational errors may have catastrophic consequences.

13.7 RECONSTRUCTION OF EVENTS

It is important to establish an overall picture of the sequence of events from initiation to the outcome of the accident. The sole aim is to establish *what* happened before one starts to deduce *why* the casualty happened. Most casualties are reported in a prose style. Although this may be acceptable to capture all pertinent information, it has its obvious shortcomings as a starting point for causal analysis. Some accidents may develop gradually over a considerable time span and involve a number of actors in terms of persons and systems. It is then vital to place the individual events in a proper context and for them to be given a certain structure and ordering.

13.7.1 STEP Diagram

The STEP diagram approach focuses on relating events to what we may call 'actors' and the time line (Hendrick and Benner, 1987). The actors may be anything from systems, vessels or individuals to organizational units. How one defines the set of actors is only a function of the nature of accidents. The actors are organized along one axis in the diagram and the time line along the other.

The approach is illustrated by the *Herald of Free Enterprise* (HFE) accident in Table 13.4. It should be emphasized that the number of actors designated in this diagram is a compromise between the need for transparency on the one hand and the need for detailed information on the other. The categories Officers and Crew might for instance have with individuals (Bosun, Master, etc.). The events given in the cells of the diagram may in the same fashion have been described in both cruder or finer terms.

13.7.2 Flowcharting

It is clear that the STEP diagram is a suitable tool for outlining the main events in an accident. On the other hand, it also has serious limitations in modelling the interaction or causality between the events. An alternative may be the use of a flowchart format where

Table 13.4. STEP diagram of main events in HFE disaster

Event No.	Management	Officers	Crew	Vessel	Contributory factors
E1				Vessel was overloaded	Inadequate control of passenger number and cargo intake Time pressure did not allow adequate control
E2	Pressure to leave port early				Delay at last port (Dover) Vessel entered this service at short notice
E3			Bow door not closed by Asst. Bosun		Assistant Bosun asleep Just relieved from cleaning and maintenance duties
E4			Bosun did not take action		Did notice that door was still open Did not see it as his duty to call Asst. Bosun to close door, or noting the bridge
E5				No indication of open door on the bridge	Requested by vessel more than once Not granted by management
E6		Chief Officer did not ensure that door was closed			Unable to check by himself; had to be on bridge 15 min before sailing. Did not seek confirmation from deck. Company standing order to accept 'negative' reporting

E7	Master did not ensure that door was closed		Did not seek positive confirmation
E8	Did not complete ballasting		Considerable mismatch between deck and ramp. High tide. Required considerable time to ballast
E9	Leaves port still trimmed nose down		High water spring tide. Considerable trim necessary in order to access deck E by ramp. Trimming not completed Inadequate seamanship
E10		Water enters through bow door on deck G	Increasing bow wave and squat as speed is picking up
E11		Inadequate capacity of scuppers to void water	Not designed for this inflow rate
E12		Free surface effect	No sectioning of car deck
E13		Progressive list to port side	Not designed for this load condition Inadequate transverse stability Top-heavy design of vessel
E14		Capsize	90 degrees heel and sunk in shallow water

the events are described as nodes and interactions by connecting arrows. We may illustrate this approach with the *Exxon Valdez* grounding (Anonymous, 1990):

> The motor tanker *Exxon Valdez* left Valdez loaded with crude oil on the night of 14 March 1989. It soon ran into difficulties due to drifting ice in the restricted fairway and inadequate manning. The mate on watch deviated from the traffic separation scheme and took the vessel back too late to avoid grounding on Bligh Reef soon after midnight. Partly due to inadequate emergency preparedness the grounding resulted in a spill of 258,000 barrels of crude. The vessel itself had no technical shortcomings and was in fact the 'pride' of the Exxon fleet. The analysis revealed a number of management issues. The state of the vessel traffic service (VTS) in the area was inadequate and had its share of the responsibility for the disaster.

The flowchart describing the main events is outlined in Figure 13.4. In order to provide some extra readability, the events are segmented in four phases:

1. *Latency:* hazardous conditions and weaknesses.
2. *Initiation:* events that initiate the accident sequence.
3. *Escalation:* despite awareness of the situation and subsequent events it is difficult to handle the failures and errors.
4. *Critical:* last opportunity to avoid the accident.

A summary of this accident is given in tabular format in Table 13.5. This description summarizes key events under each phase, the functional error that explains why they took place, and potential preventive measures revealed as a result of the analysis process.

13.8 THE LOSS CAUSATION MODEL

13.8.1 Background

The Loss Causation Model (LCM) has been developed by the International Loss Control Institute Inc. (ILCI) and is the core of the safety management approach promoted by Det Norske Veritas. It is primarily based on experience from land-based industry in the US (Bird and Germain, 1992) and follows the principles of the domino theory described earlier (Heinrich, 1950). In its present version it constitutes a fairly complete system for the management, planning and control of industrial safety.

The rationale of the LCM is that losses and safety problems can be tracked back to a lack of control in the organization. By losses are understood production problems, environmental pollution, property damage, personal injuries, employee health, etc.

13.8.2 Procedure

The application of LCM may best be described according to the elements of the model outlined in Figure 13.5 (see page 378).

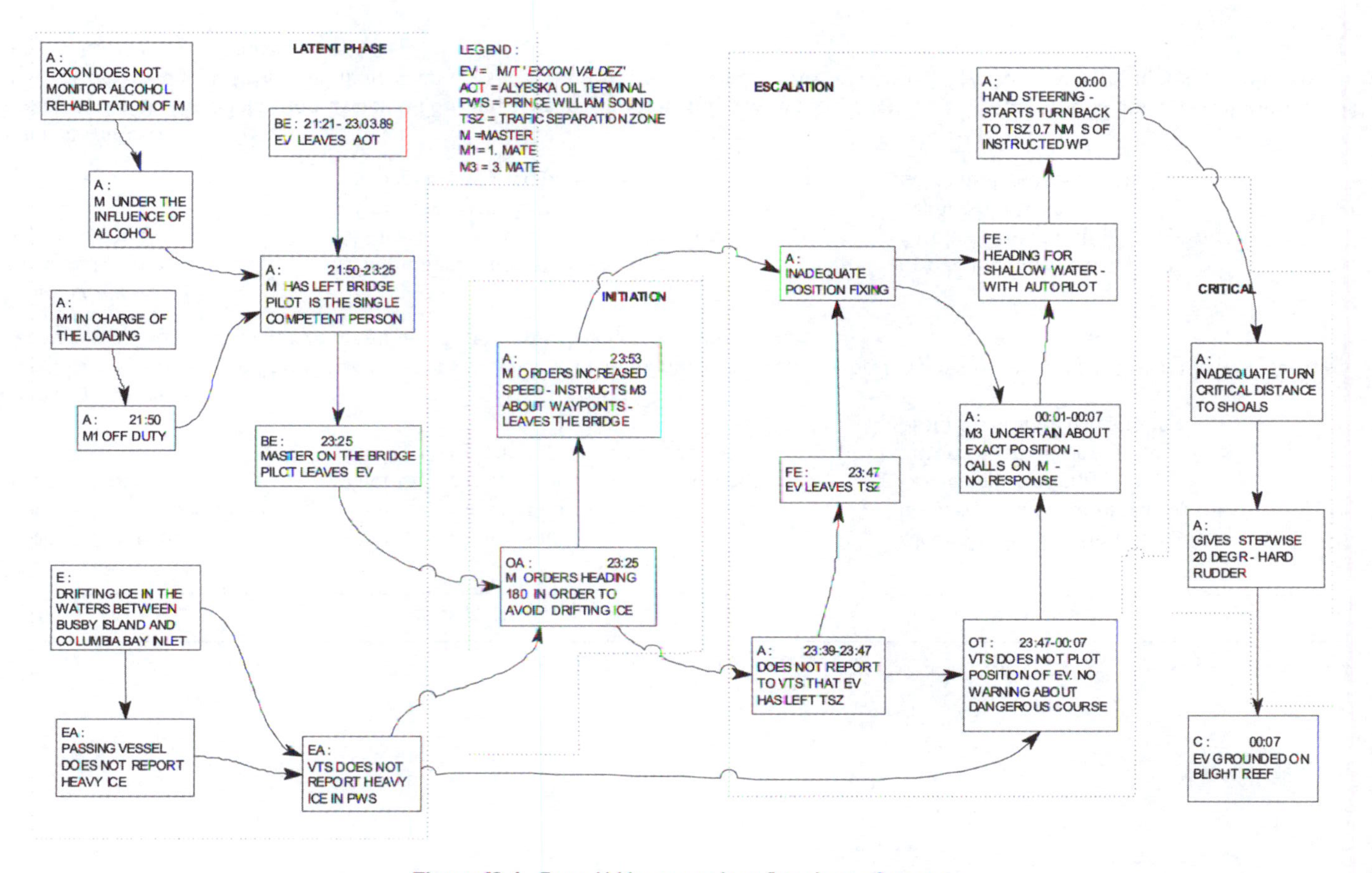

Figure 13.4. *Exxon Valdez* grounding: flowchart of events.

Table 13.5. *Exxon Valdez* grounding

	Functional error	Preventive measures
Latent phase		
Inadequately manned relative to the operational requirements	Inadequate leadership Inadequate manning of the bridge Pilot did not report manning irregularity Inadequate planning of the voyage	Alcohol rehabilitation programme Improve quality of management Leadership training Enforcement of port state control Modification of behaviour
Summary of events		
Vessel leaves port fully loaded. Master under influence of alcohol and away from the bridge most of the time until pilot leaves the vessel. Voyage planning and adequate manning of the bridge has not been taken care of.		
Initiating phase		
Inadequate bridge watchkeeping	Unqualified mate on watch Watch lacks a second officer required in Prince William Sound Inadequate voyage planning	Bridge management training Alcohol rehabilitation programme Leadership training Behaviour modification
Summary of events		
Vessel leaves Traffic Separation System (TSS) in order to avoid drifting ice. Watch is left to 3rd Mate who is not checked out for this part of the voyage in Prince William Sound. Exxon requires two qualified officers on watch. Master gives an imprecise sailing order before he leaves the bridge.		

Escalating phase		
Navigation without full control and consequent delayed turn back to safe waters (TSS)	Inadequate assessment and selection of waypoints No plotting in chart Increasing and too high speed Disturbed by drifting ice VTC did not monitor vessel	Training in coastal navigation Electronic chart display system (ECDIS) Upgrading of VTC facilities Training in VTC operations
Summary of events		
Vessel stays on a hazardous course for too long. Mate is unable to keep an updated position fix relative to the shallow water at same time as he must observe the ice-infested waters. Starts turn too late.		
Critical phase		
Too late adjustment of turning radius (rate of turn)	Too late switching to hand steering Ordered too little rudder Maloperation of rudder control	Training in coastal navigation Training in ship handling Improved ergonomics of control systems
Summary of events		
The return to the TSS starts too late and the turning of the vessel is executed with too little rudder. There are some unclarified points with respect to whether the rudder is operated wrong or too little rudder is ordered. Vessel hits Bligh Reef.		

Source: Kristiansen (1995).

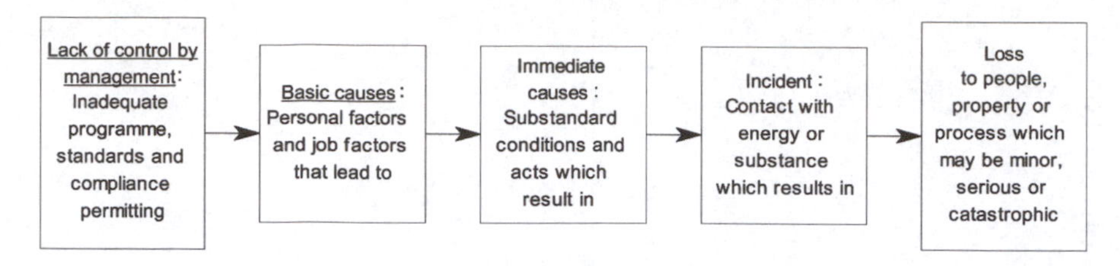

Figure 13.5. Loss Causation Model (after Bird and Germain, 1992).

Immediate causes

Immediate causes are defined as those circumstances that immediately precede the accident. They could also be labelled as unsafe acts or practices and unsafe conditions. Instead of 'error', the word 'substandard' is preferred by DNV to avoid questions of blame.

Basic causes

Basis causes are defined as those factors underlying the immediate causes. They are also referred to as root causes. The following taxonomy is used for basic causes: *personnel factors* (motivation, lack of skill, lack of knowledge, stress) and *job factors* (inadequate leadership or supervision, inadequate maintenance, unsuitable design or purchasing of equipment and tools, unclear procedures, etc.).

Lack of control

Finally, lack of control is operationalized as inadequate and/or improper regulations. These are at the start of the event sequence and produce the conditions for basic causes. Control is traditionally seen as one of the basic management functions next to planning, organizing, and leading/directing. Lack of control may originate from:

- Inadequate safety control systems
- Inadequate performance standards
- Inadequate compliance with standards

Managers are encouraged to identify hazards using an appropriate technique such as HAZOP, fault tree analysis, etc. As well as hazards, means to control them should be identified. Once identified, a prioritization can be made. This implies that decisions must be made on the basis of risk and cost estimates. Again, several tools may be used such as quantitative risk analysis, reliability analysis, etc.

13.8.3 LCM Database

The LCM approach to accident investigation and reporting puts emphasis on systematic coding. This allows the user to study trends and analyse correlations between accident

Table 13.6. Items in the LCM investigation report

Section	Description
Identifying information	Company, department, persons, date of incident Kind of injury and nature of loss, investigator identif.
Risk	Evaluation of loss potential: probability and seriousness
Description	Describe how the event occurred (text)
Cause analysis	Immediate and basic causes (text)
Action plan	What has and/or should be done to control the causes listed (text)
Immediate causes	Coding of: Substandard actions Substandard conditions
Basic causes	Coding of: Personal factors Job factors
Personal injury	Coding of: Type of contact Contact with energy, substance or people
Review	Assessment of the investigation and report
Site	Sketch of site

factors. The risk manager may thereby monitor the risk picture of the operation. The approach assumes that the incident is reported both in terms of free text information and coded data. The main items are summarized in Table 13.6.

The Loss Causation Model in its original form was designed to deal with land-based industry and did mainly focus on personal injury (work accidents) and economic loss. It can also be noticed that the database, apart from concrete findings, also contains an estimate of recurrence and potential seriousness of the kind of accident reported. Although this is merely subjective data, it is useful for presentation of the risk picture for the system or operation. This may for example be done in a matrix format showing probability versus seriousness.

13.8.4 Taxonomy

A more detailed picture of the code structure is shown in Table 13.7. The different elements in the taxonomy will be commented upon.

Losses are used to denote the consequences of an accident and focus on personal injury and property/process damage. Both the object involved and the degree of seriousness are indicated.

Incidents deal only with personal injury. It provides a classification of the kind of event or accident that directly affects the persons involved. It does not provide a description of the type of event that is related to the property or process involved. This weakness becomes evident if different accident types happen to be relevant for the system (property/process) and person. A fire in a building may, for example, lead to a fall accident to a worker involved. The fall accident may be coded (incident) but there is no coding option for the fire accident.

Table 13.7. Incident analysis summary

Losses		Incidents		Inadequate control			
Type	No.	Type	No.	Programme elements	P	S	C
Injury/illness:		Struck against		Leadership and administration			
First aid		Struck by		Management training			
Medical treatment		Fall to lower level		Planned inspections			
Lost work day		Fall on same level		Task analysis and procedures			
Fatal		Caught in		Accident/incident investigation			
Part of body harmed:		Caught between		Task observations			
Head		Overexertion		Emergency preparedness			
Eye		Overstress		Organizational rules			
Hearing		Contact with:		Accident/incident analysis			
Respiratory		Heat		Employee training			
Trunk		Cold		Personal protective equipment			
Digestive tract		Fire		Health control			
Arm		Electricity		Programme evaluation system			
Hand		Chemical–caustic		Engineering controls			
Finger		Chemical–toxic		Personal communications			
Leg		Noise		Group meetings			
Knee		Pressure		General promotion			
Ankle		Radiation		Hiring and placement			
Foot				Purchasing controls			
Toe				Off-the-job safety			
Skin							
Property or process:							
Minor (less than $100)							
Serious ($100–$999)							
Major ($1000–$9999)							
Catastrophic (over $10,000)							
Type of property damaged:							
Building				Legend:			
Fixed equipment				P = Inadequate programme			
Motor vehicle				S = Inadequate standards			
Tools				C = Inadequate compliance			

Materials
Materials handling equipment

Immediate causes		Basic causes	
Type	No.	Type	No.
Substandard practices:		**Personal factors:**	
Operating without authority		Physical incapacity	
Failure to warn		Mental incapacity	
Failure to secure		Lack of knowledge	
Improper speed		Lack of skill	
Made safety device inoperable		Physical stress	
Used defective equipment		Psychological stress	
Used equipment improperly		Improper motivation	
Did not use protective equipment			
Serviced equipment in operation		**Job factors:**	
Adjusted equipment in operation		Inadequate leadership/supervision	
Horseplay		Inadequate engineering	
Under drug/alcohol influence		Inadequate purchasing	
		Inadequate maintenance	
Substandard conditions:		Inadequate tools/equipment/materials	
Inadequate guards		Inadequate work standards	
Inadequate protection		Wear and tear	
Defective equipment		Abuse and misuse	
Congestion			
Inadequate warning system			
Fire hazard			
Explosion hazard			
Reactive chemical			
Hazardous atmosphere			
Noise			
Radiation			
Illumination			
Ventilation			
Poor housekeeping			

The explanation of why the incident took place is based on two sets of factors, namely 'immediate causes' and 'basic causes'. The immediate causes are grouped into two:

- *Substandard practices:* Errors in terms of actions or omissions directly related to the process involved. These have 12 different categories of a fairly crude nature and does not indicate very precisely what kind of functions or tasks were affected.
- *Substandard conditions:* Denote inadequate work conditions, extreme environmental conditions, or acute events that affected the person or operator.

The basic causes are also structured into two subsets as follows:

- *Personal factors:* Factors that make the operator less competent for the critical task or function. It may be a permanent or temporary disability or limitation. Another kind of explanation is lack of competence due to inadequate training. Wrong attitude may also be relevant.
- *Job factors:* Shortcomings or weaknesses related to work organization and management. This set of factors is mainly structured by management function, namely leadership, engineering, purchasing, etc.

The last element in the causal analysis chain is to pinpoint what the authors have denoted 'inadequate control'. This was a fairly innovative idea in the sense that the method put more weight on the management part than had traditionally been the case. It is, however, somewhat difficult to see the distinction between job factors and inadequate control. The main difference is that the controls are more detailed and specific. Another interesting aspect is the option to make a distinction between different kinds of inadequacies: inadequate programme, standards and compliance.

13.8.5 Data Storage and Application

ILCI has been quite concerned about the user aspects of the Loss Causation Model (LCM). In the publication *Practical Loss Control Leadership* (Bird and Germain, 1992), incident reporting is put in a wider scope as an element in safety management. Incident reporting is seen as one approach that must be supported by other functions such as conventional inspections and safety management audits. Within the scope of management it also outlines risk control, task analysis, employee training and occupational health.

The preparation for incident investigation is discussed thoroughly in terms of organization, procedure and techniques. Important requirements for performing an adequate investigation are outlined and the most frequent pitfalls are pointed out. The LCM is also prepared from a practical point of view by giving forms for reporting, announcement and major incident review. The report form contains identification data, textual description of events and causes, and coding of key parameters.

The method also offers a set of checklists for identifying basic causes termed SCAT (Systematic Causal Analysis Technique). Such factors are not readily visible in

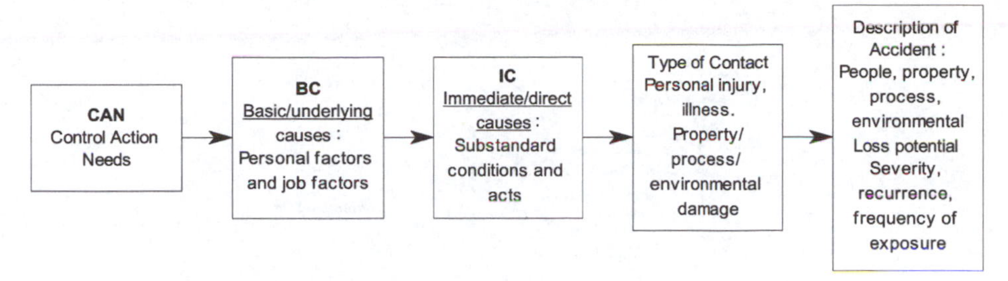

Figure 13.6. Marine Systematic Cause Analysis Technique (M-SCAT).

an accident investigation and require considerable knowledge and experience in work psychology and management. The checklist is seen as a help in identifying relevant basic causes.

13.8.6 M-SCAT

DNV took over ILCI, and thereby the services related to the LCM method which have been described in previous sections. As DNV has their main activity in the marine and maritime business, there soon emerged a need for an adaptation of the LCM to these areas.

As the critical part in accident investigation is often to identify causal factors, it was decided to develop a taxonomy for marine operations which was called M-SCAT (marine version of the original SCAT scheme). The overall structure of the coding scheme was slightly modified. The new version is shown in Figure 13.6. Apart from tailoring the framework to marine accidents, M-SCAT also eliminated some of the shortcomings of the LCM taxonomy.

The M-Scat taxonomy has the following basic classes:

- *Type of contact:* Indicates the accident type and now covers both personal injury and property/process damage.
- *Immediate/direct causes:* Substandard acts are directly related to maritime tasks. Substandard conditions are directly related to ship systems and functions. Both categories are thereby less generic.

The rest of the taxonomy, 'basic causes' and 'control action needs', is more or less unaltered. This should also be expected, as they cover fundamental human and organizational behaviour according to LCM.

13.9 ALTERNATIVE ACCIDENT MODELS

13.9.1 Systems Orientation

An important and characteristic aspect of maritime casualties is the involvement of quite large systems. The vessel in itself is a major technical system, the crew may be in the order

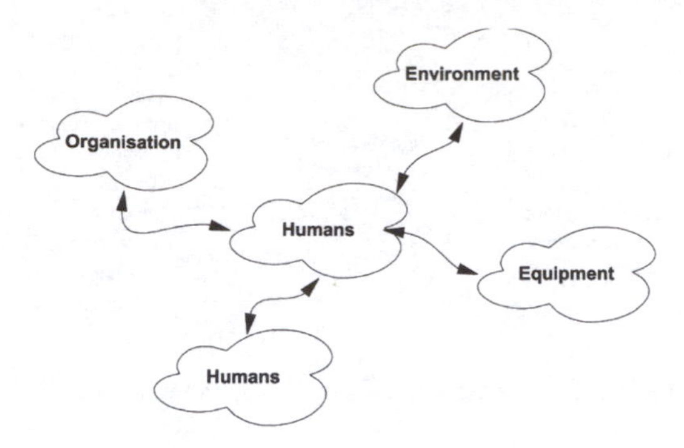

Figure 13.7. System elements.

of 15 persons or more, and the supporting land organization even much larger. Another typical trait is the physical interaction with the environment that may have consequences for both navigation and ship handling. This aspect should therefore be incorporated in the casualty analysis process.

A simple way to state this is that the taxonomy should incorporate the main systems elements or resources in line with the SHEL model (Hawkins, 1987). The generic elements are (Figure 13.7):

- Humans – 'liveware' (L)
- Equipment – hardware (H)
- Organization – software (S)
- Environment (E)

Secondly, it is important to acknowledge critical interfaces or interactions between these elements. Typical maritime problem areas are for instance:

- Seasickness: E on L
- Workplace and ergonomics: H on L
- Lack of planning: S on L
- Lack of co-operation: L on L
- Disturbance on systems: E on H, and so on.

In principle, each system element may have an adverse effect on the performance on any of the other ones and vice versa.

13.9.2 The Process Aspect

A third productive viewpoint is the information processing aspect of a maritime system. Humans can be viewed as information processors just as equipment and organizations can. The main information processing functions of the operator are detection, perception, analysis, decision, action and communication, as illustrated in Figure 13.8. A more

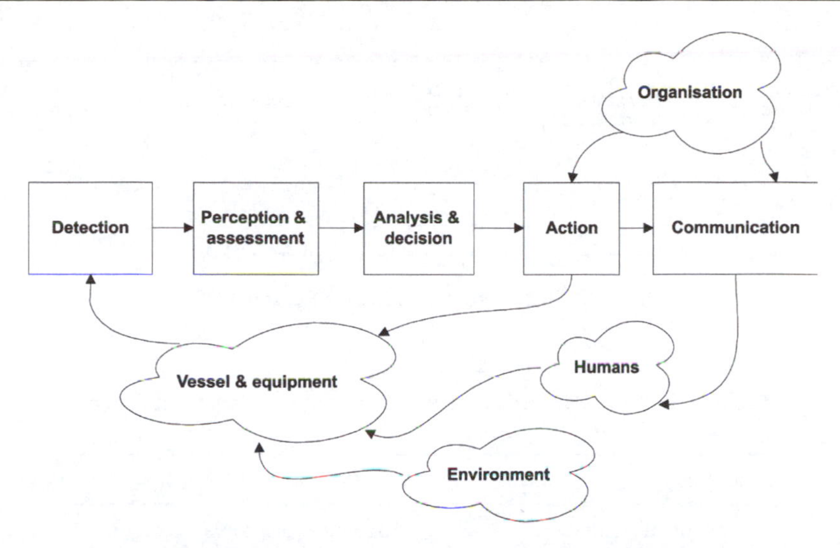

Figure 13.8. Maritime operation as an information processing system.

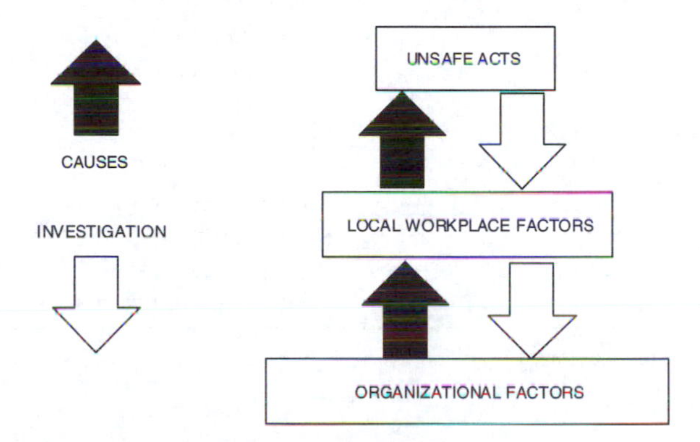

Figure 13.9. Organizational accidents (after Reason, 1997).

complete discussion of this aspect is found in the report on methodological approaches (Koster et al., 1999).

13.9.3 Organizational Factors

The research of J. Reason (1990, 1997) has had a profound influence in present thinking on why accidents happen and how they should be managed. A key point is that the observed unsafe acts should be traced back to workplace factors and organizational factors, as indicated in Figure 13.9. Reason (1990) has also pointed out that the management of accidents may be based on alternative feedback loops related to the hierarchical levels in the operation (see Figure 13.10).

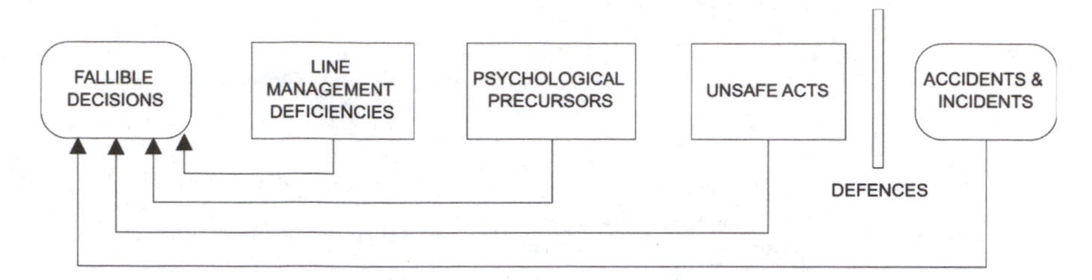

Figure 13.10. Feedback loops (after Reason, 1990).

Table 13.8. The relationship between basic systemic processes and general failure types

Processes	GFTs
Statement of goals	Incompatible goals
Organization	Organizational deficiencies
Management	Poor communications
Design	Design failures
	Poor defences
Build	Hardware failures
	Poor defences
Operate	Poor training
	Poor procedures
	Poor housekeeping
Maintain	Poor training
	Poor procedures
	Poor maintenance management

Source: after Reason (1997).

Another key point in Reason's understanding is the focusing on so-called latent failures that are difficult to detect before the accident is a fact. These factors are termed general failure types (GFTs) and can be related to the organizational processes as illustrated in Table 13.8. It is a primary objective of inspections and audits to detect the GFTs before they cause harm.

13.10 OFFICIAL ACCIDENT INVESTIGATION

All Flag States investigate, analyse and report maritime accidents. The degree of effort and procedures may, however vary considerably. The responsibility is often given to a body or agency separate from the maritime authority in order to secure independence and legitimacy. The principles for the investigation are laid down in laws and regulations. It is beyond the scope of the present discussion to elaborate on all the aspects of an official investigation. A few common aspects will, however, be highlighted (Caridis et al., 1999):

- In case of an accident the Master is obliged to report to the relevant authority.
- Which accidents to report is determined mainly by the degree of seriousness.
- In certain countries the first witness testimonies are given in a formal Sea Court proceeding.
- Further investigation is undertaken by a single inspector, a team of inspectors or an ad hoc appointed commission.
- The scope of the investigation is determined by the seriousness and potential of learning.
- Both the investigation and the report may either be open to the public or confidential.
- Legal proceedings (criminal prosecution) may be a part of the investigation or it may be performed separately by another body.
- All countries keep files on closed investigations and a few produce statistics and keep database records.

13.10.1 The Investigation Process

Maritime accidents may be investigated by the authorities, the ship manager, and other affected parties such as the insurance. The official investigation of accidents is undertaken by the national maritime administration or an affiliated agency and the conduct is regulated by law. Caridis et al. (1999) have documented a significant variation in practices within the EU.

The criteria for undertaking an official investigation are primarily related to:

- Certain casualty types that are specified
- Loss of life
- Serious damage and loss of vessel
- Environmental consequences
- Expectation that the findings will contribute to improved understanding

These criteria also vary from one country to another. The investigation may be undertaken by a single investigator or a team:

- A simplified or administrative investigation mainly based on documentation
- An investigation by a professional investigator – either with nautical or engineering background
- By a multidisciplinary team
- Or by an ad hoc appointed commission

The investigative body in the UK, the Marine Accident Investigation Branch, have published their policy and main procedures (MAIB, 1998).

The emerging view that accidents are rooted in human and organizational errors (HOE) has led to a greater focus on work psychology, ergonomics, social aspects and company policies. It has been pointed out that our present knowledge of the HOE factors is limited due to lack of focus in accident investigations. The Transportation Safety Board of Canada (TSB) has been concerned about the effect of fatigue and has recommended that the following factors should be determined in the investigation:

- The time of day of the incident
- Whether the operator's normal circadian rhythm was disrupted
- The number of hours the operator was awake
- Whether the sleep history 72 hours prior to the incident suggests a sleep debt

It has been raised as a problem that interference from police and prosecution may hamper the search for the causes of an accident. Some nations have therefore separated the two tasks, for instance the US, where both the US Coast Guard (USCG) and the National Transportation Safety Board (NTSB) are involved.

Considerable experience has now been accumulated on how such investigations should be undertaken (Ferry, 1981; Bird and Germain, 1985). The key activities are:

- Planning and organization of team and activities
- Witnesses
- Investigation of vessel and systems involved
- Medical examination
- Background information and documentation
- Examine records for the operation
- Test of material and analysis of technical failure mechanisms
- Analysis and documentation of what happened
- Analysis of human error
- Analysis of basic or root causes
- Summary of lessons to be learned
- Reporting, comments and feedback

The US Coast Guard (USCG) has developed a comprehensive set of procedures and techniques for accident investigation (USCG, 1998a, 1998b). Despite this, Hill and Byers (1991) made a critical assessment of the accident reporting system of the USCG. They found a number of weaknesses as to the reporting of so-called human errors, and the following explanations were suggested:

- Inexperience of the investigators with human error concepts
- Tendency to carry mental models which leads to over-representation of favourite causes
- Conflict with other investigation objectives
- Lack of adequate models

13.10.2 The Accident Report

The investigation process will usually be documented in a report. The typical sections or chapters are indicated in Table 13.9. The report may also be supplemented by:

- Photos and drawings
- Source documentation, logs
- Special studies
- Transcripts of testimonies, hearings, court proceedings, etc.

It is also common to have a standardized form for reporting key information about the circumstances and immediate outcome of the accident in terms of accident type, degree of seriousness and immediate consequences. The form will have an even broader scope given that the accident will be entered into a computerized database. The matter of coding will be discussed later in relation to accident models and taxonomies.

Official investigations have to some degree been under attack for not being up to date in terms of organization, technical approach and understanding of causal mechanisms:

- Use of court proceedings
- Lack of teams with broad competence and specialization in modern investigation techniques
- Narrow focus on technical and nautical aspects
- Lack of competence on human and organizational factors
- Conflict with objectives of criminal prosecution
- Bias: mental models lead to over-representation of favourite causes

13.10.3 Marine Accident Investigation Branch (MAIB)

In the UK, the MAIB has been very active since the late 1980s in developing better investigation routines and reporting formats. In 2001 they also revised their computerized accident database. The investigation policy is outlined in a memorandum (MAIB, 1998). Based on certain criteria, MAIB may undertake:

- *Administrative inquiry* for less serious cases.
- *Inspector's investigation* of cases that need full and detailed investigation in the field.
- *Inspector's inquiry* for major accidents. This will normally involve a team of inspectors and the report will be published.

The coding of accidents is based on the record structure outlined in Figure 13.11 (MAIB, 2001). The coding starts with identifying data about the case vessel and systems involved. The systems are further broken down (what), located (where) and injured persons listed (casualty). The explanation of system failures is explained at two levels: causal factor and subfactor. It may also be noticed that recommendations from the investigators are coded.

Table 13.9. The accident report structure

Section	Outline	
Summary	Short presentation of situation, key facts, events, causes and lessons to be learned	
Sources of information	Persons and parties interviewed or given testimony, transcripts Investigation technical reports	
Accident	Accident type, vessel damage, injuries and fatalities, environmental pollution	
Vessel	Technical particulars, general arrangement, photos, etc. Identification Shipbuilder, year of delivery Ownership, management, registration, classification History of operation, incidents, deficiencies, detention	
Crew	Certificates, competence and experience, nationalities, language used Watch system, hours on watch before accident	
Physical location	Day, hour Geographical location, aids to navigation Seaway type, docks, moorings Port administration, tug service	Weather conditions Visual conditions Sea state Pilot service, VTMS
Narrative events	Description of what happened from initiation through escalation to accidental outcome Description of rescue and re-floating, salvage or damage control operations Is given in chronological order and supplemented with charts, diagrams, etc.	
Analysis	Assessment of facts and evidence Technical failures, environmental impact phenomena Description of workplace, instruments, tools and ergonomic factors Discussion of causal factors related to individual behaviour and organization Human factors problems: fatigue, cognitive problems, human–machine interaction, etc. Bridge resource management The role of managing company, port authorities or other key parties involved	
Conclusions	Summary of critical events and causal factors	
Submissions		
Recommendations	Point-by-point proposal of controls or safety measures	
Appendix	Vessel documentation, hull strength analysis, weather and sea state data Background information, logs Technical reports, analysis in detail of critical items or events	

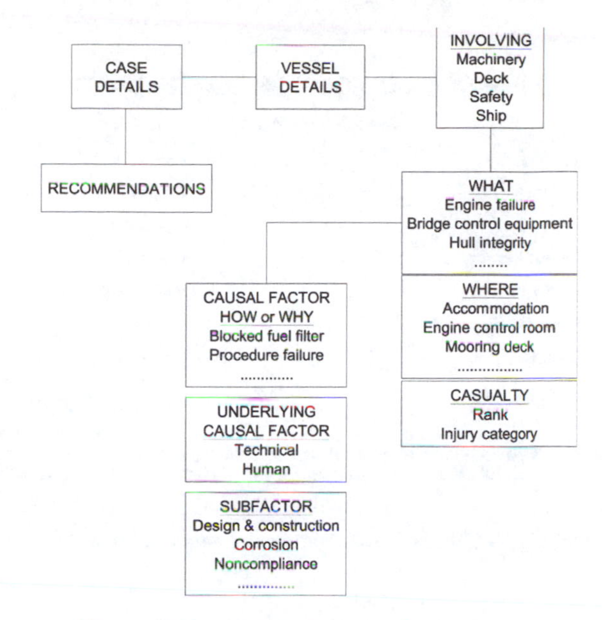

Figure 13.11. MAIB accident database structure.

MAIB has a homepage where open publications are accessible:

http://www.dft.gov.uk/stellent/groups/dft_control/documents/contentservertemplate/dft_index.hcst?n=5481&l=2

The following reports are published:

- *Annual reports*
- *Safety digests:* accident summaries and lessons learned
- *Safety bulletins:* safety preventive recommendations on the basis of a specific accident
- *Safety studies:* report on a specific safety problem based on experience gained from a number of accidents
- *Investigation reports*

13.10.4 Official Homepages

There is now an increasing trend by the maritime authorities to publish accident information on the Internet. The US Coast Guard Office of Investigation and Analysis has the following homepage on accident investigation:

http://www.uscg.mil/hq/g-m/moa/casualty.htm

It covers a broad set of information such as accident statistics, lessons learned, safety alerts, etc., quite similar to that of MAIB (Figure 13.12).

A similar homepage is published by the Australian Transport Safety Board (ATSB). Figure 13.13 shows a window from the homepage. The accident reports are presented

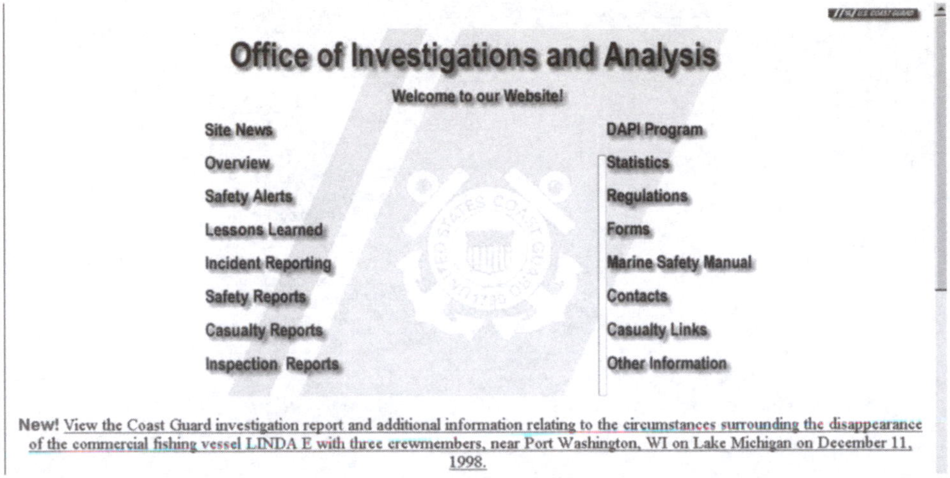

Office of Investigations and Analysis

Welcome to our Website!

Site News
Overview
Safety Alerts
Lessons Learned
Incident Reporting
Safety Reports
Casualty Reports
Inspection Reports

DAPI Program
Statistics
Regulations
Forms
Marine Safety Manual
Contacts
Casualty Links
Other Information

New! View the Coast Guard investigation report and additional information relating to the circumstances surrounding the disappearance of the commercial fishing vessel LINDA E with three crewmembers, near Port Washington, WI on Lake Michigan on December 11, 1998.

Figure 13.12. Homepage of USCG Office of Investigation and Analysis (http://www.uscg.mil/hq/g-m/moa/casualty.htm).

chronologically on the left side whereas the latest reports are accessible directly on the right side of the window. In the lower section is shown how the search for specific reports is facilitated by short text abstracts. The homepage also has basic information about ATSB, reporting routines and safety bulletins.

It is clear that this policy makes a vast information potential available to the maritime industry, research and other official agencies.

13.11 ACCIDENT ANALYSIS SOFTWARE

The fact that the ISM Code specifies that companies shall undertake systematic analysis and feedback of experience from non-conformities has opened a market for computer software that can support this function. Here we will briefly comment on two systems that are offered on a commercial basis.

13.11.1 SYNERGI

The SYNERGI system was originally developed for continental shelf operations but has also won acceptance in the shipping sector. The coding structure is to a large degree based on the Loss Causation Model (LCM) which has been outlined earlier.

Figure 13.14 shows the main window of the PC-based system. It contains mainly factual information and a short text summary (Description of Event). A separate window has been designed for coding of immediate and basic causal factors (see Figure 13.15).

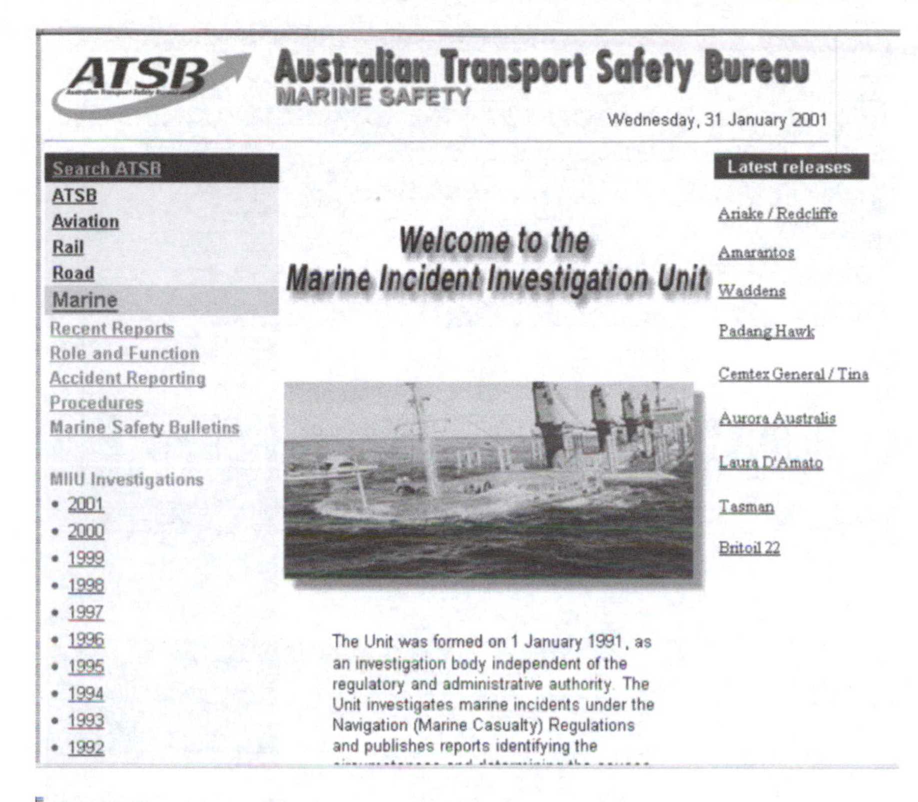

Search ATSB
ATSB
Aviation
Rail
Road
Marine
Recent Reports
Role and Function
Accident Reporting
Procedures
Marine Safety Bulletins

MIIU Investigations
• 2001

Recent Reports

Ariake/ Redcliffe	A collision occurred between the container ship *Ariake* and the tug *Redcliffe* as the *Ariake* approached its berth at Fisherman's Islands container terminal, Brisbane on 24 January 2000.
Waddens	Accidental release of lifeboat during hoisting.
Padang Hawk	While carrying a full cargo of nickel ore from New Caledonia, the bulk carrier Padang Hawk encountered rough seas, rolled heavily and developed a 15° list to port. An examination showed that in 4 out of 5 holds, the cargo had settled and appeared to have liquefied.
Cemtex General/Tina	A failure to observe the International Regulations for Preventing Collision at Sea, contributed to the collision between two vessels off Noosa Head on 2 February 1999, according to the report of the

Figure 13.13. Details from the homepage of the Australian Transport Safety Board (http://www.atsb.go v.au/marine/indxf/index.cfm).

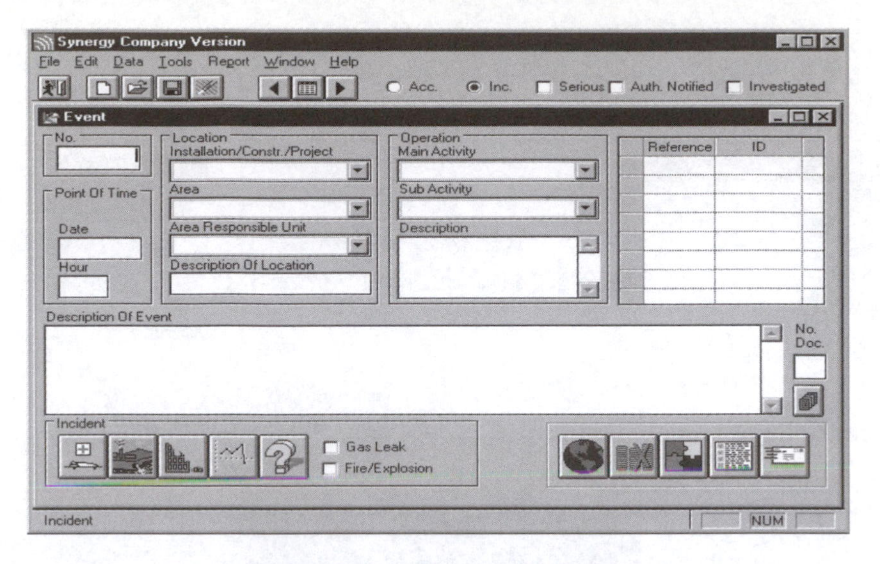

Figure 13.14. SYNERGI main window.

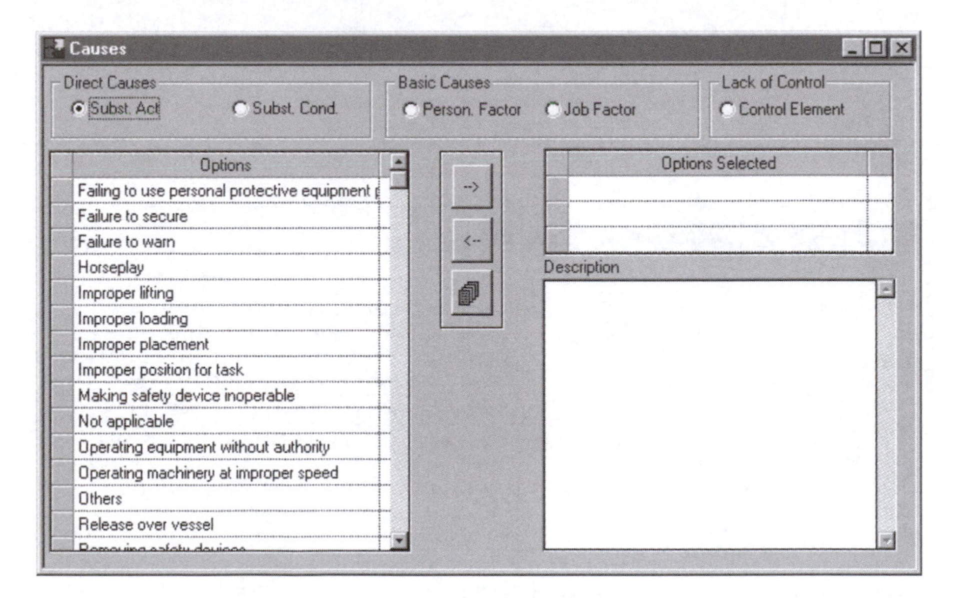

Figure 13.15. SYNERGI: data in Causes window.

13.12 HUMAN FACTORS ANALYSIS

13.12.1 What is Human Error?

The term *human error* is often heard as the explanation of accidents. On the other hand it is often misused and, as already discussed, not a very clear concept. It might be useful first to comment on some key findings of Jens Rasmussen cited by Reason (1997) on human performance as outlined in Table 13.10:

- Skill-based: highly trained tasks that are executed almost automatically and unconsciously.
- Rule-based: certain pre-trained responses to a set of known situations (*if* situation *then* response).
- Knowledge-based: basically what we understand by problem solving. The new situation requires analysis or trial and error.

Reason (1990) has further shown that the basic forms of error can be understood in terms of these performance levels. The basic error types are outlined in Figure 13.16. First, one should make a distinction between unintentional and intentional error. The

Table 13.10. Performance levels

Performance	Situation	Control mode
Skill-based	Routine	Automatic
Rule-based	Trained for problems	Mixed
Knowledge-based	Novel problems	Conscious

Source: after Reason (1997).

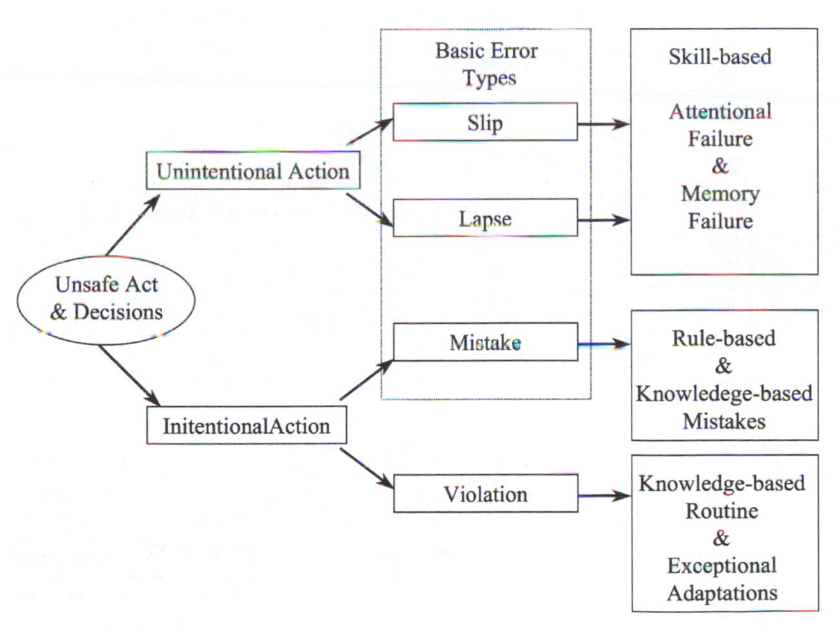

Figure 13.16. Error versus performance level (adapted from Reason, 1990).

first category may lead to slip or lapse whereas the other may result in mistake or violation.

Rockwell and Griffen (1987) describes an error classification that covers airplane flight and that may seem of relevance to ship operation (Table 13.11). The errors are related to specific tasks on the flight-deck rather than using the generic terms of Reason.

Hollnagel (1998) gives a very extensive discussion of human error models or taxonomies. It might be relevant to comment briefly on some of the approaches he mentions. Hollnagel concludes, after an extensive study of alternative accident models, that error (phenotype) is caused by man, technology and organization (Figure 13.17). He also postulates that erroneous action has certain generic modes as outlined in Figure 13.18.

13.12.2 A Practical Approach

As pointed out earlier in this chapter, the starting point for the accident analysis process is to establish the critical event sequences or, in other words, the question to answer: *what* happened? The next and very challenging task is to explain *why* it happened or, in

Table 13.11. Critical in-flight event model

1. Inadequate pre-flight checks	2. Fails to recognize early warnings of problems
3. Fails to do sequence checks	4. Decides to fly despite system discrepancies
5. Fails to recognize early warnings	6. Fails to monitor instrument readings
7. Fails to notice small discrepancies in flight sensations	8. Fails to notice lack of agreement of related instruments
9. Diagnostic error	10. Error in estimation of urgency
11. Improper corrective action	

Source: Rockwell and Griffen (1987).

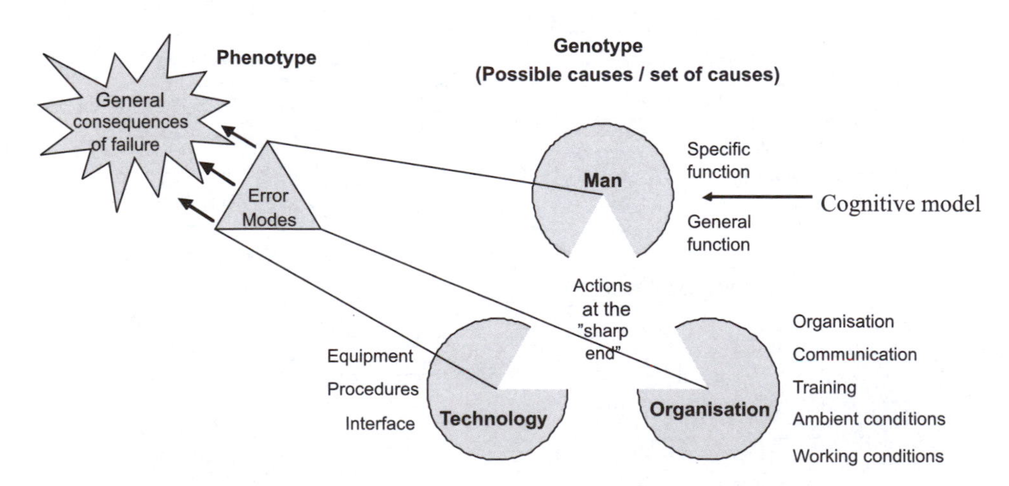

Figure 13.17. The CREAM model (adapted from Hollnagel, 1998).

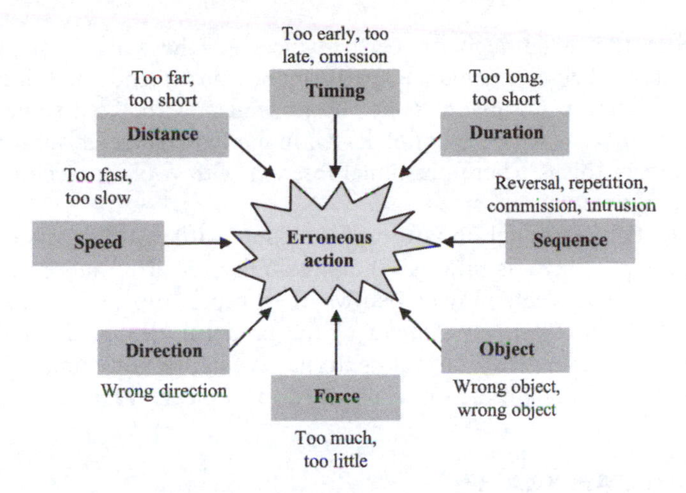

Figure 13.18. Error modes (adapted from Hollnagel, 1998).

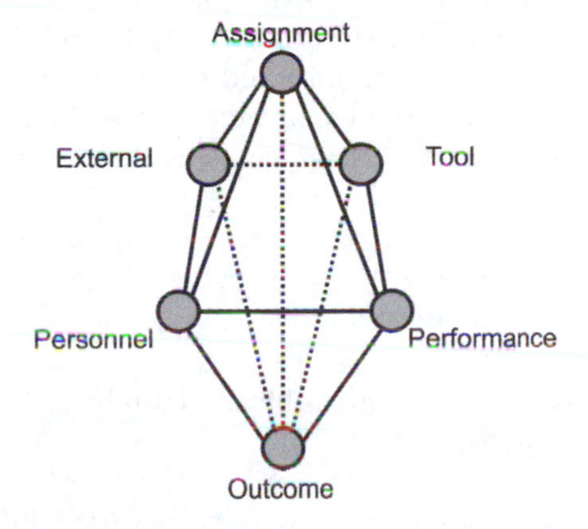

Figure 13.19. Human factors interaction model.

technical terms: what were the main causal or contributing factors? This is in essence a diagnostic process where no precise or universal approach can be prescribed. The CASMET project did, in spite of this, attempt to outline certain principles (Koster et al., 1999). The model was based on the premise that there is an interaction between the mission, humans, systems and the environment as conceptualized in Figure 13.19.

An analysis of the Human and Organizational Error (HOE) element in maritime accidents should start with the recognition that maritime operations typically take place because a certain *mission* or *assignment* has to be fulfilled (see Figure 13.19). To fulfil the assignment, a certain performance is required from two entities: the *personnel* aboard, and the *tools* with which they are equipped, including the ship itself. Thus, there is a demand-pull from the assignment on performance. At the same time, personnel and hardware interact in their performance (performance-push) and in doing so they may cause accidents, i.e., the performance is not in agreement with the assignment. Not very surprisingly, the first place where one notices an immediate cause of an accident is at the level of performance: this is where personnel interact with tools to accomplish the tasks defined by the assignment.

In addition to the personnel aboard and the tools with which they are equipped to fulfil the assignment, there is always the possibility of external events influencing the interaction. External events may be bad weather, other ships, or events preceding the one under consideration. The last element of the model is the result of the interaction between assignment, personnel, tools, and external events: the outcome of an event. The checklist for analysis of human factors is outlined in Table 13.12.

13.13 THE CASMET APPROACH

13.13.1 Overview

The CASMET approach rests on two pillars: an analytical method and a structure for coding information in a database (Kristiansen et al., 1999). The analytical method answers the question how the information should be obtained. The question of how the information obtained should be represented in a database will be dealt with by the outline of the coding and database structure. The main steps that both pillars adhere to can be outlined as follows:

1. Initial data collection.
2. Identification and reconstruction of events.
3. Human factors analysis.
4. Systems, hazardous materials and environmental analysis.
5. Summary of causal relations.

The relation between the analysis process and the resulting information to be coded, structured and stored in a database is represented in Figure 13.20.

The CASMET method has four basic levels for representing a maritime casualty (Figure 13.21); namely:

- Casualty events
- Accidental events
- Basic causal factors relating to daily operations on board
- Basic causal factors relating to management and allocation of resources.

The casualty and accidental events should be viewed in the *time domain* whereas the causal factors are *logically* linked to each other and to the events. The intention with this representation format is to keep the process character of events and the logical links between what it is possible to observe in order to understand the causal nature.

Table 13.12. Human factors checklist

External	Performance	Causal mode		
		Personnel	Tool	Assignment
Previous event(s)	**Detection:**	**Lack of knowledge** due to	**HMI/design**	**Task characteristic:**
	Technical failure	lack of experience,	Compatibility	Ambiguous task
	Personnel factor	lack of orientation,	Consistency	Habit-ignoring task
Other ships	Lack of support	inadequate training,	Context	Distracters in task
		information overload,	Structure and systematics	Inadvisable rules
		lack of information	Feedback	Error-enforcing task
Bad weather	**Assessment:**			
	Technical failure	**Lack of skills** due to	Workload	**Staffing characteristic:**
	Personnel factor	inadequate instruction,	User-directed flexibility	Personnel selection
Criminal acts	Lack of support	inadequate training,		Work schedule
		infrequent practice,	**Technical problems** due to	Workload
		lack of coaching	poor construction,	Understaffing
	Decision:		poor maintenance,	Poor training
	Technical failure		unavailable equipment	Poor motivation
	Personnel factor			
	Lack of support	**Intoxication** due to		
		alcohol use,		
		drug use,		**Poor procedures:**
		medicine use,	**Damage** due to	Operating procedures
		fumes and gases	wear-out,	Housekeeping procedures
	Action:		fire/explosion,	Maintenance procedures
	Technical failure	**Fatigue/stress** due to	physical intrusion,	Communication procedures,
	Personnel factor	task load or duration,	radiation,	Emergency procedures
	Lack of support	lack of rest,	electromagnetism	
		sensory overload,		**Incompatible goals:**
		information overload,		Time pressure
		climate,		Budget
	Causal group	time stress		
			Causal factor	**Poor communication:**
		Reduced ability due to		Ambiguous information
		physical condition,		Language problems
		mental condition,		Lack of information
		emotional condition		Too much information

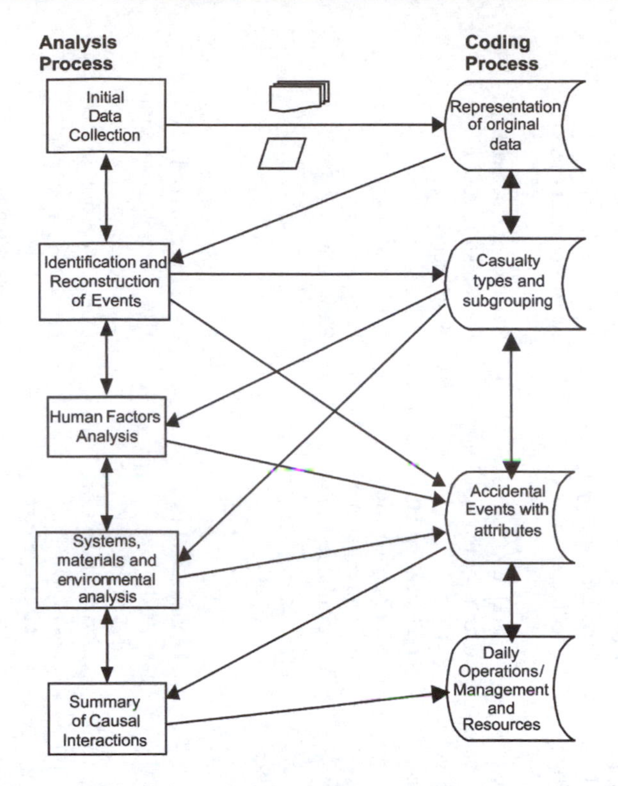

Figure 13.20. Relation between analysis and database structure.

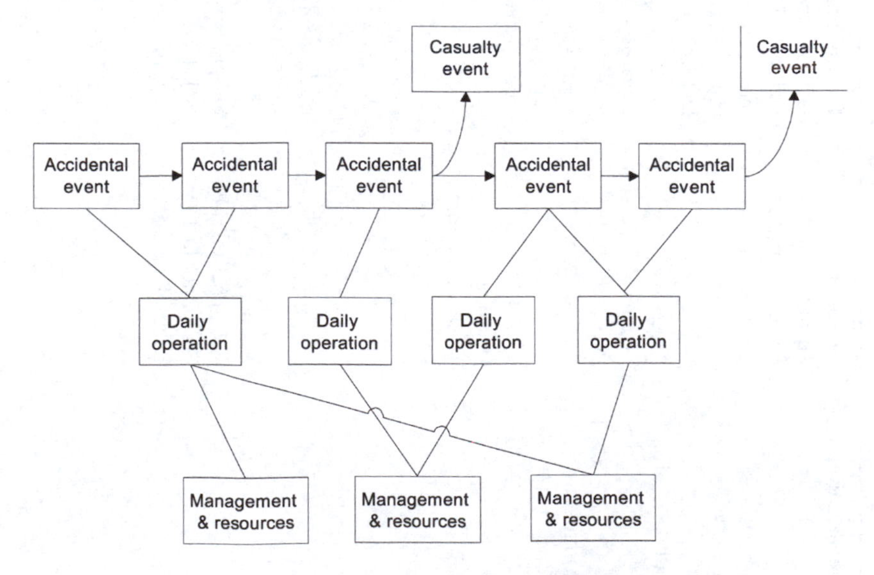

Figure 13.21. Casualty representation levels.

13.13.2 Identification of Casualty

The starting point in an accident analysis is the factual or objective information about the vessel, the subjects involved, the circumstances, and the consequences of the casualty (Table 13.13). The third of these fact groups – operation – is coded in accordance with this (Table 13.14). A distinction is made between operation of the vessel and operations onboard the vessel.

13.13.3 Casualty Summary

In order to give an introductory outline of the casualty, a summary in prose style should be given. The CASMET project found that a summary structure consisting of four segments was feasible (Table 13.15). There should not be any formal requirements for the description other than a limitation on the length of the text. The main objective of the summary is to give a synopsis of the casualty that captures all the main aspects of the accident and the lessons learned.

As an illustration, the example on pages 369–370 above shows how the summary may be written for the *Herald of Free Enterprise* accident.

Table 13.13. Fact sheet structure

Fact group	Facts
Identification	Case identification no. Vessel name Terminal casualty (final casualty event in a sequence) Date of casualty Geographical position
Vessel	Vessel type Deadweight or GRT Service speed Main dimensions (L_{pp}, B, D) Cargo intake, draft (T) Main engine type, propulsion system Yard, country, year of build Owner, flag Classification society
Operation	Vessel operation phase Operation onboard
Environmental conditions	Weather conditions, visibility Beaufort no., current speed
Manning	Number of officers and crew Nationalities Experience of key personnel
Consequences	Damage to people, vessel and environment Economic consequences

Table 13.14. Coding of operation group

Parameter	Coding		
Vessel operation phase	Sailing Manoeuvring Under tow	Anchored Enter/leave port	In port At repair yard
Operation onboard	Normal watch Mooring Loading Unloading	Cargo transfer Tank cleaning Gas freeing Bunkering	Ballasting Repair Idle, off-hire

Table 13.15. Casualty summary

Summary	Key elements
What happened	Short account of casualty Vessel, place, time, outcome
Circumstances, contributing factors	Latent technical and operational factors Environmental conditions Other hazardous conditions putting vessel at risk
Immediate causal factors	What happened onboard Events, failures and errors Individuals, systems involved
Basic causal factors	Why did it happen Individual conditions and shortcomings Inadequacies of management and workplace Lack of resources

13.13.4 Event Description

It is important to establish an overall picture of the sequence of events from initiation of the accident to the outcome. This is solely so that one can establish *what* happened before one starts to deduce *why* the casualty happened. Most casualties are reported in prose style. Although this may be acceptable to capture certain information, it has obvious shortcomings. Some accidents may develop gradually over a considerable time span and involve a number of actors in terms of persons and systems.

In the earlier sections of this chapter it has been shown how the event description can be given some structure or formalism with either the STEP diagram (Table 13.4) or a flowchart (Figure 13.4).

13.13.5 Casualty Event

A casualty case may involve one or more *casualty events* and *accidental* events. The latter are errors and failures that contribute to the casualty but do not necessarily lead to damage and serious consequences in themselves. Accidental events will be described in the next section. A casualty event, on the other hand, involves release or transformation of energy and is usually understood as the outcome of the casualty. Casualty events are listed in Table 13.16.

The taxonomy also addresses so-called maloperations (Table 13.17). These events are a group of accidents where the consequences are less critical for the vessel itself but of greater consequence for the environment. It is mainly a matter of spill and pollution of oil.

Table 13.16. Vessel-related casualty events

Casualty event	Casualty event subgroup
Allision	Ramming of buoy, marker Ramming of quay Collision with floating objects
Grounding	Powered grounding Intentional grounding Drift grounding
Collision	With other vessel With multiple vessels
Fire/explosion	Fire Explosion – incindiary Explosion – pressure vessel
Flooding (Founder)	Sinking Capsize
Structural failure	
Loss of control	Loss of electrical power Loss of propulsion power Loss of directional control

Table 13.17. Maloperation casualty event

Casualty event	Casualty event subgroup
Polluting cargo	Polluting cargo operations Cargo lost overboard Atmospheric pollution
Ballast	Unclean ballast pumped overboard Failure in ballast system
Engine room operations	Spill/dumping of bunker and sludge Polluting engine room operations

Table 13.18. Maloperation casualty class and state attributes

Casualty type	Class	State
Polluting cargo	Intentional Unintentional	Small and momentous Intermittent over time Continuous spill over time Progressive Uncontrolled
Ballast	Intentional Unintentional	
Engine room operations		

The casualty events are qualified with two different attributes, namely class and state. These attributes express certain aspects of the situation when the casualty event took place. The semantics may, however, vary for the individual casualty events. To give an example, the collision took place in a *crossing situation* and the vessel was disturbed by the *channel effect*. The definition of the attributes is given in Table 13.18 for maloperation and in Table 13.19 for casualty events.

13.13.6 Accidental Event

Based on the preparatory steps described in the preceding chapters, the basis for classification and coding of accidental events is established. It is important to keep in mind that accidental events are strictly related to the casualty case sequence as it is observed or reconstructed – or, in other words, only what did happen.

The casualty may be initiated and escalated by one or a series of accidental events. We make a distinction between the following types:

- Equipment failure
- Human error
- Hazardous material
- Environmental effect
- Other agent or vessel

This event categorization is defined further in Table 13.20 (p. 406). It is important to note that the description of events has no causal implication. This is especially vital when we deal with 'human error'. This event simply expresses that a wrong or inadequate behaviour of an individual was observed. However, it does not *explain* the behaviour. This is a later topic in the analysis process.

Similar to casualty events, we have defined certain attribute classes for each accident event type as shown in Table 13.21 (p. 406). The aim is to produce a 'richer'

Table 13.19. Coding of casualty class and state attributes

Casualty event	Class	State
Allision	Crossing	Current
	Meeting	Wind
Collision	Overtaking	Wave drift
	Multi-approach	Channel effect
	Uncontrolled	No external effect
Grounding	Straight course	
	In turn	
	Multiple course changes	
	Stopping	
	Backing	
	Uncontrolled	
Fire	Burned out	Bow
	Controlled	Cargo space
	Extinguished	Weather deck area
	Initial	Living space
	Uncontrolled	Navigation space
		Engine/pump room
		Stern
Explosion	Initial	
	Secondary	
Flooding	Controlled	
	Dewatered	
	Initial	
	Progressive	
	Uncontrolled	
Structural failure	Class 1	Deck
	Class 2	Frame
	Class 3	Hull
	Structural damage	Keel
		Longitudinal
		Watertight bulkhead
Loss of control	Restricted fairway	Current
	Heavy traffic	Wind
	Both	Wave drift
		Channel effect
		No external effect

description of the events. We will comment briefly on the attributes for each accidental event type.

Physical processes involving release or conversion of energy constitute an important aspect of maritime casualties. Events involving hazardous material will therefore be coded

Table 13.20. Definition of accidental event

Event type	Definition/explanation
Hazardous material	Critical events are associated with the presence of explosive, flammable or toxic material. The main sources are cargo and fuel.
Environmental effect	Factors like wind, waves and current may have significant effect on the behaviour of the vessel. These factors may not necessarily show extreme strength in order to explain the accidental process.
Equipment failure	A system module or component does not function as intended due to some sort of breakdown. Loss of function may also be the result of operating outside the specified performance (overload, overcapacity).
Human error	Operator performs in conflict with intended procedures or in a less than adequate way. Main forms are omission, commission, wrong timing or sequence. All kinds of negligence will also fall under this category.
Other agent or vessel	Lack of or inadequate support of other vessels, agents or infrastructure. This group should be applied for phenomena that are not a part of the investigation or an external influence.

Table 13.21. Attributes for accidental events

Accidental event type	Attributes	
Hazardous material	Material	Hazard
	Location	Failure type
Environmental effect	Phenomenon	Impact
Equipment failure	System involved	Failure type
	Location	Immediate physical cause
Human error	Position	Performance mode
	Task affected	Error type
Other agent or vessel	Role	Performance mode
	Task affected	Error type

irrespective of how they were initiated – by system failure or operator error. The attribute values are outlined in Table 13.22.

The description of an *equipment failure* needs certain attributes in order to be meaningful in a coded format (Table 13.23). The most immediate characteristics would be the system involved and where it is located. A number of other factors may also be relevant, but it is limited here to the kind of failure and the immediate physical cause.

Human error is another dominant accidental event type. In the same manner as for equipment failure, certain attributes seemed to be obvious, namely position and task affected by the error. Secondly, it seemed feasible to characterize the

Table 13.22. Hazardous material attributes

Attribute	Code values		
Material	Crude oil	White spirit	Lubricating oil
	Diesel oil	Gasoline	Chemical
Location	In the water	Engine room	Ballast tanks
	On land (near vessel)	Fuel tanks	Void, cofferdam
	Bridge	Deck stores, paint locker	Fore peak
	Living spaces, offices	Deck machinery room	Bow
	Galley	Engine stores	Open deck
	Engine control room	Cargo holds	Deck house, mast, etc.
	Pump room	Vehicle deck	Aft area, after peak
		Cargo tanks	
Hazard	Empty tank – not gas freed	Leak	Loaded tank
			Vents
Failure type	Explosive mixture	Reactivity	Toxic fumes
	Flammable mixture	Self-reaction	Pollution

kind of behaviour (performance) and error type. The attributes are outlined in Table 13.24 (p. 409).

Environmental effects, as described in Table 13.25 (p. 409), have two subgroups, namely physical loading on the vessel and limited visual conditions. The second group may primarily influence the visual conditions for the bridge watch and to a certain degree also the instrument sensors.

The sequence of events for a casualty may be influenced by the intervention of external parties. This effect may be adequate for ameliorating the consequences but may in some instances also have a negative effect on the casualty. External agents may either be systems or assisting vessels or human/organizational resources. They may basically fail in the same way as the vessel itself by equipment failure or human error. The coding will therefore be similar and so is not described here.

13.13.7 A Case on Event Representation

A casualty may involve one or a number of events. The idea behind the CASMET model is that it should handle any number of events and their sequence of appearance. The example in Figure 13.22 (p. 410) illustrates the interaction between casualty and accidental events. A broken fuel line sets off an engine room fire that is put out too late due to delayed mobilization of the firefighting team. This results in loss of propulsion power and, as the attempt to anchor is also unsuccessful, subsequently leads to drift grounding. Typical attribute values are indicated for the events.

A simplified summary of the coding structure can be found in Figure 13.23. Note that it is incomplete and therefore only shows the principles of the coding scheme. The two upper sections address the events representation. The lower part shows the coding structure for causal factors, which will be discussed in the next section.

Table 13.23. Equipment failure parameters

Attribute	Code values		
System involved	Ballast, stability Bilge, drain Boiler Cargo Deck machinery Dry cargo Electrical	Firefighting Door Hatch Bulkhead Fuel General safety Habitation	Hull Engine Life-saving Navigation Propulsion Steering
Location	Bridge Living spaces, offices Galley Engine control room Pump room Engine room	Fuel tanks Engine stores Deck stores, paint locker Deck machinery room Cargo holds Vehicle deck	Cargo tanks Ballast tanks Void, cofferdam Fore peak Bow Open deck Deck house, mast, etc. Aft area, after peak
Failure type	Bent, buckled Burst Fractured Loose, parted Missing	Pitted Penetrated, holed Torn Not appropriate Insufficient	Out of range Outdated Unclean Worn Inaccessible
Immediate physical cause		Not installed Not in operation Accident damage Corrosion Erosion Material defect	Incorrect loading Normal wear Overload Fatigue Flooded

13.13.8 Causal Factors

The analysis process has so far concentrated on identifying the accidental events. The next step is now to give a diagnosis or answer as to *why* the casualty took place. Causal factors may be seen as conditions or actions *not directly* associated with the accidental event sequence and therefore will have taken place or arisen prior to the casualty. In other words, causal factors are the factors that put the vessel at risk.

Causal factors are related to the following two decision levels:

1. Daily operations on-board.
2. Management and resource-oriented decisions ashore.

The first level focuses on working conditions and organization of work on board. The main responsibility rests with the Master, department leaders and other supervisors.

Table 13.24. Human error parameters

Attribute	Code values		
Position	Master	Engineer	Travel repairman
	Mate	Electrician	Steward department
	Bosun	Engine crew	Longshoreman
	Deck crew	Pilot	Passenger
			Visitor
Task affected	Anchoring	Lubrication	Lookout
	Towing operation	Main engine operation	Monitor radar, ARPA
	Mooring	Aux. engine operation	Assess collision risk
	Cargo handling	Engine maintenance	Monitor instruments
	Cargo pumping	Deck maintenance	Set speed, r.p.m.
	Load plan	Cargo space maintenance	Set heading, rudder
	Stability control	Vessel command	Ship handling
	Close door	Trip planning	Give navigation order
	Secure hatch	Position fixing	Communicate
	Ballasting	Monitor wind, current	Radio communication
	Bunkering		Give signal
Performance mode	Detection	Analysis	Manual control
	Identification	Decision-making	Communication
	Perception	Activation	Ordering
Error type	Attempted	Inadequate	Omission
	By-passed	Exceeded, excessive	Ignored
	Commission	Improper	Overestimated
	Delayed	Imprudent	Underestimated
	Disregarded, ignored	Ineffective	Wrong timing

Table 13.25. Environmental effect attributes

Attribute	Code values		
Phenomenon	Wind	Channel effect	Fog, haze, smoke
	Wave	Hydrostatic head	Rain, snow, hail
	Current	Light	Icing
	Shallow water	Whiteout	Debris
Impact	Drift, set	Reduced steering ability	Reduced visibility
	Impact	Unstable course	Inadequate light
	Roll, heave, etc.	Green/white water	Inflow, flooding
	Squat	List	Outflow

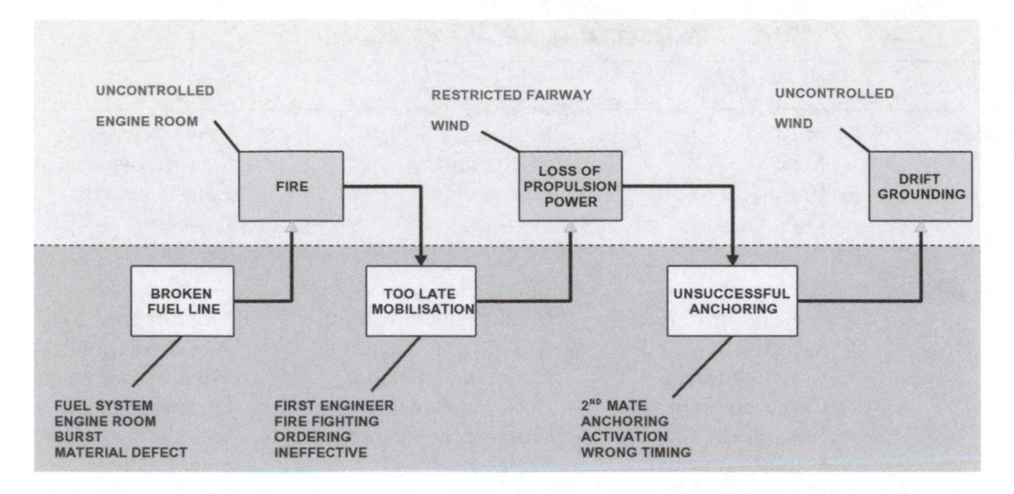

Figure 13.22. Modelling of accidental and casualty events.

The second level reflects the fact that safety critical decisions are also taken by the owning and managing company. They are responsible for setting up the organization and the management system and also for taking key decisions relating to the hiring of personnel and the acquisition of vessels.

Causal factors grouping related to conditions on board is outlined in Table 13.26 (p. 412). The causal factors associated with the supporting shore organization are summarized in the same manner in Table 13.27 (p. 413).

13.13.9 A Case on Causal Factors Representation

The analysis of casualty and accidental events was illustrated by a case example in Figure 13.22. We will develop this further here by showing how causal factors might be related to the accidental events (see Figure 13.24, p. 414). Let us take the first event, namely 'broken fuel line'. The analysis revealed that the failure could be traced back to 'lack of repair skills' and 'lack of instruction' of the crew member who did the repair task. It was further assessed that the faulty repair job was not detected due to an 'insufficient inspection programme'. The latter cause was traced back to an inadequate maintenance policy in the company. Similarly we found causal factors for the other two accidental events.

13.14 CASE-ORIENTED ANALYSIS

The focus on coding and statistical analysis of accident findings has been subject to criticism. Johnson (2000) argues that the learning process might be improved by focusing more on the individual report and emphasizing more the logic in the causal

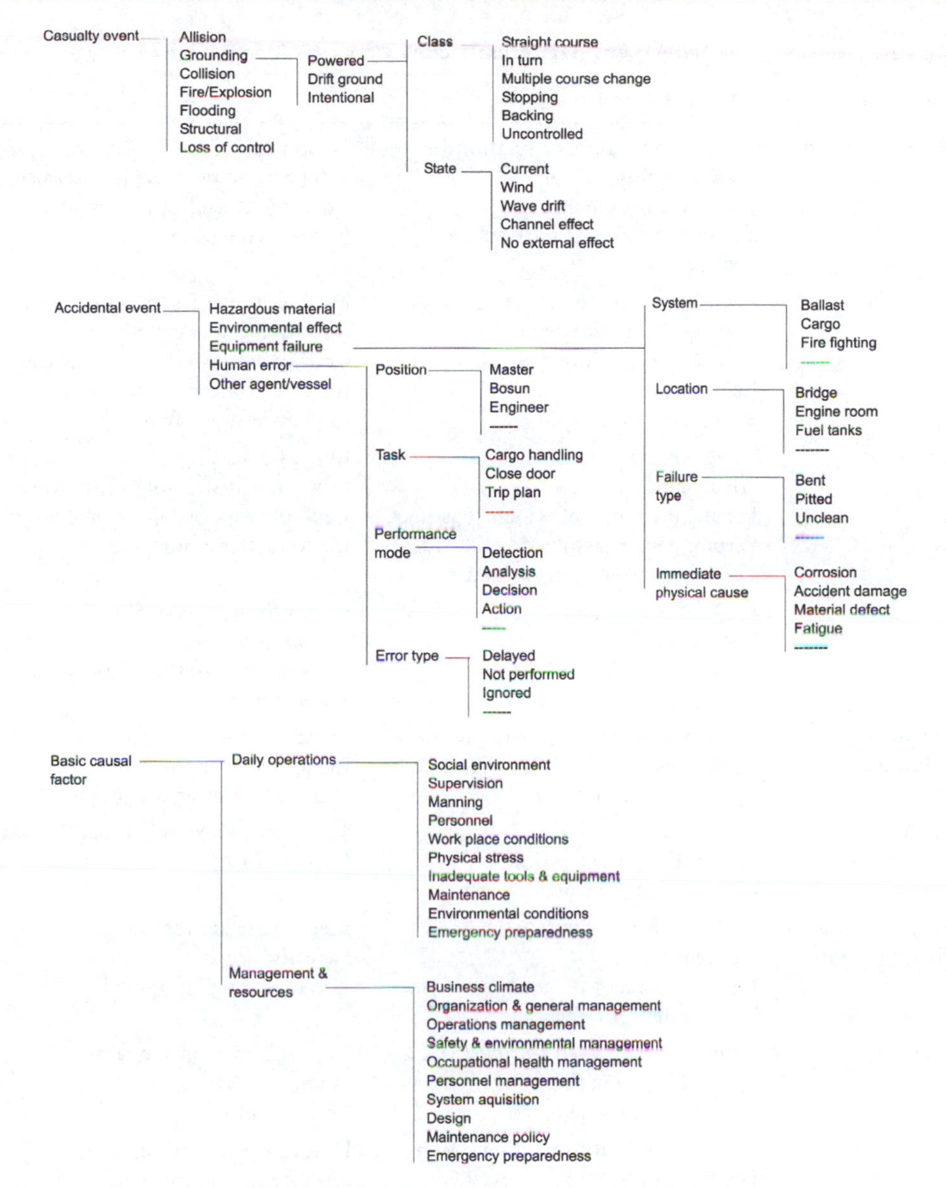

Figure 13.23. Coding taxonomy (rough outline).

analysis. He argues that each causal factor should be supported in a structured manner as follows:

- *Conclusion:* statement of the causal explanation.
- *Analysis:* analysis of factors and lines of argument that support the conclusion.
- *Evidence:* facts and observations that support the analysis.

Table 13.26. Basic causal factors, daily operation

Causal group	Causal factors	
Social environment	Labour–management relations LTA communication Language problem Social and cultural barriers and conflicts	Person-to-person conflict/animosity Safety awareness, cutting corners Cowboy attitudes, horseplay Resistance to change
Supervision	Lack of co-ordination of tasks Inadequate work preparation Inadequate briefing, instruction Lack of resources Supervisors not in touch	Expectations of supervisor is unclear Inadequate work method Conflicting orders, cross-pressure Inappropriate peer pressure Improper supervisory example
Manning	Long working periods, much overtime Frequent change of watch schedule Wrong person assigned Too high/too low work load	Idleness, waiting Low job satisfaction, monotony Lack of responsibility for own job Inadequate manning
Personnel	Lack of motivation Lack of skill Lack of knowledge	LTA physical/physiological capability LTA mental and psychological state Sea sickness
Workplace conditions	Anthropometric factors/dimensions Lack of information, inadequately presented information	Display design, controls Inadequate illumination Hazardous/messy workplace
Physical stress	Noise, vibration Sea motion, acceleration Climate, temperature	Toxic substance, other health hazards Lack of oxygen
Inadequate tools and equipment	Right tools and equipment unavailable LTA assessment of needs and risks Inadequate tool or aid	Inadequate standards or specifications Use of wrong equipment
Maintenance	Failure not detected during IMR Lack of maintenance Inadequate maintenance	Improper performance of maintenance/repair System out of operation
Environmental conditions	Too low visibility for observation Traffic density hinders vessel control	Hindrances in the seaway Restricted fairway
Emergency preparedness	Contingency plans not updated Training ignored Lacks initiative to deal with emergencies	Inadequate control of life-saving equipment Lack of leadership Lack of information to passengers

Table 13.27. Causal factors related to management and resources

Causal group	Causal factors	
Business climate	Economic conditions Market change	Bad relations with other organization Extreme competition
Organization and general management	Policy, ethical values Focus on liability and punishment Communicate policy Set standard by example Company loyalty and commitment Response to feedback from employees Vessel undermanned	Support from land organization Too wide control span Authoritarian command style Unclear roles and responsibility Cross-pressure from schedule and economy Lack of communication and coordination
Operations management	Pressure to keep schedule and costs Inadequate procedures and checklists	No review of critical tasks/operations Management training
SE management	Critical system and cargo documentation Inspection Follow-up of non-conformities Incident reporting, analysis, improvement Work instruction	Concern for quality improvement Inadequate promotion of safety LTA safety plan and program LTA formal safety assessment, risk analysis
Occupational health management	Information about health risks Personal protective equipment Health control of personnel Workplace inspections	Substandard hygiene onboard LTA medical services provided Follow-up of programmes and plans No off-the-job safety policy
Personnel management	Hiring and selection policy Inadequate training programme Selection/training of officers	Control by use of overtime Opportunity for advancement High turnover, lack of continuity
System acquisition	Substandard components Substandard contractors Control of contractors	Verification of contract requirements Inadequate testing
Design	Deviation from standards/ specifications Inappropriate regulations Design error	LTA design verification LTA system review and evaluation LTA change management
Maintenance policy	Lack of priority to IMR Lack of competent repair personnel	LTA planning Lack of follow-up and compliance
Emergency preparedness	Emergency plans Emergency procedures Management training Crisis handling Maintenance of life-saving equipment	Inadequate fire-fighting equipment Emergency training programme Life-saving equipment Lack of decision support Lack of warning systems

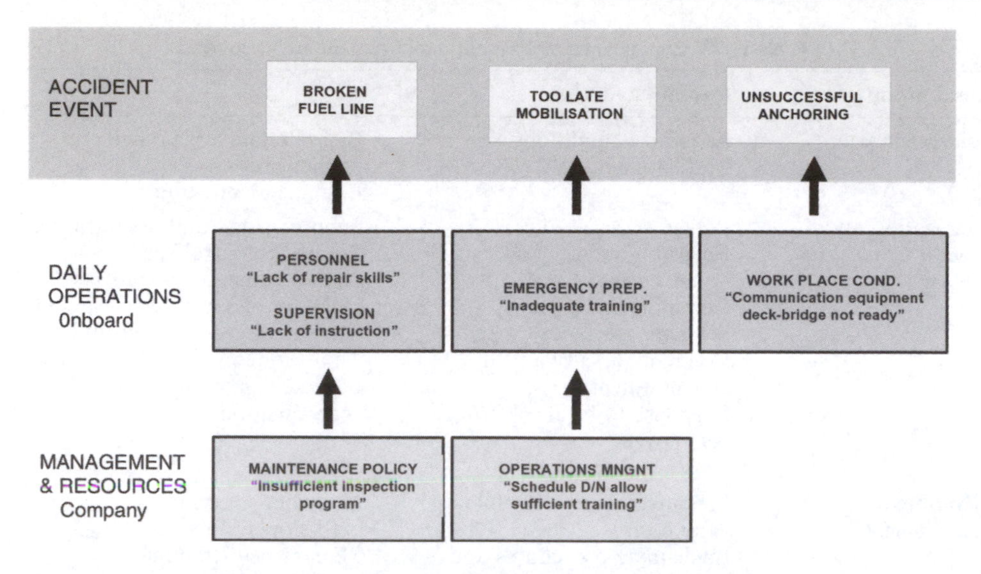

Figure 13.24. Causal factors related to accidental events.

In order to visualize the analysis, Johnson (2000) propose a so-called CAE diagram format. Figure 13.25 shows an example of a machine room engine fire. The cause of the fire was a result of technical failures and lax maintenance procedure (C2). This assessment was based on four lines of argument:

A6: Diesel oil sprayed from a shut-off cock.
A7: The spindle bonnet of the shut-off cock had vibrated loose.
A8: The handle of the shut-off cock had been released and resulted in removal of the locking function.
A9: Despite having spare assemblies the missing handles had not been replaced.

Each of these analysis elements is supported by evidence documented in the right-hand diagram blocks. Apart from bringing an overview of the analysis in the accident report, the CAE diagram is also a useful medium for integrating supplementary or contradictory evidence from other sources. Figure 13.26 shows how findings from the Chief Engineer's workbook indicate that fuel links had been detected earlier but not dealt with.

13.15 INCIDENT REPORTING

13.15.1 Motivation

There is a growing understanding that accidents represent a limited source for learning by the fact that they happen fairly infrequently. This raises the question of the potential of incidents and other non-conformities. A second argument for the study and analysis of less serious events is the very fact that the ISM Code states that all kinds of incidents and non-conformities should be used in the learning process.

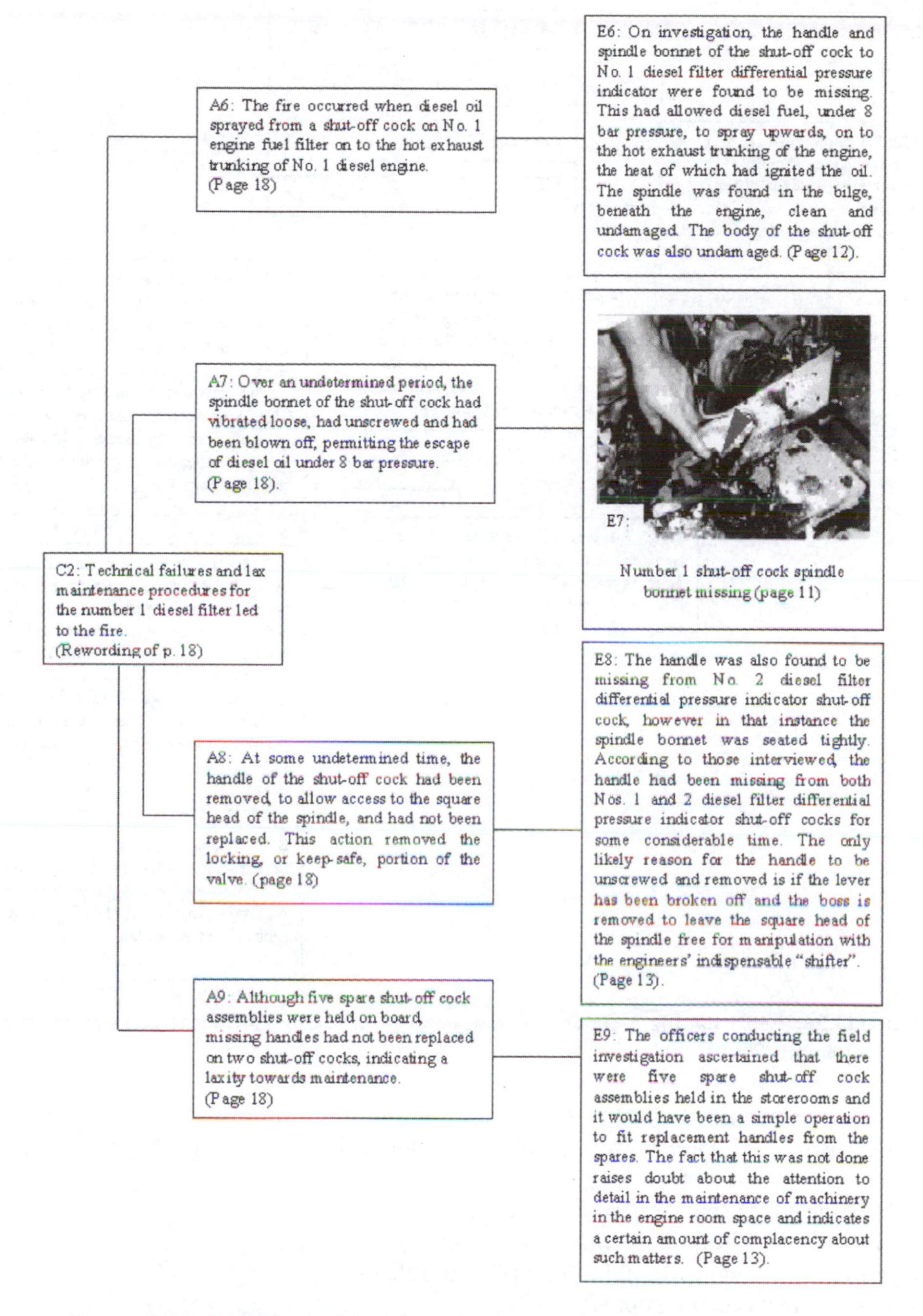

E6: On investigation, the handle and spindle bonnet of the shut-off cock to No. 1 diesel filter differential pressure indicator were found to be missing. This had allowed diesel fuel, under 8 bar pressure, to spray upwards, on to the hot exhaust trunking of the engine, the heat of which had ignited the oil. The spindle was found in the bilge, beneath the engine, clean and undamaged. The body of the shut-off cock was also undamaged. (Page 12).
A6: The fire occurred when diesel oil sprayed from a shut-off cock on No. 1 engine fuel filter on to the hot exhaust trunking of No. 1 diesel engine. (Page 18)
A7: Over an undetermined period, the spindle bonnet of the shut-off cock had vibrated loose, had unscrewed and had been blown off, permitting the escape of diesel oil under 8 bar pressure. (Page 18).
E7:
Number 1 shut-off cock spindle bonnet missing (page 11)
C2: Technical failures and lax maintenance procedures for the number 1 diesel filter led to the fire. (Rewording of p. 18)
E8: The handle was also found to be missing from No. 2 diesel filter differential pressure indicator shut-off cock, however in that instance the spindle bonnet was seated tightly. According to those interviewed, the handle had been missing from both Nos. 1 and 2 diesel filter differential pressure indicator shut-off cocks for some considerable time. The only likely reason for the handle to be unscrewed and removed is if the lever has been broken off and the boss is removed to leave the square head of the spindle free for manipulation with the engineers' indispensable "shifter". (Page 13).
A8: At some undetermined time, the handle of the shut-off cock had been removed, to allow access to the square head of the spindle, and had not been replaced. This action removed the locking, or keep-safe, portion of the valve. (page 18)
A9: Although five spare shut-off cock assemblies were held on board, missing handles had not been replaced on two shut-off cocks, indicating a laxity towards maintenance. (Page 18)
E9: The officers conducting the field investigation ascertained that there were five spare shut-off cock assemblies held in the storerooms and it would have been a simple operation to fit replacement handles from the spares. The fact that this was not done raises doubt about the attention to detail in the maintenance of machinery in the engine room space and indicates a certain amount of complacency about such matters. (Page 13).

Figure 13.25. CAE diagram of the causes of the fire. (Source: Johnson, 2000.)

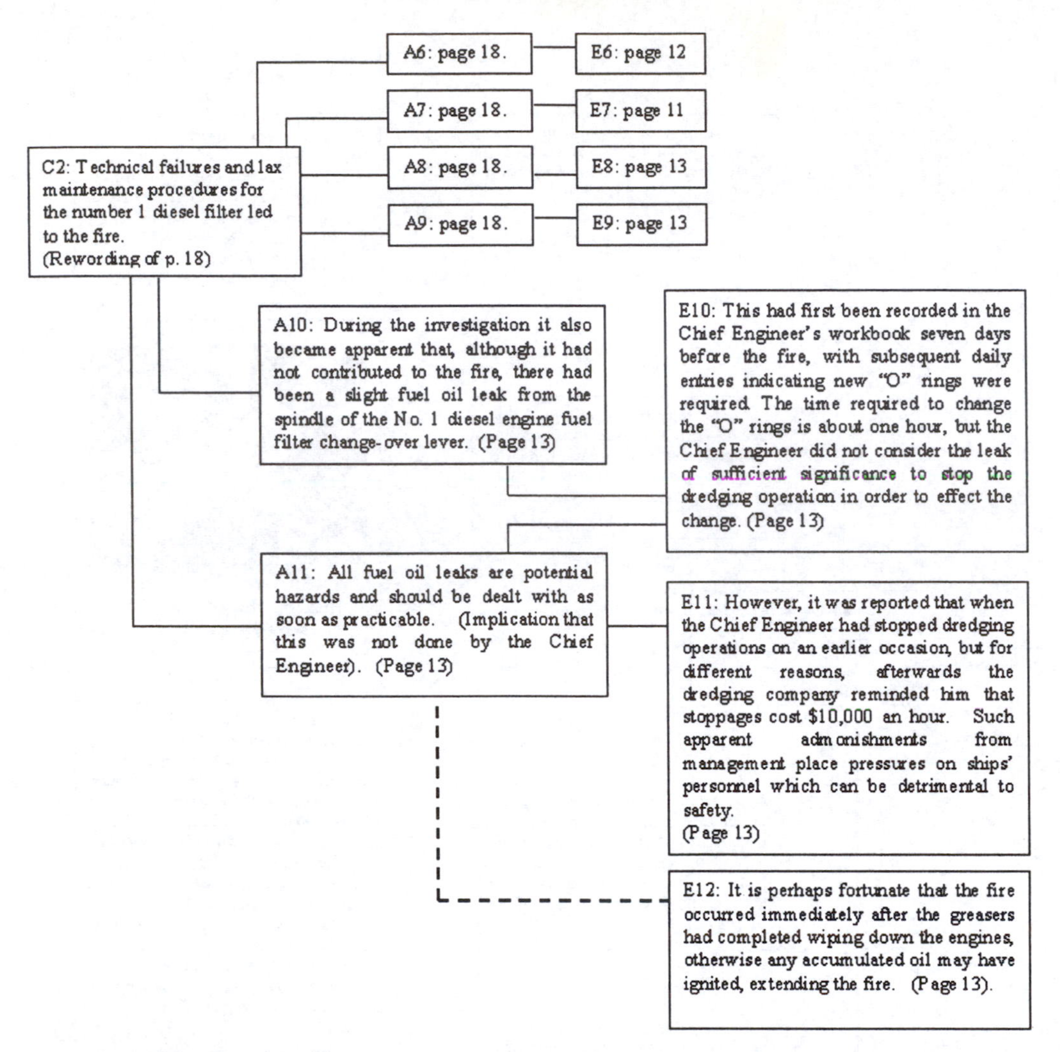

Figure 13.26. Representing secondary findings and contradictory evidence in an accident report. (Source: Johnson, 2000.)

Ferguson et al. (1999) lists the following alternative sources of information:

- Near-accident (near-miss) occurrences
- Hazardous situations/events
- Human or organizational relationship problems
- Lessons learned on safety
- Perceived safety problems or issues

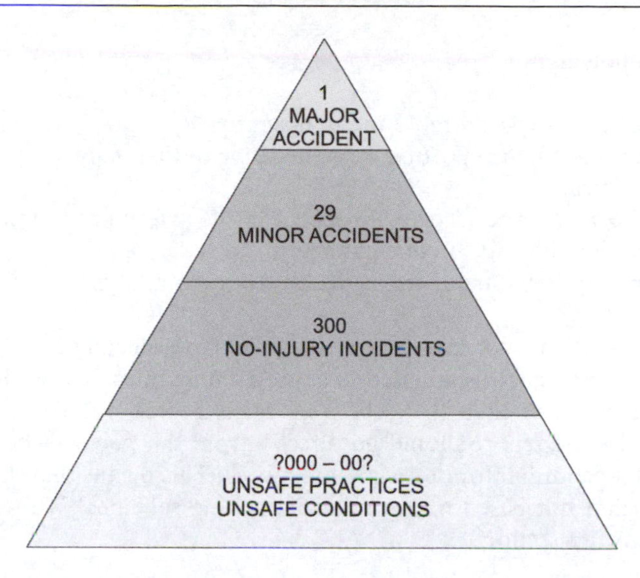

Figure 13.27. Iceberg theory (according to Heinrich, 1950).

There are different opinions about the frequency of incidents relative to accidents. Heinrich (1950) postulated that there are 300 no-injury accidents for every major accident (Figure 13.27). Ferguson et al. (1999) proposes that there are 100 incidents and 1000 safety situations for every accident. The figures in themselves are not critical as there will be different conditions in the different sectors of industry and transport, and there are also different definitions of incidents and situations.

A fundamental assumption is that incidents are basically an identical phenomenon to accidents. The only difference is that the sequence of events is interrupted before it leads to damage for incidents. This opinion has been challenged because the very difference is that incidents do not end as accidents. Incidents are in other word different phenomena and therefore have limited learning potential.

Incident reporting may be organized in two different forms:

- Open reporting within the shipping organization
- Confidential reporting by an external and independent organization (CHIRP)

13.15.2 Open Reporting

There is obvious potential for safety improvement in operating an incident reporting programme within the company. The main challenge is to motivate the crew to report every relevant incident or situation in a candid manner. This requires that the company arrange for the following:

- Top management have to promote the philosophy.
- Assure the employees that reporting is not negative to their career and does not lead to sanctions or reprisals.
- Demonstrate how the experience will be used in the safety programme.
- Incidents reported will lead to concrete action.
- Ensure that the reporters are given feedback from the analysis of the incident.

It is presently a fact that companies have trouble in establishing a culture of reporting. Indications from even quality-oriented shipping companies show that reporting is infrequent and much lower than the iceberg model indicates (Figure 13.27). In order to function adequately, the report should go directly from the person who experienced the incident to the safety coordinator or review committee. This means that the reporting form should be simple and not based on complicated coding schemes. We will return to this matter in the following section.

13.15.3 CHIRP

The Confidential Hazardous Incident Reporting Program (CHIRP) has been practised in the aviation industry for many years. As the name indicates, the identity of the reporter is kept confidential from the public. It represents a vast information source by the very fact that it has also involved international exchange of data. The UK CHIRP has been operated since 1982 and is organized as an independent charitable company (CHIRP, 2003). The system receives reports from pilots, air traffic controllers, licensed engineers and approved maintenance organizations. The findings are reported in the newsletter *Feedback* which has a circulation of 30,000.

The Marine Reporting Scheme (MARS) was set up in 1992 and has been operated by Captain R. Beedel. It is a voluntary and non-official system and has primarily engaged the maritime community in the UK. Reports are published in *Seaways*, the journal of the Nautical Institute. The main lessons learned are (Beedel, 1999):

- Inadequate bridge watchkeeping
- Did not know Collision Regulations
- Over-reliance on instruments
- Lack of cooperation between Master and pilot
- Poor bridge and control room design

13.15.4 Reporting Form

The example on page 419 indicates how an incident might be reported by means of a form. Notice that coding is kept to a minimum and that the description of the incident is given primarily in plain text. The layout of the form proposed by a US initiative is given by Ferguson et al. (1999).

Incident Report

1. This report is about (cross out):
    - ❑ Near accident
    - ❑ Dangerous condition, potential hazard
    - ❑ Observation, proposal for improvement
2. My position on-board:
    - ❑ Master
    - ❑ Chief mate
    - ❑ Engine officer
    - ❑ Engine crew
    - ❑ Deck crew
    - ❑ Combined position
    - ❑ Other position
3. The incident might have resulted in:
    - ❑ Allision
    - ❑ Collision
    - ❑ Grounding, stranding
    - ❑ Fire, explosion
    - ❑ Founder
    - ❑ Engine breakdown
    - ❑ Pollution
    - ❑ Fatality, injury
    - ❑ Other
4. When did it occur: Time and day
5. What kind of vessel:
6. Where did it take place (fairway category, place on-board)?
7. What kind of activity was going on (vessel phase, activity on-board)?
8. Describe incident or potential hazard:
    (What initiated the incident and what happened next? Indicate failures and errors made with reference to persons (role/position) and systems involved)
9. Describe particular physical circumstances:
    (Influenced by external conditions, visibility, sea state, traffic. Conditions on-board: Noise, vibration, climate)
10. What caused the accident?
    (Technical failure, lack of equipment. Human factors: Did not see, detect, wrong assessment of situation, and wrong action, and miscommunication. State or skills of individuals: Lack of motivation, fatigue, and competence. Organization: Watch system, supervision, work load, social and cultural factors, etc.)
11. What went right?
    (How was the potential accident avoided: Critical action or system function)
12. How can similar accidents be avoided?
    (Indicate technical, personnel-related or organizational measures. Who should address these measures?)

Processing of report

Information on where to send the report and its confidential treatment.
How to contact the reporter for supplementary information.

Table 13.28. Comparison of the planned maritime CHIRPs

Activity	US	UK	Norway	Finland
Planning of the system	Industry-based working group under SNAME	MAIB	MARINTEK	VTT
Funding	Industry and government	Government first 5 years Industry	Government	Government
Management	Maritime administration		Independent organization	Independent organization
Permanent staff	7 employees			3 employees
Reporting	Paper form	Paper form	Paper form	Paper form
Analyses	Trend analyses	Trend analyses Narrative text	Classification of incidents Possible causes Possible consequences Possible frequencies	Narrative text Classification of incidents Trend analyses
Status	Pending	Starting 2003	Pending	Pending

Source: Kristiansen et al. (2003).

13.15.5 Organization

A number of countries are presently in the process of organizing national systems for maritime CHIRP (Kristiansen et al., 2003; Ferguson et al., 1999). Table 13.28 gives a summary of the characteristics of some of the systems that are in the planning phase.

We will mention only a few factors that are vital for its success:

- Existence of general positive attitude to CHIRP within the industry: shipowners, managers, shipyards, authorities and crew organizations.
- Secure independence of the industry itself.
- The CHIRP organization has the necessary resources to make the system known and operate efficiently.
- Guarantee confidentiality and legal protection against criminal prosecution as a result of a report.
- Personnel acknowledge the fact that information about incidents may contribute to the prevention of serious accidents.
- Personnel have the motivation to take the time to report.
- Individuals are in a position to report.
- Being aware of the existence of the CHIRP system.
- Professional processing and analysis of reports.
- Relevant feedback in the form of reports, newsletter and statistics.

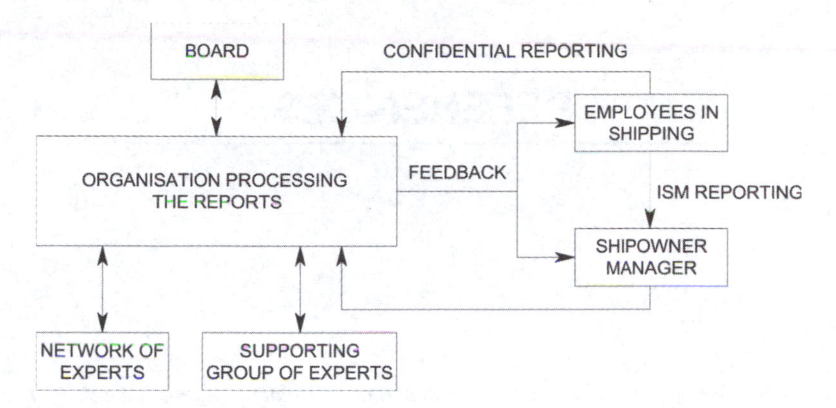

Figure 13.28. A schematic picture of the proposed organization for the Finnish confidential incident reporting system. (Source: Kristiansen et al., 2003.)

Figure 13.28 shows how it is envisaged that the Finnish maritime CHIRP organization will process the reports. The operation will be monitored by a board. The staff will be supported by experts in various fields due to the fact that it has to address a broad set of incidents and related phenomena.

REFERENCES

Anonymous, 1990, *Grounding of the US Tankship Exxon Valdez on Blight Reef, Prince William Sound near Valdez, Alaska.* National Transportation Safety Board, PB 90-916405.

Beedel, R., 1999, International Marine Accident Reporting Scheme MARS – Lessons Learned from MARS. Presentation at International Conference on Learning from Marine Incidents, 20–21 October, London. Royal Institution of Naval Architects, London.

Bird, F. E. and Germain, G. L., 1992, *Practical Loss Control Leadership.* International Loss Control Institute, Inc., Loganville, GA.

Brown, I. D., 1990, Accident reporting and analysis. In Wilson, J. R. and Corlett, E. N. (eds.), *Evaluation of Human Work: A Practical Ergonomics Methodology.* Taylor and Francis, London.

Caridis, P. A. et al., 1999, *State-of-the-Art in Marine Casualty Reporting, Data Processing and Analysis in EU Member States, the IMO and the United States.* CASMET Project Report C01.D.001.1.1. National Technical University of Athens, Department of Naval Architecture and Marine Engineering, Athens.

CHIRP, 2003, Homepage: www.chirp.co.uk

Dand, I. W., 1988, *Hydrodynamic Aspects of the Sinking of the Ferry 'Herald of Free Enterprise'.* RINA Transactions, London.

DOT, 1987, *MV Herald of Free Enterprise – Report of Court No. 8074 – Formal Investigation.* HMSO, London.

Ferguson, S. J. et al., 1999, IMISS – An International Maritime Information Safety System – The Next Frontier. Presentation at International Conference on Learning from Marine Incidents, 20–21 October, London. Royal Institution of Naval Architects, London.

Ferry, T. S., 1981, *Modern Accident Investigation and Analysis: An Executive Guide.* John Wiley, New York.

Hawkins, F. H., 1987, *Human Factors in Flight.* Gower Technical Press, Aldershot, UK.

Heinrich, H. W., 1950, *Industrial Accident Prevention.* McGraw-Hill, New York.

Hendrick, K. and Benner, L., 1987, *Investigating Accidents with STEP.* Marcel Dekker, New York.

Hill, S. G. and Byers, J. C., 1991, *Evaluation of the U.S. Coast Guard Taxonomy for Human-related Causes of Marine Accidents.* Interim Report, Human Factors Research Unit, Idaho National Engineering Laboratory, Idaho Falls.

Hollnagel, E., 1998, *Cognitive Reliability and Error Analysis Method – CREAM.* Elsevier Science, Oxford, UK.

Johnson, C. W., 2000, Viewpoints and bias in accident and incident reporting. In Cottam et al. (eds), *Foresight and Precaution: Proceedings of European Safety and Reliability Conference ESREL 2000*, Balkema, Rotterdam.

Kjellén, U. and Tinmannsvik, R. K., 1989, *SMORT – Säkerhetsanalys av industriell organisation* [In Swedish: Safety analysis of an industrial organization]. Arbetarskyddsnämnden, Stockholm.

Koster, E. et al., 1999, *CASMET – Casualty Analysis Methodology for Maritime Operations: A Review of Accident Analysis Methods and Database Structures*. Report No. ID C05.D.004, TNO Human Factors Research Institute, Soesterberg.

Kristiansen, S., 1995, An approach to systematic learning from accidents. *IMAS'95: Management and Operation of Ships – Practical Techniques for Today and Tomorrow*, 24–25 May, Institute of Marine Engineers, London.

Kristiansen, S. et al., 1999, A new methodology for marine casualty analysis accounting for human and organisational factors. Presentation at International Conference on Learning from Marine Incidents, 20–21 October. Royal Institution of Naval Architects, London.

Kristiansen, S. et al., 2003, *Final Report on Accident Data Base Implementation, CHIRP, Voyage Data Recorder and Inspection Information*. THEMES – Thematic Network for Safety Assessment of Waterborne Transport, Deliverable D3.4. European Community 'Competitive and Sustainable Growth' Programme, Brussels.

MAIB, 1998, *Memorandum on the Investigation of Marine Accidents*. Report, Marine Accident Investigation Branch, Southampton.

MAIB, 2001, *Database Taxonomy*. Report, Marine Accident Investigation Branch, Southampton.

NMD, 1996, *DAMA – Description of Coding Elements* [in Norwegian: *DAMA – Databank til sikring av maritime operasjon. Element beskrivelse. 1996. Sjøfartsdirektoratet*]. The Norwegian Maritime Directorate, Oslo.

Reason, J., 1990, *Human Error*. Cambridge University Press, Cambridge, UK.

Reason, J., 1997, *Managing the Risks of Organizational Accidents*. Ashgate, Aldershot, UK.

Rockwell, T. H. and Griffen, W. C., 1987, General aviation pilot error modelling – again? Proceedings of the Fourth International Symposium on Aviation Psychology, Columbus, Ohio.

USCG, 1998a, *Marine Safety Manual*, Vol. 1: *Administration and Management*. Chapter 12: Information and data systems. United States Coast Guard, Internet Home Page http://www.uscg.mil/hq/g-m/nmc/pubs/msm

USCG, 1998b, *Marine Safety Manual*, Vol. 5: *Investigations*. Chapter 3: Marine casualty investigations. United States Coast Guard, Internet Home Page http://www.uscg.mil/hq/g-m/nmc/pubs/msm

14

EMERGENCY PREPAREDNESS

Hell is the place where everything test perfectly and nothing works
("Campbell's Maxim")

14.1 INTRODUCTION

A substantial part of this book has so far focused on how to improve the safety of maritime activities, primarily through the implementation of risk-reducing measures. If the safety work is effective the number of accidents will be reduced, but the chance of an accident occurring will always exist as no activity or system is 100% reliable and safe, mainly because of human involvement. The role of emergency preparedness, also known as contingency planning, is that of being ready to take the necessary and correct actions in the undesirable event that an unforeseen accident occurs. Emergency preparedness also includes being ready to initiate proper mitigating actions in the event that incidents threaten to escalate into a major accident with serious consequences for people, property and the environment. One thing is definitely certain: many accidents could have had less serious consequences if the right actions had been made at the right time and in the right place. Emergency preparedness requires that one makes the necessary planning and training proactively, i.e. before something undesirable happens.

The lack of emergency preparedness was, for instance, devastatingly evident in the grounding of the oil tanker *Exxon Valdez* in Prince William Sound (Alaska) in 1989. The accident took place in protected water and under favourable weather conditions, and the catastrophic environmental and economic consequences of the accident were largely the result of an inadequate handling of the situation. The mobilization of necessary resources for the clean-up operation was seriously delayed, and in addition the co-ordination of the containment and clean-up operation was poor. The result is now a tragic chapter in maritime history: what was initially only a moderate-sized spill resulted in one of the worst man-made environmental disasters of modern times.

This chapter will examine some key aspects related to emergency preparedness. After a brief presentation of a few maritime accidents in which improved and appropriate emergency preparedness could have reduced the consequences, the focus will be on the

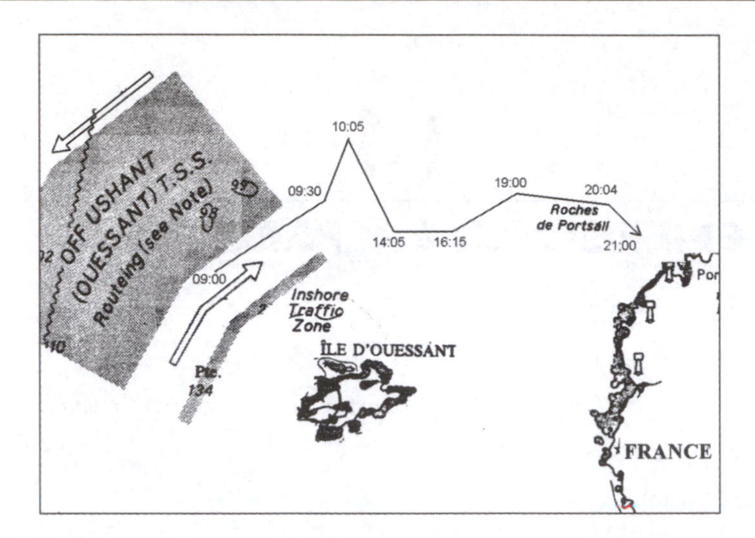

Figure 14.1. The drift route of the *Amoco Cadiz* from when it lost control of its rudder until it grounded off Portsall.

following main topics:

- Emergency and life-saving regulations (i.e. SOLAS, the ISM Code and STCW)
- Emergency preparedness activities and functions
- Human behaviour in catastrophes
- Evacuation risk and the importance of emergency preparedness in evacuations
- Evacuation simulation
- Pollution emergency planning

14.2 EXAMPLES OF ACCIDENTS

14.2.1 *Amoco Cadiz*

The VLCC (i.e. very large crude carrier) *Amoco Cadiz* was on a laden voyage when it lost its rudder control 10 miles off the coast of Brittany (France) on 16 March 1978. The loss of control happened at 09:46 and was due to failure of the steering engine (Hooke, 1989). The weather was harsh and the vessel immediately started drifting towards the shore. The Master of the *Amoco Cadiz* was not prepared for this very undesirable event, and over the course of the next couple of hours he made a number of fatal decisions that in the end contributed to the grounding off the village of Portsall almost 12 hours after the initial failure. Subsequently the vessel lost its integrity and broke up, resulting in the entire cargo of 223,000 tonnes of crude oil being spilt. Figure 14.1 illustrates the drift route of the *Amoco Cadiz* from when it lost control of its rudder until it grounded off Portsall. Figure 14.2 shows the loss of the *Amoco Cadiz*.

After the failure of the steering engine, an unsuccessful attempt was made to repair the engine. Given the harsh weather conditions, it was obvious that a salvage operation would

Figure 14.2. Loss of the *Amoco Cadiz*.
(Source: http://response.restoration.noaa.gov/photos/ships/images/cadiz.jpeg)

be complicated. By pure chance, the radio traffic from the *Amoco Cadiz* was intercepted by the ocean tug *Pacific*, which immediately started to steam towards the disabled vessel. After some delay, for reasons described below, the tanker was taken under tow at 14:25, but due to the hard weather the tow broke at 17:19. A second tow started at 20:35, but it was not able to take control of the drifting vessel and the vessel went aground.

After the steering engine failure the captain took a number of poor and directly unwise decisions that contributed to the loss of the vessel. These included the following:

- It took 1 hour and 45 minutes before a call for tug assistance was sent out.
- The main engine was stopped.
- It took 1 hour and 30 minutes to negotiate a towing contract. The Master of the *Amoco Cadiz* wanted to avoid Lloyd's Open Form, and the negotiations were complicated by language problems on both sides.
- The initiation of the second tow was inadequately co-ordinated.

Through research, initiated as a result of the accident, it was found that the vessel would have been easier to control had the propulsion and forward speed been maintained. Having the superstructure at the aft, *Amoco Cadiz* could have sailed into the wind, which was blowing towards land, and thereby kept the vessel offshore. It was also a tragedy that valuable time was lost as the Master was reluctant to accept the Lloyd's Open Form, primarily for economic reasons.

Having a single propeller and a single rudder, *Amoco Cadiz* was obviously at risk of the hazard of steering engine failure becoming reality. The tragic fact was that the preparedness for the emergency situation of steering engine failure was inadequate, and that such preparedness could have resulted in a much more desirable outcome.

14.2.2 *Capitaine Tasman*

One of the key reasons why incidents sometimes lead to serious accidents is that the seemingly non-serious initiating event is not handled with the necessary determination. This happened in the engine room fire onboard the cargo ship *Capitaine Tasman* (Cowley, 1994). The key events of this accident are outlined in Figure 14.3. It was the motorman who first detected smoke from the fuel oil heater. An attempt was made to extinguish the fire with the use of a powder unit, but when this failed the chief engineer was called upon. The chief sounded the general alarm and various measures were then taken to isolate the heater electrically, but these mainly failed. 45 minutes after the smoke was first detected, the fire was put out by a party of four firefighters in SCBA (i.e. self-contained breathing apparatus) by means of powder and foam. However, the fire re-ignited and the fire-fighting team had to return to fight the fire using water. The fire hose was left with spraying water in order to prevent further re-ignition. It was then observed that the fire had spread to the workshop above the heater, and only then was it decided to activate the CO_2 flooding system. Being without power and with empty SCBA bottles, it was finally decided to request tug assistance. Seven hours after the initial smoke detection a hose party quenched the local hot spots.

The response to the fire was inadequate in a number of ways:

- The general alarm was not sounded immediately
- Oil supply to the heater was not shut off
- Failure to isolate the fuel oil heater electrically

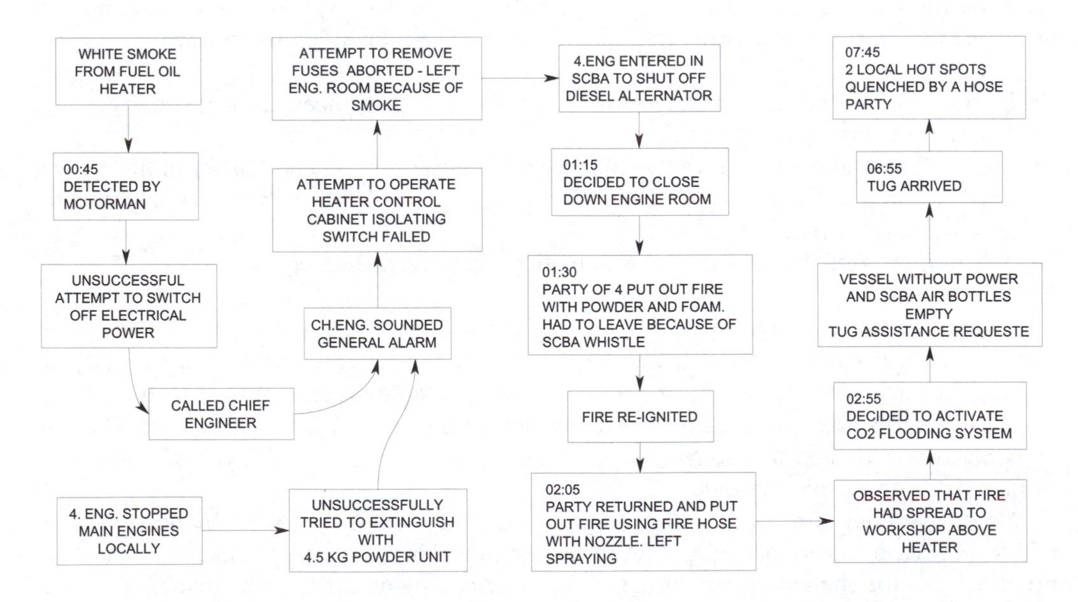

Figure 14.3. Sequence of events in the engine room fire onboard *Capitaine Tasman*.

- Delayed start of fire pumps
- Persistent use of portable equipment
- Hot spots and secondary fires were not detected

It can be concluded that the crew was never in control of the situation during the fire-fighting operation.

14.2.3 HSC *Sleipner*

The HSC *Sleipner* (HSC = high-speed craft) stranded on a small rock/shoal on the west coast of Norway on a dark autumn evening in 1999 (Anon., 2000). 45 minutes after the impact the seriously damaged vessel floated/slid off the rock, disintegrated, and sunk to about 150 metres depth. The evacuation of the 85 passengers and crew on board was totally out of control, and 69 people had to jump into the water when the vessel sank. The majority of these were picked up by nearby vessels, but a total of 16 persons perished. This accident demonstrated that inadequate certification of life-saving appliances and lack of emergency training can result in fatal consequences when an unexpected accident occurs. Although the accident was rooted in navigational failure, the dramatic nature of the consequences was, in addition to poor emergency preparedness and improper life-saving equipment, to a large degree a result of the heavy damage to the hull. In light of the potential impact forces and the extent of damage in accidents involving high-speed craft, the present design requirements for high-speed craft should be questioned.

The main events in the grounding/stranding of HSC *Sleipner* were as follows:

1. Stranding:
   - Damage to the bottom on both hulls
   - Water ingress, also in engine room
   - Progressive list
   - Starboard generator stopped
   - Port generator started but stopped almost immediately
   - Loss of internal communication
   - Transitional emergency power did not function
2. Attempt to release starboard liferafts:
   - Raft containers under the waterline
   - Did not release due to lack of hydrostatic release units
   - Manual release system did not function
   - The release system was fairly complicated
   - Lack of training in use of the system
3. Attempt to release port liferafts:
   - Fore unit did not release upon activation. It eventually released but overturned in the sea.
   - Aft unit released, but the container was filled with water
   - The release line was tangled and did not function

The failure to attain a safe evacuation was also related to the following factors:

1. The lifejackets were stowed in enclosed recesses.
2. The liferaft release system was brand new and inadequately tested.
3. The organization of the evacuation was chaotic due to shock and lack of training.

In the investigation that was initiated after the accident, critical remarks were made regarding the lifejackets. It was found that the lifejackets were difficult to put on, did not have a good fit, and had a tendency to slip off over the head. They also had limited buoyancy and thermal protection. In the aftermath of the *Sleipner* accident there have been discussions within the Norwegian maritime administration of whether the lifejackets should have been approved, despite the fact that the approval process was formally in order. HSC *Sleipner* was equipped with immersion suits for the crew, but the crew lacked knowledge of their existence, as well as training in the use of these immersion suits, and only a few succeeded in putting them on.

Given the fact that HSC *Sleipner* was certified to carry 380 passengers, it is not difficult to envisage the potential for a major catastrophe under these circumstances. After the accident the shipowner and operator of HSC *Sleipner* were heavily criticized for inadequate safety management and lack of emergency preparedness. Without doubt, proper execution of these activities would have reduced the consequences of the accident.

14.3 EMERGENCY AND LIFE-SAVING REGULATIONS

14.3.1 SOLAS

A key section of the SOLAS (i.e. the International Convention for the Safety of Life at Sea) regulation is Chapter III: Life-saving appliances and arrangements. This chapter is organized into three parts (IMO, 2001a):

A. General
B. Ship requirements
C. Life-saving appliance requirements

The content of Chapter III is outlined in Table 14.1. The regulation has special requirements for passenger ships on top of the general provisions for cargo ships. The regulation focuses on two main aspects, namely design requirements and guidelines for operation.

As for IMO's regulations in general, the main criticism of SOLAS has focused on the following aspects:

- Too much concern about the technical details of life-saving appliances and systems.
- Too little focus on the overall performance of life-saving appliances, i.e. the ability to save people.
- Unrealistic testing conditions – primarily in calm weather in protected waters.

Table 14.1. SOLAS Chapter III: Life-saving appliances and arrangements

Part	Content	
A. General	Application of the regulation Definitions Approval	Evaluation, testing Exemptions Production tests
B. Ship requirements • Section I – All ships • Section II – Passenger ships	Communications Personal life-saving appliances Survival craft, rescue boats Stowage of craft Rescue boat embarkation Manning of survival craft Muster stations Muster lists	Launching stations Launching arrangements Line-throwing appliances Operating instructions Abandon ship training Operational readiness Maintenance Inspections
C. Life-saving appliance requirements	Lifebuoys Lifejackets Immersion suits Thermal protective aids Requirements for liferafts Requirements for lifeboats Rescue boats Flares, smoke signals	General emergency alarm system Training manual Instruction for onboard maintenance Muster list and emergency Instructions

These aspects have led to some ambivalence among seafarers. They know that the risk of evacuation is high, but on the other hand they see no point in training with inadequate systems under unrealistic conditions. In addition, the average passenger seems to have an unrealistic perception of the effectives of evacuation and life-saving appliances and systems. This is shown by the shock and anger found among the general public in the aftermath of maritime catastrophes such as the loss of *Herald of Free Enterprise*, *Scandinavian Star* and *Estonia*.

The increasing number of high-speed ferries and high-capacity cruise vessels has brought to the surface the problem of inadequate approaches for the verification of evacuation systems. There are obvious ethical problems related to realistic full-scale testing of such systems, most importantly the significant risk of injury in such testing. In this context the computer simulation approach to testing has aroused considerable interest. The simulation approach is based on models of the vessel arrangement and the flow of people towards mustering and lifeboat stations. The approach is very much dependent on the ability to model and simulate both individuals and crowd behaviour in emergency situations. As a response to this situation, IMO has introduced regulations that address the use of simulation approaches in the assessment of life-saving effectiveness. We will return to this later in the chapter.

14.3.2 ISM Code: Emergency Preparedness

The International Safety Management (ISM) Code focuses primarily on the implementation of systematic safety management but also has a chapter on emergency preparedness (IMO, 1994). The ISM Code is now incorporated as Chapter IX in the SOLAS Convention (IMO, 2001a). The requirements of the ISM Code were adopted by the IMO in 1993 through Resolution A.741(18). Guidelines on the implementation of ISM are found in Resolution A.788(19) (IMO, 1995).

Chapter 8, 'Emergency Preparedness', in the ISM Code states the following:

- 8.1: The Company should establish procedures to identify, describe and respond to potential emergency shipboard situations.
- 8.2: The Company should establish programmes for drills and exercises to prepare for emergency actions.
- 8.3: The safety management system should provide for measures ensuring that the Company's organization can respond at any time to hazards, accidents and emergency situations involving its ships.

It is clear that these requirements go much further than the SOLAS regulations in the sense that the company has to identify potential emergency situations and respond to these, and not only equip its vessels in accordance with certain standardized (prescriptive) requirements. These regulations also indicate that a shipping company or manager should undertake contingency planning in terms of the following aspects (ICS, 1994):

- Duties of personnel
- Procedures and checklists
- Lists of contacts, reporting methods
- Actions to be taken in different situations
- Emergency drills

The ISM Code introduces proactive safety management with regard to emergency preparedness. The Code is examined in greater detail elsewhere in this book.

14.3.3 STCW Requirements

STCW is short for the International Convention on Standards of Training, Certification and Watchkeeping for Seafarers. Chapter VI of the STCW Code specifies 'standards regarding emergency, occupational safety, medical care and survival functions' for crew members (IMO, 2002a). Key elements in securing minimum emergency preparedness are:

- Familiarization training:
  - Communicate with other persons onboard on elementary safety matters
  - Ensure understanding of safety information symbols, signs and alarm signals
  - Know what to do if:
    - a person falls overboard

    - fire or smoke is detected
    - the fire or abandon ship alarm is sounded
  - Identify muster and embarkation stations and emergency escape routes
  - Locate and learn how to use lifejackets
  - Learn how to raise the alarm
  - Have basic knowledge of the use of portable fire extinguishers
  - Learn to take immediate action upon encountering an accident or other medical emergencies before seeking further medical assistance onboard
  - Identify the location of fire- and watertight doors fitted in the particular ship

- Basic training for crew with designated safety or pollution prevention duties:
  - Personal survival techniques
  - Fire prevention and fire-fighting
  - Elementary first aid
  - Personal safety and social responsibilities

- Crew competence requirements:
  - Competence to undertake defined tasks, duties and responsibilities
  - Competence evaluation in accordance with accepted methods and criteria
  - Examination or continuous assessment as part of an approved training programme

14.4 EMERGENCY PREPAREDNESS ACTIVITIES AND FUNCTIONS

14.4.1 Planning

The ISM Code requires that emergency preparedness should be based on an identification of hazards, estimation of risks, and the introduction of safety (or risk reduction) measures. This requirement has obvious implications for how a company plans and prepares for emergency situations. Key activities in the planning process include the following:

- Risk assessment:
  - Identify/locate hazards
  - Outline accident scenarios
  - Estimate probabilities and consequences

- Establish resources:
  - Ship arrangement
  - Safety-related equipment and systems
  - Manning and safety functions

- Outline emergency plan objectives:
  - Evacuation
  - Safeguard people
  - Mobilization of rescue operations
  - Control and mitigation of incidents

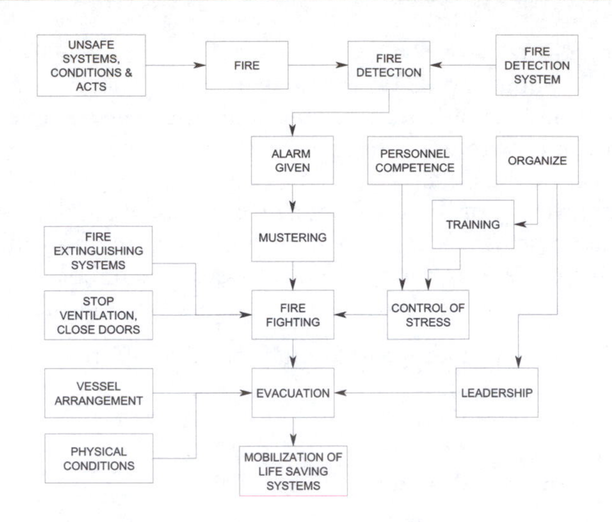

Figure 14.4. Key elements of a fire accident scenario.

  - Salvage of vessel
  - Rehabilitation of conditions onboard

- Maintain plan:
  - Train
  - Arrange drills/exercises
  - Audit/review plan

Risk assessment may be undertaken using well-established risk analysis techniques and accumulated experience. The description of the likely and relevant accident scenarios should emphasize both the development of events and the role of equipment and human resources. Figure 14.4 shows how the key elements of a fire accident scenario may be outlined as a basis for the planning process.

The emergency plan should cover all the main accident scenarios, which may include the following:

- Fire
- Explosion
- Collision
- Grounding/stranding
- Engine breakdown
- Disabled vessel, loss of power and control
- Cargo-related accidents
- Person overboard
- Emergency assistance to other ships
- First aid
- Unlawful acts threatening safety and security

With regard to emergency preparedness plans, it is necessary to focus on realistic scenarios that comprise many different event variations. The plans should be wide-ranging in scope, because focus on all possible variations at a detailed level may result in too many and complicated plans. Emergency preparedness plans must be based on a realistic time frame, and should take into consideration factors such as the likely speed of escalation, how damage may propagate, and possible energy releases. The key elements in an emergency plan are as follows:

1. Preface
2. Safety systems, life-saving equipment
3. Information systems, decision support systems
4. Organization of emergency teams, job descriptions
5. Distress signals
6. Information to crew and passengers
7. For each accident scenario:
   - Situation assessment
     - Category 1: Minor accident
     - Category 2: Alert situation
     - Category 3: Distress situation
   - Decision criteria
   - Defence and containment measures
   - External resources
8. Whom to contact depending on situation assessment
9. Evacuation plans:
   - Muster plan, boat stations
   - Evacuation routes
   - Information systems, control
   - Lifeboat/liferaft manning
10. Training:
    - Familiarization, basic training
    - Specialist training: fire-fighting, lifeboat coxswains, first aid
    - Drills
11. Revision of plan

14.4.2 Land Support

Experience has shown that the engagement of shipowner and manager is vital in the case of a serious accident. They should serve as a support and co-ordinator for the crew onboard and supply information on a continuous basis to families, official agencies and the press. The psychological effect of prompt and truthful information should not be underestimated. In an otherwise difficult or even tragic situation, this may have a considerable positive impact on the company's goodwill. An example of how the managing company sets up the organization ashore is given in Figure 14.5.

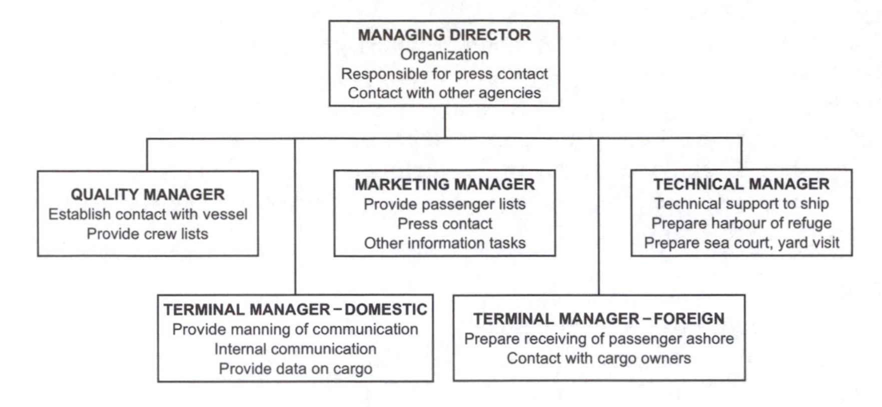

Figure 14.5. Land support team.

14.4.3 Decision Support

Regulation 29 in Chapter III of SOLAS states that a decision support system for emergency management shall be provided on the navigation bridge of passenger vessels. The system shall, as a minimum, consist of printed emergency plans. All foreseeable emergency situations shall be identified in the emergency plans. In addition to the printed emergency plans one may also accept the use of a computer-based decision support system (DSS) on the navigation bridge. A DSS provides all the information contained in the emergency plans, procedures and checklists. The DSS should also be able to present a list of recommended actions to be carried out in foreseeable emergencies. The main objectives of a DSS include:

- Issue warnings of dangerous situations and damage
- Detect critical trends
- Give a quick presentation of critical information
- Presentation of contingencies
- Enhance the overall understanding of the emergency situation

The emergency DSS shall have a uniform structure and be easy to use, and the following data from sensors and alarm systems might be presented in time series:

- Draught, heel, trim, freeboard
- Water level in tanks and compartments
- Status of watertight doors and fire doors
- Temperature and smoke concentration
- Status of all emergency systems

An emergency DSS may also integrate input from operators, external sources and static information such as hydrostatic calculations (curve sheet and stability). Figure 14.6 outlines the data structure for an integrated fire-fighting system (IFFS), which may be one

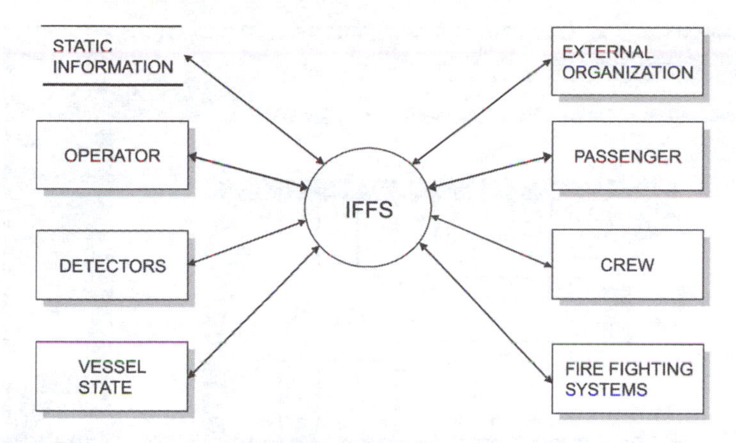

Figure 14.6. The data structure for an integrated fire-fighting system. (Source: Kristiansen, 1994.)

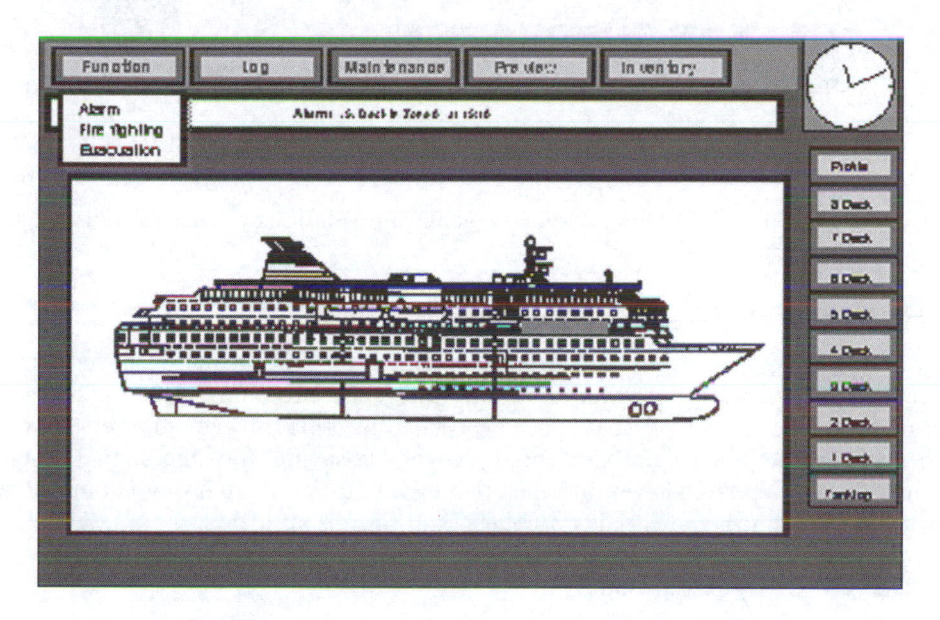

Figure 14.7. IFFS – localization of fire on deck plan. (Source: Kristiansen, 1994.)

module within a more comprehensive decision support system (Kristiansen, 1994). The IFFS may, for example, support the crew, provide information to passengers and external parties, and be used to control remotely operated firefighting systems.

A very important requirement for all decision support systems is fast and user-friendly input and output of information. The use of graphical interfaces in the DSS is therefore highly recommended. Two examples of how such graphical interfaces might be configured are given in Figures 14.7 and 14.8. In addition to presenting the instantaneous fire situation, the integrated firefighting system shown in the figures may also be used for

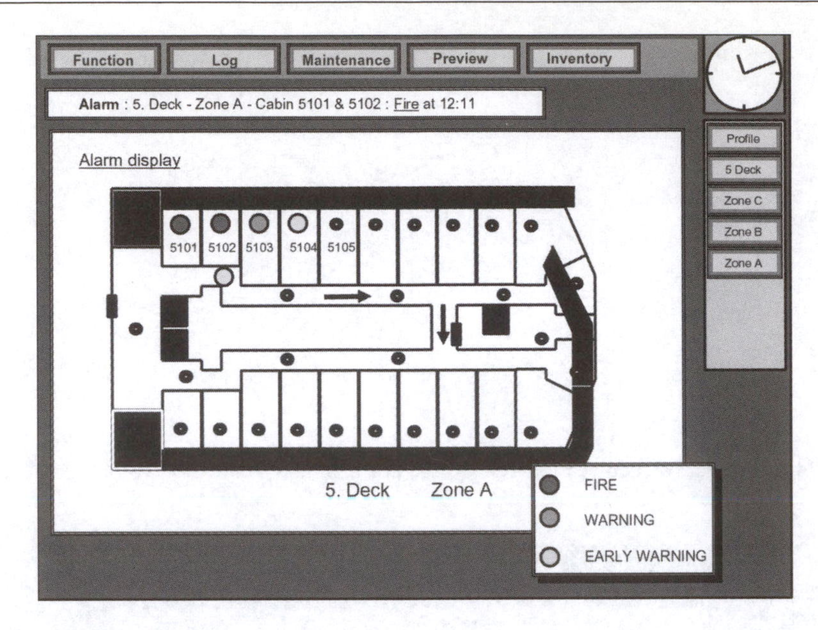

Figure 14.8. IFFS – localization of fire in section. (Source: Kristiansen, 1994.)

preview (prognosis) and maintenance of emergency information systems, as well as logging of events and developments.

14.5 HUMAN BEHAVIOUR IN CATASTROPHES

As a result of investigation of and research on catastrophic accidents we have improved our understanding of how humans react in emergency situations. This enables us to make realistic assessments of what can be expected in terms of evacuation effectiveness, which again gives us a better basis for design of life-saving appliances. In particular, investigations and research have confirmed that evacuation training is of great importance in terms of reducing the potential consequences of serious accidents.

14.5.1 General Characterization

Maritime accidents are normally of a very dramatic nature. Some important characteristics of emergency situations are presented in Table 14.2. The emergency situation is a function of the physical nature of the accident, the dramatic and uncontrolled development (or escalation) of events, and the perceived threat to people's own safety. The degree of drama in maritime accidents is further compounded by the degree of isolation that is experienced on a ship and the limited availability of assistance from external resources (e.g. other ships). The degree of rescue and salvage help is often limited or delayed, and there is nowhere to flee other than evacuating into the sea.

The degree of drama that can be experienced onboard a ship may be illustrated by the loss of the Ro-Ro passenger ferry *Estonia* (JAIC, 1997). The vessel sailed at a speed of

Table 14.2. Characteristics of emergency situations

Parameter	Characteristics
Accident nature	Degree of immediate threat to own life How fast the events develop or situations change
Lack of warning	People are unprepared for the next event High degree of uncertainty Influenced by rumours and 'hearsay'
Time pressure	Quickly changing situations
Drama	Degree of injury and number of fatalities Despair, fear and other stress reactions
Physical chaos	Trapped in enclosed areas, moving objects Blocking of escape routes Darkness, smoke
Vessel state	Fire, explosion, water ingress, sea motion Heeling, sinking
Threat to own life	Heat, lack of oxygen, drowning Impact from explosion, structural failure Extreme weather
Lack of control	Strong feeling of anxiety Impaired by own stress reactions and trauma Lack of information Lack of leadership, team spirit and solidarity Influenced by reactions of other people
Isolation	The vessel is an 'island' in the ocean No or limited assistance from other vessels or land-based resources Critical delay in rescue and salvage No safe haven: forced to evacuate into the sea

about 14 knots in bow seas, with a significant wave height of 4.3 metres, when the bow visor's three locking devices failed. When the bow door fell off the locks on the inner ramp failed, allowing water to flow into the vehicle deck. It has been estimated that the water inflow might have been in the order of 300–600 tonnes per minute. Because of free-surface effect the ship heeled to 30° within minutes, and simulations indicate that the vessel reached a heel of 60° in only 16 minutes. At 40° heel the water reached the windows on deck 4, probably resulting in progressive flooding. Owing to the rapid development of this event, no alarm was given nor was any evacuation organized from the bridge. Later studies indicated some local attempts to assist passengers, and that some passengers managed to reach the boat deck, although without being able to launch any lifeboats. Many of these saved their lives when they managed to get into the liferafts that were released as the ship sank. Almost immediately after the vessel started to heel, people had problems with leaving their cabins and movement in the narrow corridors (1.2 m wide) was

Table 14.3. Stress reactions in different phases of an emergency event/situation

Phase	Stress reactions
Pre-accident	Denial: 'This will not happen to me'
Warning	Denial, illusion of being invulnerable
Acute	Shock and stress reactions: alarm, psychosomatic, passiveness, uncontrolled behaviour
Intermediate	Development of syndromes: emotionally unstable, depression, guilt, isolation, over-reaction
Post-accident	Post-traumatic disturbance such as stress and neurosis, etc. Regaining emotional stability, control and good health Continued need of treatment

Source: based on Sund (1985).

difficult. Many passengers were trapped inside their cabins, and many of those that were able to get out of their cabins got stuck in staircases. Loose furniture and large objects also hampered movement in public spaces. It has been confirmed by many of those that survived that the people onboard were struck by well-known emergency reactions, ranging from panic to apathy, despite early attempts to take responsibility and assist each other. Of the 989 people onboard it is judged that only 300 reached the outer decks, and only 160 of these succeeded in boarding a liferaft or lifeboat when the vessel sank. In the end, helicopters or vessels picked up 138 people, giving a survival rate of only 14%.

Studies of stress reactions under emergency situations show that it is feasible to make a distinction between four phases of an emergency or catastrophic event/situation (Sund, 1985). These phases and corresponding stress reactions are presented in Table 14.3. The early phases of 'pre-accident' and 'warning' are characterized by denial and/or a feeling of being invulnerable. This may lead to a critical delay in the necessary response or fighting of the accident. In the 'acute' phase people are typically subject to more dramatic reactions such as shock, panic or becoming paralysed. If these reactions strike the majority of the crew and passengers onboard a ship, the consequences may be severely worsened. It is important to note that there is a risk of developing so-called post-traumatic reactions. This knowledge has led to a greater focus on treatment and counselling in the aftermath of accidents and catastrophes.

Research has shown that persons involved in emergencies have a limited ability to deal with challenges related to evacuation and salvage operations. As indicated in Table 14.4, people in emergencies have a tendency to become narrow-minded and stereotypic, and become unable to deal adequately with complicated problems. An immediate lesson of this fact should be to design simple evacuation systems and other life-saving appliances. For example, in a number of emergency situations there have been accidents related to the release of lifeboats, such as inability to activate the system and premature release leading to uncontrolled fall.

Sund (1985) has also given indicative numbers on the relative distribution of how people manage emergencies. These are presented in Table 14.5. The group that behaves

Table 14.4. Negative stress reactions

Behaviour	Characteristics
Sensing	Narrow-minded, selective focus 'Everything or nothing behaviour'
Cognition	Stereotypic, 'frozen attitude' Short-term oriented, loss of perspective/overview Unable to solve complicated problems
Reaction	Limited search for information Stereotypic behaviour Perseverance/persistence Impulsive or lamed

Table 14.5. Stress reactions in emergencies: distribution within a group

Part of population	Characteristic behaviour
10–30%	Behaviour balanced Realistic perception and assessment of situation Helps others in the group, able to co-operate Takes leadership, demonstrates initiative
50–75%	Light psychic lameness or apathy Slightly puzzled or confused Becomes active under leadership No need for medical help
10–25%	Strong psychic reactions needing medical treatment
1–3%	Loss of mental control Symptoms of serious nervous breakdown Acute mental disorder or panic

Source: based on Sund (1985).

optimally and takes leadership may be from 10% to 30% of the total. A larger group, of about 50–75% of the total, will be slightly reduced but will function reasonably well with adequate leadership. Strong psychic reactions can be seen for as much 25% of a group, while between 1% and 3% will lose mental control and/or experience nervous breakdown. In addition to the characteristics of the emergency situation under consideration, the following background factors may determine the degree of adequate/balanced behaviour:

- Earlier experience with emergencies
- Personality type and psychic health
- Duration of employment and age
- Leadership experience
- Intelligence

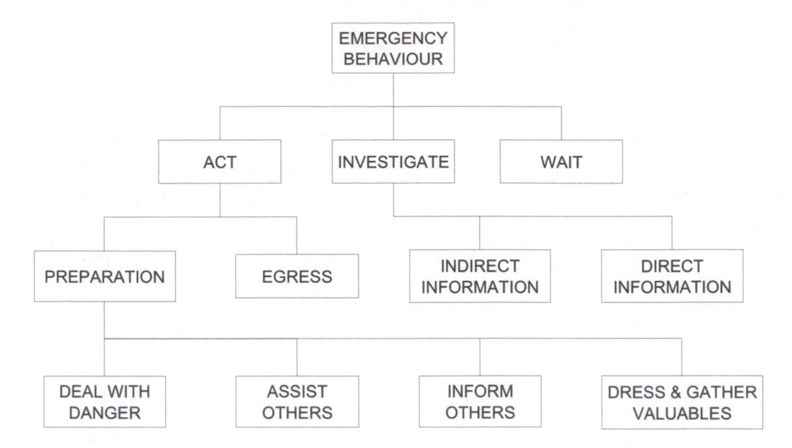

Figure 14.9. Hierarchical task structure of an emergency. (Source: Reisser-Weston, 1996.)

14.5.2 Emergency Behaviour

Considerable knowledge has been accumulated over time about concrete behaviour in emergency situations. Reisser-Weston (1996) proposes that one should see the total evacuation time as a function of the following phases:

- Detection
- Decision
- Non-evacuation behaviour
- Physical evacuation

The author has further proposed a task structure in emergency situations, which is outlined in Figure 14.9. In the event of an alarm one has three basic options: act, investigate or wait. Reisser-Weston further points out that the emergency behaviour is influenced or determined by a set of four so-called performance-shaping factors (PSF), which are presented in Table 14.6. Performance-shaping factors are factors assumed to have an effect on human behaviour.

Reisser-Weston has also summarized the results of a number of studies of human behaviour in emergency situations, primarily relating to fires in office buildings. Some of the findings of this study are briefly presented in Table 14.7.

14.6 EVACUATION RISK

One of the first investigations of the risks associated with evacuation from ships was undertaken by Pyman and Lyon (1985). The main findings of their investigation are summarized in Table 14.8. It was found that the average probability of a fatality (or several fatalities) occurring during an evacuation was in the order of 35%. The effect of weather conditions was, not surprisingly, found to be considerable. According to this study, the probability of fatalities occurring during evacuation is more than three times

Table 14.6. Performance-shaping factors (PSF) in an emergency situation

PSF	Description
Structural	Organization of the workplace Physical characteristics, rules
Effective	Emotional, cultural and social factors Behaviour is affected by stress, perceived risk, trust and cultural norms
Informational	Direct information Warning cues and information about escape routes Communication and advice
Task and resource characteristics	Possible conflict between current task or job function and the need for evacuation

Source: Reisser-Weston (1996).

Table 14.7. Human behaviour in emergency situations: summary of research findings for fires in office buildings

PSF	Findings
Informational	High-rise office building: 14% interpreted the alarm as genuine Informative warning system: 81% responded Many have to observe a fire directly in order to respond Tendency to investigate ambiguous signals further 45% were unable to differentiate fire alarm from other signals False alarms desensitize people
Effective	Investigated alarm signal: men: 15%, women: 6% Called fire department: men: 6%, women: 11% Got family together: men: 3%, women: 11% Women will warn others and evacuate, whereas men have a tendency to deal with the danger
Structural	Hospitals are hierarchical: individuals respond adequately in accordance with their position Persons with authority are critical for mobilizing large crowds Time to initiate evacuation without direction from staff: 8 min 15 sec Time to initiate evacuation when directed by staff: 2 min 15 sec In public places people are slower to break out of the normal routine

Source: Reisser-Weston (1996).

as high during hard weather conditions compared to calm or moderate weather. The relative number of fatalities during evacuation averages 14%, i.e. one in seven trying to evacuate does not survive. The effect of hard weather conditions on the number of fatalities is even more dramatic: approximately five times as high in hard weather compared to calm/moderate weather.

Table 14.8. Evacuation risk worldwide, 1970–80

	Merchant vessels (%)	Fishing vessels (%)
Accidents with fatality during evacuation	37	32
Fatalities among those who attempted evacuation	13	15
Hard weather accidents with fatality	78	73
Fatalities in hard weather evacuation	35	36
Calm/moderate weather accidents with fatality	16	26
Fatalities in calm/moderate weather evacuation	5	8
Fatalities among those in a fast sinking accident	86	—

Source: Pyman and Lyon (1985).

Table 14.9. Evacuation risk for bulk carriers

Type of accidental event	No. of events	Evacuation frequency (per ship year)	Fatalities	No. on board	Probability of fatality (%)
Collision	14	$3.1 \cdot 10^{-4}$	116	332	35
Contact	5	$1.1 \cdot 10^{-4}$	54	119	45
Fire/explosion	16	$3.6 \cdot 10^{-4}$	6	379	2
Foundered	51	$1.1 \cdot 10^{-3}$	618	1209	51
Hull failure	5	$1.1 \cdot 10^{-4}$	0	119	0
Machinery failure	1	$2.2 \cdot 10^{-5}$	0	24	0
Wrecked/stranded	23	$5.1 \cdot 10^{-4}$	0	545	0
Total	115	$2.6 \cdot 10^{-3}$	794	2727	29

Source: DNV (2001).

There has for quite some time been an increasing concern for the safety of bulk carriers due to an alarmingly high number of losses and many crew fatalities. At the 70th session of IMO's Maritime Safety Committee (MSC), the topic of life-saving appliances (LSAs) for bulk carriers was discussed, and it was decided to include LSAs as part of the formal safety assessment (FSA) process for these vessels (see Chapter 10). As a response to this, a comprehensive FSA project on LSAs for bulk carriers was performed in Norway (DNV, 2001), and as a part of this study the evacuation risk was estimated for this particular ship type based on reported evacuations for the period 1991–98. The results of this study are presented in Table 14.9.

The evacuation frequency for bulk carriers is $2.6 \cdot 10^{-3}$ per ship year. The average probability of fatality in evacuation is 29%, which is defined as the ratio of fatalities to the number of crew at risk. This figure is much higher than the earlier cited estimate for merchant ships in calm/moderate weather evacuation, but more comparable with the hard

Table 14.10. Evacuation of bulk carriers: probability of fatality for different evacuation methods

Evacuation method	Number of events	Fatalities	No. to be evacuated	Probability of fatality
Transferred by helicopter	8	17	219	0.078
Transferred to vessel	8	1	201	0.005
Lifeboat	4	57	112	0.509
then picked up by helicopter	0	—	—	—
then picked up by vessel	3	24	79	0.304
unknown further salvage	1	33	33	1.000
Liferaft	3	10	68	0.147
then picked up by helicopter	0	—	—	—
then picked up by vessel	3	10	68	0.147
Both lifeboat and liferaft	13	81	310	0.261
then picked up by helicopter	1	21	25	0.840
then picked up by vessel	9	4	209	0.019
then by helicopter and vessel	1	5	25	0.200
unknown further salvage	2	51	51	1.000
Direct into sea (wet evacuation)	6	99	127	0.780
then picked up by helicopter	0	—	—	—
then picked up by vessel	4	63	90	0.700
unknown further salvage	2	36	37	0.973
Transferred to helicopter and evacuation to survival craft	3	1	78	0.013
Transferred to helicopter and picked up by vessel	1	0	25	0.000

Source: DNV (2001).

weather figure (Pyman and Lyon, 1985). This supports the assessment that bulk carrier losses often happen under dramatic conditions such as hard weather and fast sinking.

The data material from which Table 14.9 was complied has also been analysed with respect to the effectiveness of different evacuation methods, and the results of this analysis are shown in Table 14.10 in terms of probability of fatality. The data indicate that direct transfer to another vessel is one of the safest evacuation means with a fatality rate of less than 1%. It is further clear that liferafts are safer than lifeboats. It should not be a big surprise that the least preferred method of evacuation is directly into the sea, i.e. so-called 'wet' evacuation.

One important piece of knowledge to be drawn from the evacuation risk data presented above is that evacuation is a very risky activity with a high probability of fatality. Given the high fatality rate in evacuation, it should be clear that appropriate emergency/evacuation training and preparedness is of essential importance for seafarers. Learned responses to critical events and situations, as well as a degree of familiarity with simulated situations (i.e. training), can be of significant importance in terms of saving lives in evacuation. This accounts for all phases of an evacuation, from calm and controlled behaviour at muster stations, to correct use of personal life-saving appliances such as lifejackets and immersion suits, proper use of lifeboats and liferafts, behaviour in these evacuation crafts, use of first aid, etc.

14.7 EVACUATION SIMULATION

SOLAS specifies the following maximum times for key evacuation phases on passenger ships:

- The maximum time from when abandon ship signal is given to when all survival craft are ready for evacuation is 30 minutes (Chapter III, Regulation 11).
- The maximum time for abandonment of mustered people is 30 minutes (Chapter III, Regulation 21.1.4).

The emerging computer simulation technology prompted IMO to develop standards for the adoption of such techniques in the assessment of evacuation effectiveness. In 2002 the Maritime Safety Committee (MSC) of IMO formally adopted the 'Interim guidelines for evacuation analysis of new and existing passenger ships including Ro-Ro' (IMO, 2002b). These guidelines only address the assembly/mustering part of the evacuation process and two scenarios are defined, namely day and night conditions. The SOLAS performance requirements are based on calm weather conditions, no list, and no effect of fire. It is evident that in harsh weather conditions, with list and/or the effects of fire, it will be much more difficult to achieve evacuation within the given performance requirements.

This section of the chapter will take a closer look at evacuation simulations. First, however, a brief introduction to crowd behaviour is given. Not considering crowd behaviour is considered to be a major shortcoming of many evacuation simulation models.

14.7.1 Crowd Behaviour

Jørgensen and May (2002) have discussed a number of important issues related to crowd behaviour. The authors point to the fact that evacuation simulation models basically estimate individual behaviour and more or less neglect so-called crowd behaviour, which according to them is a major shortcoming of these models. Jørgensen and May have defined the concept of group-binding, which expresses the fact that people both rationally and emotionally have an interest in finding their relatives before being evacuated. The crew ideally manages the mustering of crowds in an emergency situation, but due to group-binding people will often be non-compliant to the instructions given by the crew and instead focus on finding their relatives. The degree of group-binding will be a function of the social composition of the passengers: singles, couples, families, and groups of friends. The effect of group-binding will also be related to the size and arrangement of the vessel as well as the time of the day. Jørgensen and May studied the social composition of the passengers on different Danish vessels, and interviewed people about their willingness to disobey crew instructions. An average of 30% of the passengers would disobey crew instructions in order to find family members and other people they felt closely connected/related to. With an estimated probability of actually being separated of 0.2, the group-binding problem would affect 6% (i.e. $0.3 \cdot 0.2 = 0.06$) of the passengers in a given situation.

Jørgensen and May (2002) also discuss panic in relation to emergency situations. As cited earlier (see Table 14.5), it has been estimated that 1–3% of a group will panic and/or lose mental control. Jørgensen and May challenge this view based on psychiatric generalizations and the fact that crowd behaviour is not taken into consideration. Panic behaviour should also be seen as a sociological phenomenon, and they refer to work by Berlonghi (1993) who makes a distinction between different crowd phenomena:

- Passive crowd (e.g. spectators)
- Active crowds
  - Hostile crowd (e.g. mobs)
  - Escape crowds (often characterized by panic)
  - Acquisitive crowds (often characterized by craze)
  - Expressive crowds (often characterized by mass hysteria)

Other aspects of the realism of evacuation simulations are also discussed by Jørgensen and May (2002). First, to verify a numerical simulation model it is necessary to run large and full-scale evacuation exercises with people in actual ships. This, however, is impossible in practice, mainly for economic reasons. Secondly, in terms of arranging such evacuation exercises, it is problematic to make the exercise fully realistic, as people will not be influenced by the perception of danger and feeling of urgency that characterizes real emergency situations. It may also be dangerous to arrange such realistic evacuation exercises as real panic may arise.

14.7.2 Modelling the Evacuation Process

The evacuation of crew and passengers is a process involving a number of phases. In terms of modelling evacuation processes the following phases can be used:

1. Detection and acknowledgement of an emergency.
2. Sound the alarm.
3. Recognition (by people) of the alarm.
4. Collection of lifejacket, orientate oneself about the situation.
5. Search for and unite with family and friends.
6. Evacuate to safe place or mustering station.
7. Prepare and deploy survival craft or escape system.
8. Board survival craft.
9. Launch craft or leave the vessel.
10. Rescue by external resource.

The simulation approach replicates the evacuation process described above in the time domain. In a time-stepping mode the behaviour of each individual is estimated, taking into consideration the physical conditions and interaction with other people. The vessel is normally incorporated into the simulation model by a two-dimensional space grid. The result of the simulation is an estimation of the total time of evacuation. As pointed out by Galea et al. (2002), the simulation must address a number of aspects:

- *Configurational:* The physical layout and arrangement of the vessel with dimensions of rooms, corridors and stairways.
- *Environmental:* Environmental factors that affect people under the evacuation, such as list, ship motion, presence of debris, heat, smoke, toxic substances, etc.
- *Procedural:* Basic rules for the phases in the evacuation process, for example related to the guidance of passengers by crew, the organization at mustering stations, etc.
- *Behavioural:* Characteristics of how individuals behave and perform. The group of people onboard should reflect a realistic composition in terms of sex, age, walking speed and ability to respond adequately. Some of these attributes may be dynamic and change value during the evacuation.

The EXODUS numerical simulation tool (Galea et al., 2002) consists of a number of interacting program modules as illustrated in Figure 14.10. The model considers the interaction of people relative to other people, physical arrangement, the state of the vessel and fire threat. The models are rule-based and the behaviour of individuals is based on heuristic rules.

The 'behaviour' module in EXODUS is critical for the realism of the simulation. It controls how people respond to the changing situation and controls the 'movement' module. It functions on two levels, globally and locally, where the former addresses the decision on escape strategy and the latter determines behavioural responses and decisions made locally by individuals. For example, through the 'behaviour' module EXODUS reflects reactions to such phenomena as congestion and group ties.

The EXODUS model has a number of output formats that visualize the development of the evacuation. A so-called footfall contour map indicates the most heavily used routes

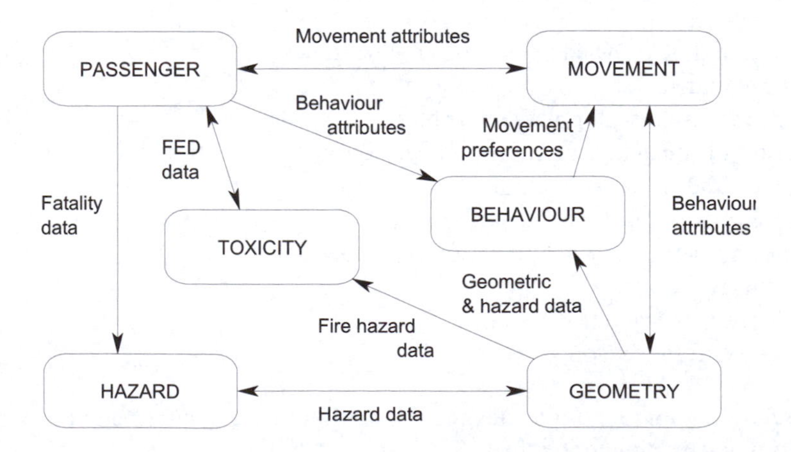

Figure 14.10. Module interactions in the EXODUS numerical simulation tool. (Source: Galea et al., 2002.)

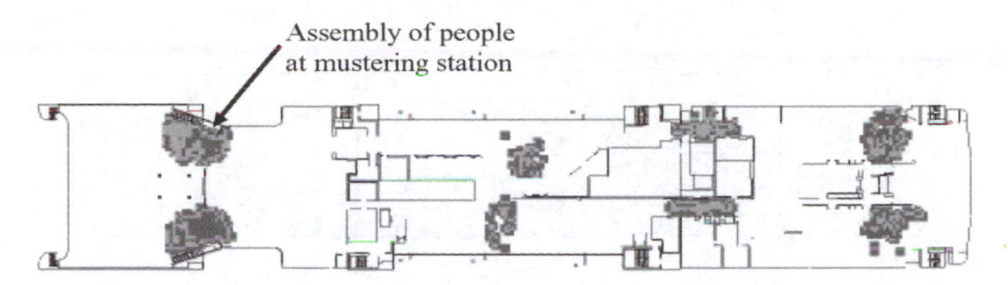

Figure 14.11. Population density at mustering stations. (Source: Galea et al., 2002.)

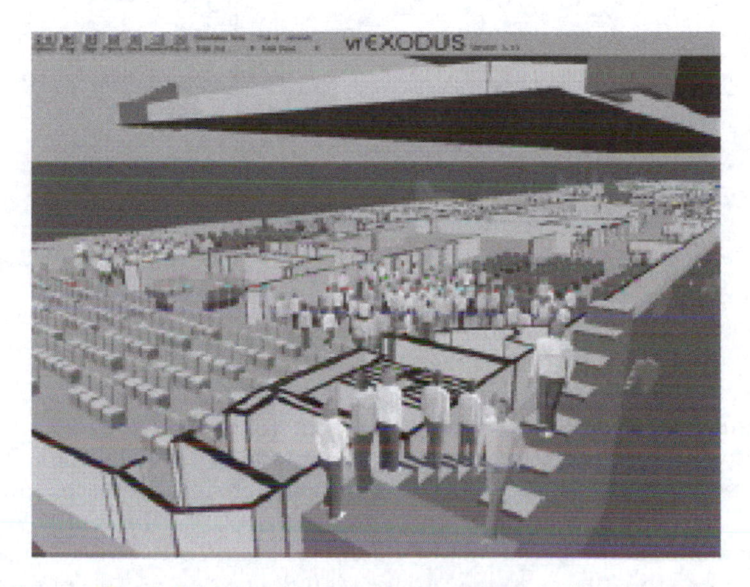

Figure 14.12. Live 3D view/presentation of the EXODUS simulation. (Source: Galea et al., 2002.)

that passengers use during the evacuation. Another format presents the final assembly and density of people at the designated mustering stations. This format is presented in Figure 14.11. The simulation package also offers an option where parts of the evacuation can be viewed 'live' in 3D as indicated in Figure 14.12, which makes it possible to study the effects of the ship's arrangement and potential 'bottlenecks'.

The effectiveness of an evacuation may be expressed by a number of variables:

- Times:
  - For individuals to muster
  - Total time used to muster and evacuate
  - Time wasted in congestions
  - Time to clear particular compartments or decks

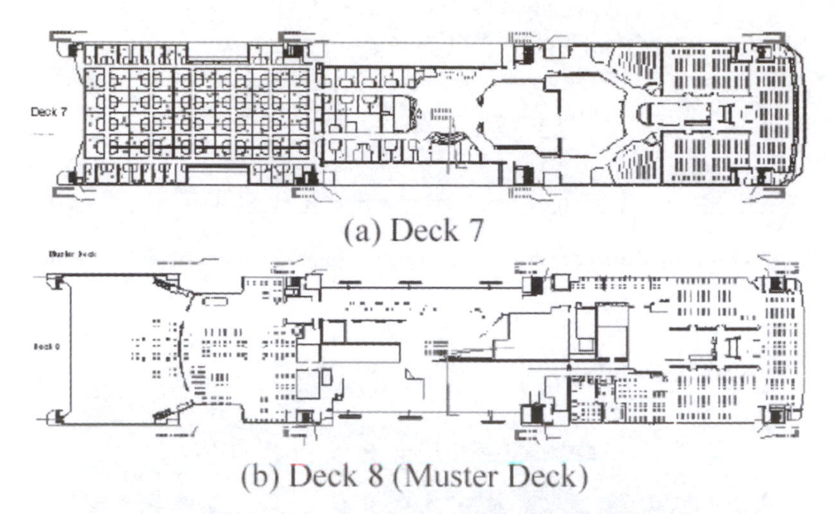

Figure 14.13. Deck arrangement. (Source: Galea et al., 2002.)

- Distance travelled by individuals
- Flow rate through doors or openings

An experimental evacuation exercise has been conducted for a so-called Thames pleasure boat with two decks (Galea et al., 2002). A total of 111 volunteers, from 16 to 65 years of age, were engaged in the exercise. 49 of these were located on the lower deck and 62 on the upper deck. For each deck there were four exits, two forward and two aft, and a twin set of staircases connected the two decks. During the experiment the vessel was moored with its starboard side to the jetty. Several evacuation tests were performed with different restrictions on access to the exits. The results from these evacuation exercises were then compared to the EXODUS simulation model where the predicted evacuation times were within 7% of the experiments.

14.7.3 A Simulation Case

The evacuation time for a large passenger ship with a total of 650 passengers has been estimated using the EXODUS simulation tool (Galea et al., 2002), and some interesting aspects of this simulation case are presented below. The vessel in question had ten decks divided into three vertical fire zones. The muster deck (i.e. deck 8) and the deck below are shown in Figure 14.13. The initial distribution of passengers within the ship before evacuation was as shown in Table 14.11. Passengers in fire zones 1 and 3 are assumed to be in their cabins. Fire zones 1 and 3 have four staircases each, with each staircase located in the far corners and only allowing a single lane of passengers. Fire zone 2 has a single centrally located staircase allowing two lanes of people to move.

Table 14.11. Initial location of passengers

Deck	Fire zone 1	Fire zone 2	Fire zone 3
6	172	28	
7	176	24	
8			
9			150
10			100

Source: Galea et al. (2002).

Table 14.12. Passenger travel speed (m/s) as specified by MSC Circular 1033 (IMO, 2002b)

	Age (years)/ impairment	Walking on flat terrain (m/s)	Walking downstairs (m/s)	Walking upstairs (m/s)
Female	< 30	1.24	0.75	0.63
	30–50	0.95	0.65	0.59
	50+	0.75	0.60	0.49
	Impaired 1	0.57	0.45	0.37
	Impaired 2	0.49	0.39	0.31
Male	< 30	1.48	1.01	0.67
	30–50	1.30	0.86	0.63
	50+	1.12	0.67	0.51
	Impaired 1	0.85	0.51	0.39
	Impaired 2	0.73	0.44	0.33

The simulation was done for night-time conditions with a response time from 7 to 13 minutes, allowing for people sleeping in their cabins to wake up and get dressed (an IMO requirement). Travel speeds are also specified by IMO, and these are summarized in Table 14.12. Depending on gender, age and degree of impairment, the travel speed in flat terrain varies by a factor of 3. The speed of walking downstairs is 30% lower than that on flat terrain, and less than 50% lower than walking upstairs. The IMO regulation specifies that the simulation should be run 50 times with random values. MSC Circular 1033 (IMO, 2002b) gives ranges of variation for the travel speeds cited in Table 14.12, and the range of variation is in the order of ±25%.

The estimated mustering times for an even keel (i.e. no heel) vessel condition are given in Table 14.13. The fact that fire zone 1 has the longest mustering time can be explained by the relatively high number of passengers, which may result in congestion, and that the evacuation includes walking up the stairs from deck 6 and 7 to the muster deck (i.e. deck 8).

IMO specifies that the dimensioning evacuation time should be taken as the highest value of four scenarios and an extra 10 minutes added to account for the assumptions and

Table 14.13. Range of mustering times with even keel vessel condition

Estimate	Fire zone 1	Fire zone 2	Fire zone 3
Minimum	14′ 59″	13′ 34″	13′ 42″
Average	15′ 32″	14′ 00″	14′ 32″
Maximum	15′ 58″	14′ 43″	15′ 24″

Source: Galea et al. (2002).

Table 14.14. Muster time in the EXODUS model for fire zone 1 at different degrees of heel

Heel	0°	10°	20°
Assembly time	15′ 32″	15′ 34″	16′ 00″

Source: Galea et al. (2002).

uncertainties in the model. The highest value in each scenario is to be taken as the 95th percentile. For the given scenario (see Table 14.13) the estimated maximum mustering time was 15 minutes and 58 seconds, and given an added safety margin of 10 minutes the regulation says that the vessel needs 25 minutes and 58 seconds to evacuate/muster.

Congestion (of people) in specific areas of the ship can be studied in the EXODUS model, and for the case described above there was congestion in the range of 2.2–3.5 persons/m^2 at the base of the staircases for 19 seconds. IMO defines a congested area as an area where there is a passenger density of 4 persons/m^2.

The effect of heel on the mustering time is not very significant in the present version of the EXODUS simulation model. Muster times for fire zone 1 are shown in Table 14.14 for 0°, 10° and 20° heel, respectively. With 10° heel the mustering time increases by only 2 seconds, and for a heel of 20° the increase in time is still marginal with 26 seconds. 20° heel is quite dramatic and makes it considerably more difficult to move around the vessel. The heel itself may, in addition, result in increased stress and even panic. The increase in mustering time will therefore most certainly be much larger than 26 seconds. Evacuation from partly capsized passenger vessels is discussed in more detail below. Table 14.14 confirms the inability of the EXODUS model to take factors such as change in human behaviour, physical chaos and potential loss of electricity because of heel into consideration. When using such simulation tools it is important always to have a clear understanding of the inherent limitations.

14.7.4 Evacuation from Partly Capsized Vessels

Planning and training for evacuation of passenger vessels normally assumes that the vessel is in an upright or only moderately heeled condition. However, experience shows that this is not necessarily the case in real emergency situations:

- *European Gateway* (1982): Collided with another ship off Felixstowe (England) as a result of confusion at a bend in the channel. The collision resulted in puncturing of the main vehicle deck and the generator room below the waterline. Because of asymmetric flooding the vessel started to heel, reaching 40° in only 3 minutes, at which point the bilge grounded. During the next 10–20 minutes the ship rolled on to its side. Six of the 70 people on board drowned.
- *Herald of Free Enterprise* (1987): Uncontrolled flooding through the open bow door resulted in rapid heel to 30° only minutes after leaving port at Zeebrugge (Belgium). Within 90 seconds the vessel heeled/capsized to 90°, at which point the side of the vessel was resting on the seabed in the shallow water. At least 193 passengers and crew died.
- *Estonia* (1994): Failure of the bow door and ramp in a severe storm in the north Baltic Sea led to rapid water ingress onto the vehicle deck. Because of free-surface effect the ship heeled to 30° within minutes, increasing to 90° only 20 minutes after the bow ramp opened. About 10 minutes later the ship sank completely, resulting in 852 fatalities.

According to Spouge (1996), the difficulty of evacuation increases dramatically when a vessel heels beyond 45°. The main causes of death in such situations are as follows:

- Falling headlong with extreme heel
- Shock of water immersion results in heart diseases or other paralysing illnesses
- Drowning due to rising water in compartments and inability to swim or escape (primarily to higher level)

After capsizing a vessel will usually come to rest in a stable position for a period of time. This will give some time for evacuation from inside the ship as well as rescue away from the ship. After a while, depending on the vessel's construction and the extent of damage, further water ingress will result in heel to 180° or sinking. Spouge (1996) has assessed the fatality risk for this accident scenario. The consequences of an evacuation of a passenger ferry, primarily consisting of large public spaces (i.e. type A), after capsize to 90° are summarized in Table 14.15. Of the estimated 45% fatality ratio, most people perished inside the vessel. The data in Table 14.15 also emphasize the importance of dry compared to wet evacuation. For a ship with cabins (i.e. type B) the fatality rate during night conditions will typically be 56%, considerably more than for type A vessels with mainly large open public spaces.

Spouge (1996) also proposed technical measures to improve the evacuation success rate. It was estimated that the survival rate could be improved by 3–7% for capsize scenarios beyond 45° if additional escape equipment and arrangement features were implemented. The following types of equipment/features were proposed: ladders, bridges, ropes, escape windows and elimination of full height partitions in public areas. In addition, limiting heel is considered very important in terms of saving lives.

Table 14.15. Fatalities and survivors on a passenger ferry for short crossings in the case of 90° capsize

Outcome	Relative number of people (%)
Killed by fall to side of compartment	5
Killed by shock of immersion	4
Drowned in rising water	26
Drowned awaiting rescue	10
Total fatalities	45
Escaped on own	3
Rescued from dry by survivors	23
Rescued from water by survivors	1
Rescued from dry by rescuers	26
Rescued from water by rescuers	2
Total surviving	55

Source: Spouge (1996).

14.8 POLLUTION EMERGENCY PLANNING

14.8.1 Contingency Planning

MARPOL, the International Convention for the Prevention of Pollution from Ships (IMO, 1997), includes requirements on pollution emergency planning. MARPOL Regulation 26 of Annex I requires that all oil tankers of 150 grt and above, and all other ships greater than 400 grt, shall carry a Shipboard Oil Pollution Emergency Plan (SOPEP) approved by the Administration (IMO, 2001b). Regulation 16 of Annex II requires all ships of 150 grt and above, certified to carry noxious liquid substances in bulk, to have onboard a pollution emergency plan. Ships to which both regulations apply may have a combined plan called a Shipboard Marine Pollution Emergency Plan (SMPEP). Such a pollution emergency plan should cover the following four elements:

- Procedures for reporting pollution incidents.
- A listing of authorities to be notified.
- A detailed description of actions to be taken by the ship's crew to reduce or control a discharge of oil or a noxious liquid substance.
- Procedures for co-ordinating shipboard activities with national and local authorities.

Some countries (coastal states) define additional measures to be taken against marine pollution (ICS, 2002). All ships operating in the territorial waters of these countries must adhere to these requirements. For example, the US requires the following additional measures:

- The ship must identify and ensure, through contract or other approved means, the availability of private firefighting, salvage, lightering and clean-up resources.
- A qualified individual with full authority to implement the response plan, including the activation and funding of contracted clean-up resources, must be identified.
- Training and drill procedures shall be described.

IMO (2001b) and the International Chamber of Shipping (ICS, 2002) have published guidelines that may assist companies in setting up a SOPEP or SMPEP. The ICS guideline gives flowcharts that support the decision-making process during an emergency. Figure 14.14 shows the guideline for the reporting of a polluting discharge. MARPOL specifies in detail what kind of information has to be reported: ship description, position, nature of discharge, vessel damage, etc. The guidelines further list whom to contact in case of a polluting discharge:

- Coastal state
- Port: terminal master, agent, port authority, etc.
- Ship interest contacts:
  - Head office
  - Charterer
  - Classification society
  - P & I Club

Another important step in the pollution emergency plan is to mobilize the vessel's pollution prevention team, which involves all key officers onboard the vessel. The emergency plan should have detailed job descriptions for each team member. For a specific spill scenario the pollution emergency plan, which gives a description of measures to be taken, should be given in both plain text and as a checklist. The plan is to be categorized according to the source of the spill and the causes. Examples of an emergency plan from the ICS guideline are given in Tables 14.16 and 14.17 in plain text and as a checklist, respectively.

14.8.2 Organization

Maritime pollution accidents may directly involve a number of parties:

- Master and crew
- Shipping company
- Salvage vessel
- Port administration
- Firefighting brigade
- Pollution prevention agency, etc.

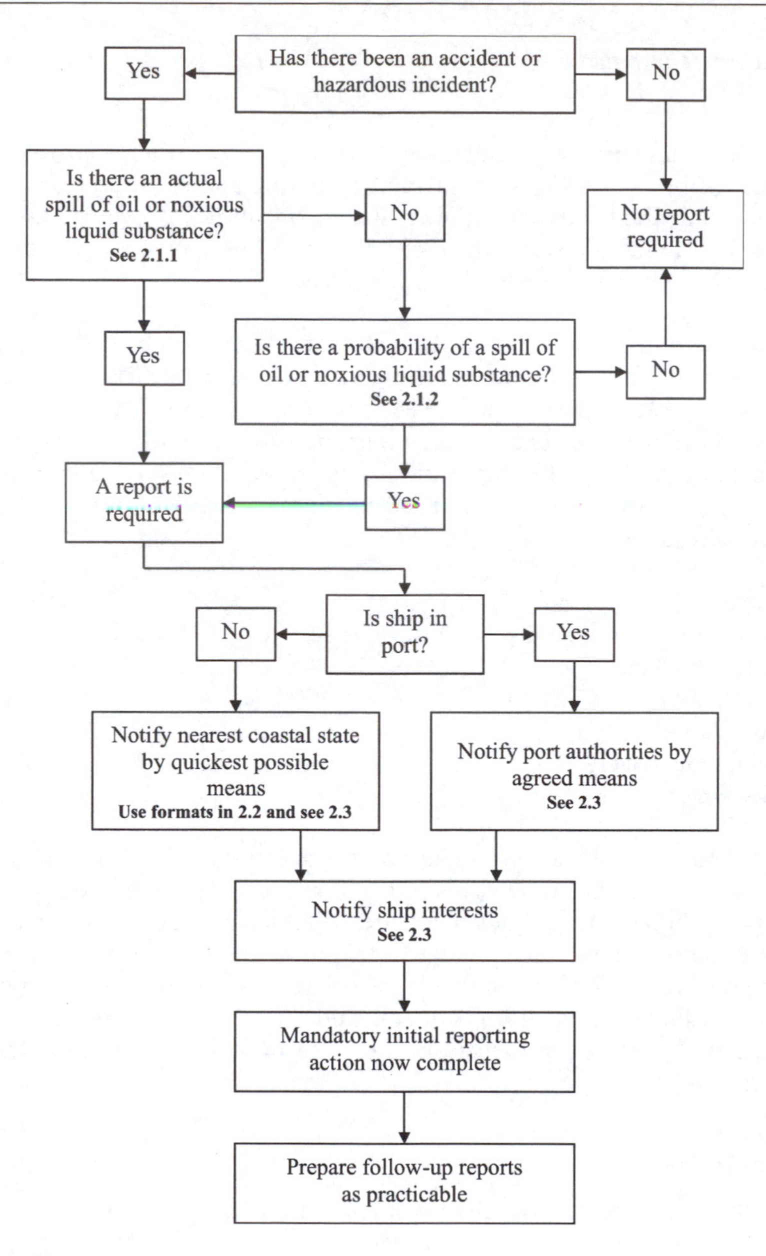

Figure 14.14. Decision-making process for the reporting of a polluting discharge. (Source: ICS, 2002.)

In addition, there are always several concerned parties in the case of maritime pollution accidents, such as the shipowner, cargo owner, local community (e.g. local fishermen) and government. Because of the interests involved, maritime pollution accidents often raise both political and legal issues with respect to overall management.

Table 14.16. Pollution emergency plan for the event of tank overflow during loading or bunkering: measures to be taken

3.1.2 Tank Overflow During Loading or Bunkering

Measures to be implemented immediately:

- Stop all cargo and bunkering operations, and close manifold valves
- Sound the emergency alarm, and initiate emergency response procedures
- Inform terminal/loading master/bunkering personnel about the incident

Further measures:

- Consider whether to stop air intake into accommodation and non-essential air intake to engine room
- In the case of a noxious liquid substance, consider what protection from vapour or liquid contact is necessary for the response team and for other crew members
- Consider mitigating activities such as decontamination of personnel who have been exposed
- Reduce the tank level by dropping cargo or bunkers into an empty or slack tank
- Prepare pumps for transfer of cargo/bunkers to shore if necessary
- Begin clean-up procedures
- Prepare portable pumps if it is possible to transfer the spilled liquid into a slack or empty tank.

If the spilled liquid is contained on board and can be handled by the pollution prevention team, then:

- Use absorbents and permissible solvents to clean up the liquid spilled on board.
- Ensure that any residues collected, and any contaminated absorbent materials used in the clean-up operation, are stored carefully prior to disposal.

Source: ICS (2002).

Table 14.17. Checklist for response to operational spill of oil or a noxious liquid substance

This checklist is intended for response guidance when dealing with a spill of oil or a noxious liquid substance during cargo or bunkering operations. Responsibility for action to deal with other emergencies which result from the liquid spill will be as laid down in existing plans, such as the Emergency Muster List.

Actions to be considered (Person responsible)	Action taken?	
	YES	NO
Immediate Action		
• Sound emergency alarm (Person discovering incident)	☐	☐
• Initiate ship's emergency response procedure (Officer on duty)	☐	☐
Initial Response		
• Stop all cargo and bunkering operations (Officer on duty)	☐	☐
• Close manifold valves (Officer on duty)	☐	☐
• Stop air intake to accommodation (Officer on duty)	☐	☐
• Stop non-essential air intake to machinery spaces (Engineer on duty)	☐	☐

(*continued*)

Table 14.17. Continued

This checklist is intended for response guidance when dealing with a spill of oil or a noxious liquid substance during cargo or bunkering operations. Responsibility for action to deal with other emergencies which result from the liquid spill will be as laid down in existing plans, such as the Emergency Muster List.

Actions to be considered (Person responsible)	Action taken?	
	YES	NO
• Locate source of leakage (Officer on duty)	☐	☐
• Close all tank valves and pipeline master valves (Officer on duty)	☐	☐
• Commence clean-up procedures using absorbents and permitted solvents (Chief Officer)	☐	☐
• Comply with reporting procedures (Responsible: Master)		
Secondary Response		
• Assess fire risk from release of flammable liquids or vapour (Chief Officer)	☐	☐
• Reduce liquid level in relevant tank by dropping into an empty or slack tank (Chief Officer)	☐	☐
• Reduce liquid levels in tanks in suspect area (Chief Officer)	☐	☐
• Drain affected pipeline to empty or slack tank (Chief Officer)	☐	☐
• Reduce inert gas pressure to zero (Chief Engineer)	☐	☐
• If leakage is at pump room sea-valve, relieve pipeline pressure (Chief Officer)	☐	☐
• Prepare pumps for transfer of liquid to other tanks or to shore or to lighter (Chief Engineer)	☐	☐
• Prepare portable pumps for transfer of spilt liquid to empty tank (Chief Engineer)	☐	☐
Further response		
• Consider mitigating activities to reduce effect of spilt liquid (Master)	☐	☐
• Pump water into leaking tank to create water cushion under oil or light chemical to prevent further loss (Chief Officer)	☐	☐
• If leakage is below waterline, arrange divers to investigate (Master)	☐	☐
• Calculate stresses and stability, requesting shore assistance if necessary (Chief Officer)	☐	☐
• Transfer cargo or bunkers to alleviate high stresses (Chief Officer)	☐	☐
• Designate stowage for residues from clean-up prior to disposal (Officer on duty)	☐	☐

Source: ICS (2002).

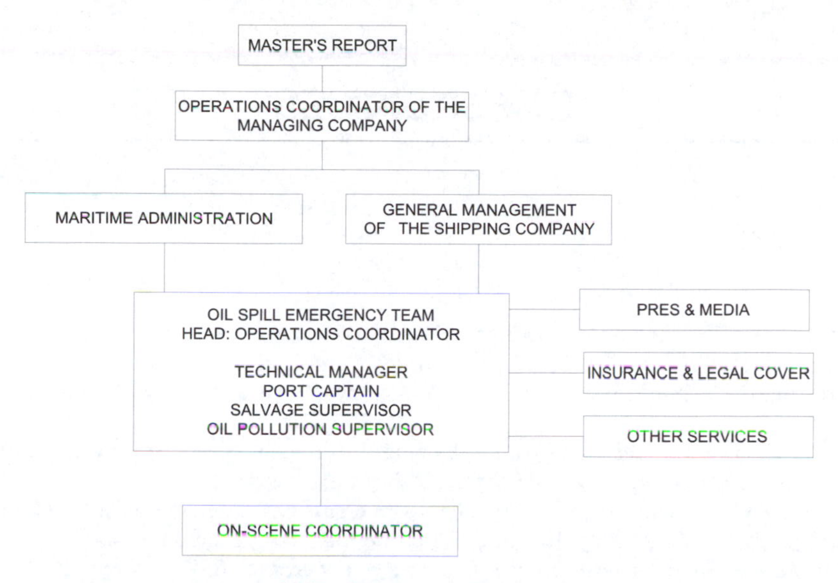

Figure 14.15. Oil spill response organization.

Figure 14.15 indicates how an oil spill accident may be organized in order to be both effective and well co-ordinated. The on-scene co-ordinator should have the necessary authority to direct the vessel and manage the available resources relating to salvage, oil spill containment and clean-up. In order to co-ordinate the mobilization of various resources, an oil spill emergency team should support the on-scene co-ordinator with representatives from the involved parties.

REFERENCES

Anonymous, 2000, *The High-Speed Craft MS Sleipner Disaster, 26 November 1999* [In Norwegian: *Hurtigbåten MS Sleipners forlis 26. November 1999. Norges offentlige utredninger NOU 2000: 31*]. Report from the Commission of Inquiry Appointed by the Royal Norwegian Ministry of Justice and Police, Oslo.

Berlonghi, A. E., 1993, Understanding and planning for different spectator crowds, in Smith and Dickie (eds.), *Engineering for Crowd Safety*. Elsevier, Amsterdam.

Cowley, J., 1994, Fire casualties and their regulatory repercussions. Conference IMAS 94, Fire Safety on Ships, London, 26–27 May. Institute of Marine Engineers, London.

DNV, 2001, *Formal Safety Assessment of Life Saving Appliances for Bulk Carriers (FSA/LSA/BC)*. Det Norske Veritas Strategic Research, Høvik, Norway.

Galea, E. R. et al., 2002, The development of an advanced ship evacuation simulation software product and associated large scale testing facility for the collection of human shipboard behaviour data. International Conference on Human Factors in Ship Design and Operation, 2–3 October. Royal Institution of Naval Architects, London.

Hooke, N., 1989, *Modern Shipping Disasters 1963–87*. Lloyd's of London Press, London.

ICS, 1994, *Guidelines on the Application of the IMO International Safety Management Code*. International Chamber of Shipping and the International Shipping Federation, London.

ICS, 2002, *Model Shipboard Marine Pollution Emergency Plan*. International Chamber of Shipping, London.

IMO, 1994, *International Safety Management Code (ISM Code): International Management Code for the Safe Operation of Ships and Pollution Prevention, 1994 Edition*. International Maritime Organization, London.

IMO, 1995, *Guidelines on Implementation of the International Safety Management (ISM) Code by Administrations*, Resolution A.788(19) adopted 23 November 1995.

IMO, 1997, *International Convention for the Prevention of Pollution from Ships 1973/1978, Consolidated Edition*. International Maritime Organization, London.

IMO, 2001a, *SOLAS Consolidated Edition 2001*. Consolidated text of the International Convention for the Safety of Life at Sea, 1974, and its Protocol of 1988: articles, annexes and certificates. International Maritime Organization, London.

IMO, 2001b, *Guidelines for the Development of Shipboard Marine Pollution Emergency Plans, 2001 Edition*. International Maritime Organization, London.

IMO, 2002a, *STCW – International Convention on Standards of Training, Certification and Watchkeeping for Seafarers*. International Maritime Organization, London.

IMO, 2002b, *MSC Circular 1033*, May.

JAIC, 1997, *Final Report on the Capsizing on 28 September 1994 in the Baltic Sea of the Ro-Ro Passenger Vessel Estonia*. The Joint Accident Investigation Commission of Estonia, Finland and Sweden, Edita, Helsinki.

Jørgensen, H. D. and May, M., 2002, Human factors management of passenger ship evacuation. International conference on Human Factors in Ship Design and Operation, 2–3 October. Royal Institution of Naval Architects, London.

Kristiansen, S., 1994, Development of an integrated fire-fighting system. Conference IMAS 94, Fire Safety on Ships, London, 26–27 May. Institute of Marine Engineers, London.

Pyman, M. A. F. and Lyon, P. R., 1985, Caualty rates in abandoning ships at sea. *Transactions of the Royal Institution of Naval Architects*, Vol. 127, 329–348.

Reisser-Weston, E., 1996, Simulating human behaviour in emergency evacuations. International Conference on Escape, Evacuation and Rescue – Design for the Future, 19–20 November. Royal Institution of Naval Architects, London.

Spouge, J., 1996, Escape from partly capsized ferries. *International Conference on Escape, Evacuation and Rescue – Design for the Future*, 19–20 November. Royal Institution of Naval Architects, London.

Sund, A., 1985, *Accidents, Catastrophes and Stress* [In Norwegian: *Ulykker, katastrofer og stress*]. Gyldendal Norsk Forlag, Oslo.

15

SAFETY MANAGEMENT

Any action for which there is no logical explanation will be deemed "company policy".

("Second Law of the Corporation")

15.1 INTRODUCTION

Broadly speaking, the objective of safety management is to ensure that activities such as ship operations, for example, are carried out safely and efficiently. Safety management should therefore be seen as an essential and integral element of the overall management system of an organization. This chapter will examine some important aspects of safety management and safety management systems. This should only be considered as an introduction to the topic, as safety management is a large and complex field of study that could itself be the subject of a whole book.

Four aspects of safety management are given special attention in this chapter. These are safety management in the context of Total Quality Management (TQM), the International Safety Management (ISM) Code utilized within shipping, the topic of auditing which is of great relevance for both TQM and the ISM Code, and finally the issue of safety cultures, which is briefly introduced. This introduction will try to establish a context for the discussion of these four important aspects of safety management. First, a brief discussion of the vulnerability of modern organizations is given. Secondly, different strategies on how accidents can be avoided or prevented are presented. Thirdly, the historic development of the maritime safety management regime is reviewed.

15.1.1 The Vulnerability of Modern Organizations

Large-scale accidents such as those at Flixborough (1974), Bhopal (1984) and Chernobyl (1986), the capsize of the Ro-Ro passenger and freight ship *Herald of Free Enterprise* (1987), and the loss of the space shuttle *Columbia* (1986), have been attributed to the increasing complexity of technology and organizations. According to Mitroff and Pauchant (1990), 'the potential for large-scale disasters seems literally to be built into the very fibre and fabric of modern civilization'. Mitroff and Pauchant illustrated this situation with the *Exxon Valdez* grounding accident, which resulted in massive oil spills and catastrophic environmental damage on the coast of Alaska. The company Exxon

Shipping was in for their judgement in deep trouble:

- Exxon Shipping (ES) was unable to make the necessary decisions to deal satisfactorily with the alcohol problem of the Master on *Exxon Valdez*.
- ES had just some weeks before the *Exxon Valdez* accident decided not to reassess their environmental preparedness plan, primarily because they had never had any major spills.
- It took almost one and a half days before ES started to respond to the *Exxon Valdez* oil spill.
- Experts said that the company was in what is called a 'restructuring blues' as the result of a 28% reduction of the workforce. Even company executives admitted that certain operations were 10% undermanned.
- An Exxon employee said: 'The more things change at Exxon ... the more they have stayed the same.'
- During the initiation of an internal inquiry into whether there was anything fundamentally wrong with the huge company, ES's president said: 'We don't believe there is ...'
- At a conference the president of ES denied that the management structure or culture was at fault. Instead he put the blame on individuals onboard the ship and the ship culture.

The analysis of Mitroff and Pauchant (1990) shows that crisis-prone companies are unable to admit in a frank and honest manner that the companies are far from perfect and professional. Instead they exhibit various defensive behaviours that are similar to those experienced by individuals in crisis situations, such as narrow-mindedness and focus on short-term survival.

15.1.2 Accident Prevention Strategies

Before discussing different ways of managing safety, it is useful to take a pragmatic look at how accidents can be avoided or prevented. Morone and Woodhouse (1986) have proposed the following five accident prevention strategies:

1. *Protect against the potential hazard:* If we accept that accidents are inevitable, a good strategy is to protect against the consequences of these accidents. For example, in order to prevent oil pollution from grounding and collision accidents with tankers, double hulls were introduced to withstand penetration of cargo oil tanks. Recently it has been discussed whether high-speed marine vessels should have better strength against impact from groundings and collisions to avoid rapid sinking resulting from such accidents. Such consequence-reducing measures are often very expensive to implement.
2. *Proceed cautiously:* One way to interpret this strategy is to prepare and plan for the worst possible accidents. This requires the use of, for example, formal safety assessment (FSA) in order to identify all hazards and accident scenarios. Another

interpretation is that new technical solutions and methods of operation should be considered dangerous until proven otherwise.

3. *Test the risks:* This strategy implies that simulations and testing should be undertaken under realistic conditions in order to assess how the system responds to certain conditions and situations. Important questions are whether realistic conditions can be prepared and whether such testing is acceptable. Evacuation and life-saving equipment/systems may be used as an example. The function and reliability of lifeboats and liferafts has been a topic of continuous discussion: Have the equipment/systems been tested under realistic conditions? Is it acceptable to test these systems with people under realistic conditions such as heavy seas, etc.?
4. *Learn from experience:* This principle has been generally accepted, but there is still some way to go before it is satisfactorily implemented. The strategy is a question about developing procedures for reporting, analysis and corrective response to non-conformities, incidents and accidents. The main challenge today is to take the least serious incidents into consideration. These incidents have the largest learning potential by being the most frequent. It is a paradox that most resources today are used on learning from infrequent accidents/catastrophes.
5. *Set priorities:* The safety manager is confronted with a large number of hazards and accident scenarios. It is a key management function to continuously assess the alternative risks the operation is confronted with and to give priority to the critical ones or those with the highest benefit-to-cost ratio. In other words, apply scarce resources where the safety improvement is highest.

On the basis of this, we may try to identify which management functions are related to these strategies. A simple categorization is shown in Table 15.1. This overview gives a simple presentation of what we mean by safety management functions.

Some will perhaps say that these strategies and safety management functions reflect too narrow an approach towards achieving safety and that we thus need a broader safety orientation in order to reach a sustainable safety level. Table 15.2 proposes some elements that constitute such a broader orientation. The key idea is that safety must be reflected in both company and human values and be visible in the so-called company culture. The culture concept may be seen as a fairly vague term, and this will be addressed in

Table 15.1. Strategies applied to prevent accidents with corresponding safety management functions

Strategies	Safety management functions
Protect against potential hazards	Engineering, Formal Safety Assessment (FSA)
Proceed cautiously	Policy formulation
Test the risks	Formal Safety Assessment (FSA), design verification and testing
Learn from experience	Accident reporting and analysis, inspections, auditing
Set priorities	Policy formulation, Formal Safety Assessment (FSA)

Table 15.2. Safety orientation dimensions

Safety orientation dimensions	Explanation
Safety view	Safety incorporated as an essential part of the business policy Proactive attitude Public responsibility Long-term perspective
Set priorities	Documented and visible policy Give safety priority and allocate the necessary resources
Culture	Credible leadership: 'Do as you say' True concern for safety at all levels of the organization Establish a set of 'symbols/heroes/rituals/values'
Human values	Genuine concern for crew, staff and middle management Emphasis on worker safety and health Personnel policies that support motivation and responsibility
Operate systematically	Implement safety plan, programmes and routines Establish a safety management system TQM approach: Plan/do/check/review cycle
Protect against potential hazards	Formal Safety Assessment (FSA), Safety Case approach Risk-based engineering Emergency preparedness
Proceed cautiously	Limit the pace of technical innovations in each project Design verification Test the risks
Learn from experience	Openness to risks and safety problems Inspections, maintenance Auditing of the safety management system

greater detail later in the chapter. True safety orientation is also visible through the priorities set by top management. Fundamentally it is a matter of how safety is balanced against time and money in daily operations.

15.1.3 The Maritime Safety Management Regime

The rules and regulations governing safety and environmental protection within the shipping industry have evolved over time through what may be described as a set of interrelated stages. The earliest and most basic stage focused primarily on the consequences of accidents resulting from failures made in relation to safety. In the context of this safety regime, major efforts were made in the aftermath of accidents to find someone to blame for personal injuries, fatalities, damage to or loss of ship and cargo, and environmental pollution. This created a *culture of punishment*, where the essential theme was to identify and allocate/apportion blame. Frequently the last people in the chain of events, i.e. the people at the sharp end of the system, were found responsible. The underlying principle of this safety management regime was that the threat of punishment should influence companies and individual behaviour to the extent that safety gained a higher priority. Although the maritime safety management regime has principally evolved away from this early stage of development, the culture of punishment is still to be found in the aftermath of accidents as well as in many maritime regulations. For example, the US Oil Pollution Act (OPA 90) gives shipowners full economic liability for oil spills in US coastal waters.

What may be described as the second stage of development regarding the maritime safety management regime involves the regulation of safety by prescription, i.e. the prescriptive regime in which the maritime industry is given sets of rules and regulations to be obeyed. For example, the provisions of ILLC 1966 (the International Convention on Load Lines), SOLAS 1974 (the International Convention for Safety of Life at Sea), MARPOL 73/78 (the International Convention for the Prevention of Pollution from Ships), COLREG 1972 (the Convention on the International Regulations for Preventing Collisions at Sea) and STCW 78/95 (the Standards for Training, Certification and Watchkeeping for Seafarers) provide the basis for the prescriptive (or external) regulatory framework for international shipping. The prescribing party is normally a country government or its legislative bodies, or an international organization in which a number of countries participate (e.g. the IMO). The prescribed rules and regulations have normally been based on past experiences and very rarely have proactive rules been included. The prescriptive regime affects and is employed in all parts of a vessel's life-cycle, i.e. design, construction, operation, modification/refurbishment, and decommissioning.

The second stage of development is an advance on the first stage (the *culture of punishment*) because it is designed to attack known points of danger before actual harm occurs. This leads to a *culture of compliance* with prescriptive rules. However, more recently there has been a growing belief that the application of prescriptive rules is not enough: rules and regulations provide the means to achieve safety, but this should not be an end in itself.

The third and most advanced stage in the evolution of the maritime safety management regime is the creation of a so-called *culture of self-regulation* of safety, where regulation goes beyond the setting of externally imposed compliance criteria as in the second stage. The culture of self-regulation concentrates on internal management and organization for safety, and encourages individual industries and companies to establish targets for safety performance. Self-regulation also emphasizes the need for every company and individual to be responsible for the actions taken to improve safety, rather than seeing them imposed from external prescriptive parties. This requires the development of company-specific and, in the case of shipping, vessel-specific safety management systems (SMS). It can be concluded that in the culture of self-regulation, safety is organized by those who are directly affected by the implications of failure.

As mentioned earlier, the regulation of safety in the shipping industry has, historically, been characterized by a culture of punishment and a culture of external compliance. IMO's adoption of the International Safety Management (ISM) Code, which is made mandatory in all member states, is an important step towards the creation of a culture of self-regulation in shipping. The increasing focus on safety management represents an important transition from the traditional principle of prescriptive regulations that dominate the maritime sector. Self-regulation is not, however, completely effective on its own. In order to achieve safer seas and environmental protection it is necessary for all three regimes/stages described above to coexist. Each regime plays a significant part in influencing company and individual behaviour. The ISM code will be reviewed in detail later in this chapter.

With regard to improving maritime safety, the following factors and aspects have been, and still are, in focus:

- Technical solutions
- Training and competence (i.e. human factors)
- Workplace conditions (i.e. ergonomics)
- Management and organization
- Risk-based planning and design

The causal factors resulting in ship accidents reveal that there is considerable potential for improvement with regard to human and organizational factors (HOF), at least relative to what can be gained today through the implementation of improved technical solutions in North European and North American flag states. In the emerging Total Quality Management (TQM) thinking there is an increasing tendency to see health, safety and environment (HSE) as elements in an integrated management approach.

15.2 TOTAL QUALITY MANAGEMENT (TQM)

15.2.1 Basic Theory

As pointed out earlier, safety has traditionally been seen in a regulatory perspective. This means that the shipowner or manager operates within the framework of prescriptive safety regulations. This view is now changing rapidly because of the so-called quality thinking that is gaining increasing acceptance and thereby blending the different aspects of quality management together. Increasingly, safety management will therefore be seen as an integral part of the overall management system of a company. Safety is only one of a number of factors that express the *quality* of a business, which may involve the following factors:

- Long-term perspective
- Customer orientation
- Leadership involvement
- Continuous improvement
- Fact-based management
- Employee involvement at all levels
- Good relations with subcontractors
- Corporate responsibility
- Good and effective health, safety, and environmental policies

Quality affects every aspect of business. A popular and emerging management philosophy that bases itself on this all-embracing notion of quality is the so-called Total Quality Management or TQM (Costin, 1998). Juran (1998) has stated the following motivations for the TQM approach:

- There is a crisis in quality.
- Our traditional ways of dealing with quality crises are inadequate.
- Quality management affects all functions and every level (of the hierarchy) of the organization, and hence requires a universal way of thinking.
- There is a need for continuous learning and improvement at all organizational levels.

The so-called 'Juran Trilogy', as illustrated by Figure 15.1, focuses on quality planning, control and improvement, as well as quality assurance. This quality model is the underlying basis of the TQM philosophy. In order to clarify the concepts in Figure 15.1, the quality of the financial function of an organization can be studied as shown in Table 15.3.

Three important characteristics and key conditions of successful quality management are a clear policy regarding quality, continuous improvement, and comprehensive management commitment. In addition to presenting the quality model (i.e. Figure 15.1), Juran (1998) also outlines the processes underlying each of the three basic quality functions in this model, i.e. quality planning, quality control and quality improvement. These processes are presented in Table 15.4.

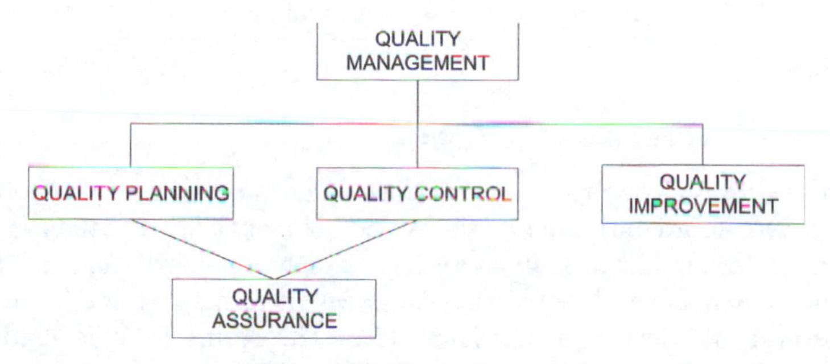

Figure 15.1. The quality model.

Table 15.3. The financial function of an organization seen in the context of the quality model

Financial	Quality
Budgeting	Quality planning
Cost/expense control	Quality control
Cost reduction, profit improvement	Quality improvement
Audit	Quality assurance

Table 15.4. The processes underlying quality planning, quality control and quality improvement

Quality planning	Identify customers, both external and internal Determine customer needs Develop product features (both goods and services) that respond to customer needs Establish quality goals that meet the needs of customers and suppliers alike, and do so at a minimum combined cost Develop a process that can produce the needed product features Prove process capability – prove that the process can meet the quality goals under operating conditions
Quality control	Choose control subjects, i.e. what to control Choose units of measurement Establish measurement Establish standards of performance Measure actual performance Interpret the difference between actual and standard performance Take action on the difference
Quality improvement	Prove the need for improvement Identify specific projects for improvement Organize to guide the projects Organize for diagnosis, i.e. for discovery of causes Diagnose to find causes Provide remedies Prove that the remedies are effective under operating conditions Provide for control to maintain gains

15.2.2 Safety Management Based on TQM

Central to the TQM philosophy is the so-called 'plan/do/check/review' loop, in which quality is improved on a continuous basis. As pointed out earlier, safety is a quality of an organization. The organizational safety level can be managed through the use of a set of basic safety management activities. These activities may be modelled in a 'safety management spiral' as illustrated in Figure 15.2. The spiral form is used to indicate that safety management within an organization should be an iterative process, which is in accordance with the important TQM principle of continuous improvement (Aitken et al., 1996).

The five basic components in the safety management system (SMS) in Figure 15.2, which is heavily influenced by TQM, are as follows:

- *Policy:* All organizations have a set of policies that are used to guide the performance of the staff so that the overall objective of the organization can be achieved in an effective way. Safety goals (i.e. the level of safety one wishes to achieve) should be included as a vital part of the policies. Policies have a tendency to be quite static documents, but to achieve a well-functioning safety management system one must

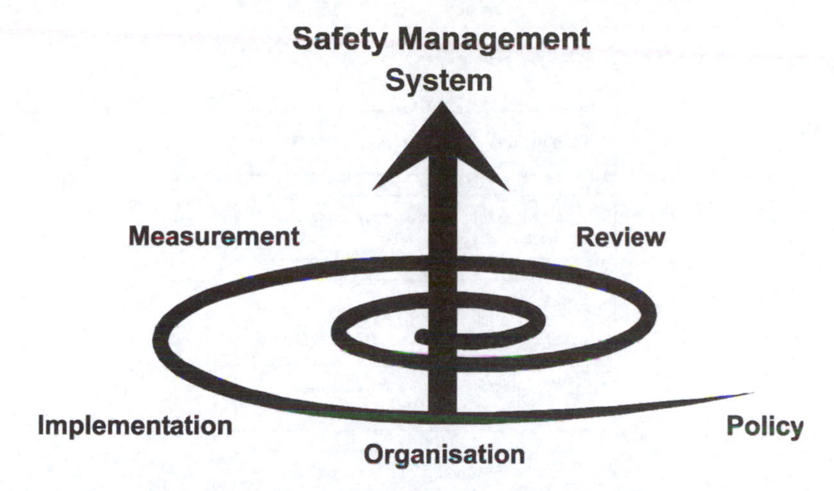

Figure 15.2. Safety management modelled as a spiral illustrating safety management as an activity of continuous improvement.

establish a culture where policies are developed and improved over time as a result of the iterating safety management process.

- *Organization:* It is of great importance that the management system establishes the responsibilities of individuals with regard to safety matters. This has to be done with an understanding of the needs of communication and co-operation between the individuals involved, and the need for proper education and competence within the organization.
- *Implementation:* The implementation phase should make sure that the policies and objectives are translated into practice. The results of this implementation are 'tested' through the application of the SMS.
- *Measurement:* The objective of the measurement task is to measure whether the implementation went as intended and its effectiveness. The information from this stage is fed into the review phase/stage of the SMS.
- *Review:* It is necessary to have a mechanism for reviewing the performance of a system and to seek ways to continuously improve it. The review phase/stage uses the information obtained by measurement to review/audit and analyse the performance of the system. Auditing is the only non-destructive way in which lessons can be learned and fed into the system for enhancement. The review phase should examine the total range of safety management activities, i.e. policies, organization, implementation and measurement.

Clement et al. (1996) have outlined how principles from TQM can be applied in the development of an integrated health, safety and environment (HSE) management system suitable for implementation into all business processes. The process of introducing TQM involves a number of steps, which can be illustrated by Figure 15.3. As can be seen from

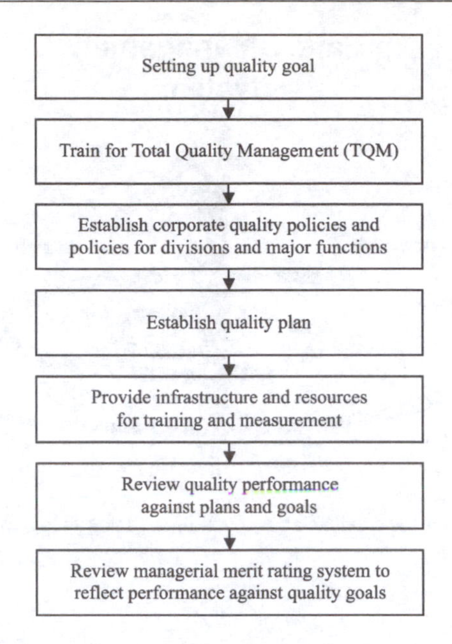

Figure 15.3. The process of introducing TQM. (Source: Clement et al., 1996.)

this figure, the basic elements of this approach are consistent with the TQM-based safety management system described above.

The integrated HSE management system that was set up put emphasis on the wide range of elements shown in Table 15.5. The HSE policy in this table may seem vague. The intention of incorporating and stating the company's policies in its vision, mission and value statements is to give the company/organization a committing and visible direction/goal for its HSE efforts. For a particular company the policies will normally be described more explicitly and may, for example, include the following (Clement et al., 1996):

- Comply with all HSE laws, regulations and industry standards, and self-regulate where there are no such prescriptive requirements.
- Exhibit socially conscious leadership and demonstrate excellent HSE performance.
- Seek to participate in developing HSE legislation, regulations and standards.
- Integrate HSE protection into every aspect of the business activities.
- Design and operate the company's facilities following industry standards. Prevent discharge of hazardous substances.
- Satisfactorily train employees and contractors, emphasizing individual responsibility for sound HSE performance.
- Conserve natural resources by prudent management of emissions and discharges, and by eliminating waste.

Table 15.5. HSE management system

Leadership and commitment	Top management shall provide strong, visible leadership and be fully involved HSE shall be on the agenda in all management meetings HSE is to be seen as a key business strategy Accountability is of great importance
Policy	HSE aspects shall be reflected or incorporated in the company's vision, mission and value statements
Organization	The organization shall evolve from being functional to becoming a team-based organization focused on asset value optimization HSE engineers/specialists shall be assigned to teams The training programme shall be improved and strengthened Emergency response programmes shall be developed Records shall be kept for documentation of plans, management system, procedures, etc.
Implementation and monitoring	Awareness and communications are key condition to success Planned inspections and preventive maintenance shall be carried out Accidents, incidents and near-misses shall be investigated Work rules and permits are to be used Personal protective equipment shall be provided and used Health hazard identification and evaluation shall be carried out at regular intervals Good change management is important Environmental issues shall be identified and action plans produced regularly Establishing and maintaining good relations with external stakeholders is of great priority
Measurement and performance	The implementation shall be measured and the performance assessed This process shall result in specific targets for the individual business units
Audits and reviews	It shall be assessed whether the HSE management system is implemented effectively and according to the plan Are the policy's principles being fulfilled? Are the objectives and performance measures achieved? Are we in compliance with rules and regulations? Establish areas for improvement

- Encourage employees to communicate within the company, as well as with the public, regarding HSE matters.
- Work to resolve any problems created by past operations and practices.
- Ensure conformity with these policies through a comprehensive compliance programme including audits.

Measurement and performance must be seen in relation to the activities of the different business units. The following list indicates typical performance measures for oil exploration and production activities:

- Number of oil spill incidents
- Total volume of oil spilled
- Number of km of road transport replaced with pipeline
- Area re-greened
- Km pipeline upgraded to present standard
- Volume of oily water and waste water discharged
- Hydrocarbon emission to air
- NO_x and SO_x emissions

The final element of the HSE management plan is auditing, which may be undertaken both internally on a corporate scale, by the different business units, or by an independent external auditor. With regard to which HSE-relevant elements of an organization should be audited/reviewed, Clement et al. (1996) refer to a scheme developed by the International Loss Control Institute (ILCI, 1995). According to this scheme the audit elements are as follows:

1. Leadership and administration
2. Leadership training
3. Planned inspections and maintenance
4. Critical task analysis and procedures
5. Accident/incident/near-miss investigation
6. Task observation
7. Emergency preparedness
8. Internal rules and work permits
9. Accident/incident/near-miss analysis
10. Knowledge and skill training
11. Personal protective equipment
12. Health and hygiene control
13. Programme evaluation
14. Engineering and change management
15. Personal communications
16. Group communications
17. General promotion
18. Hiring and placement
19. Materials and contractor management
20. Off-the-job HSE awareness
21. Environmental issue identification
22. Environmental action plan
23. Environmental performance monitoring and assessment

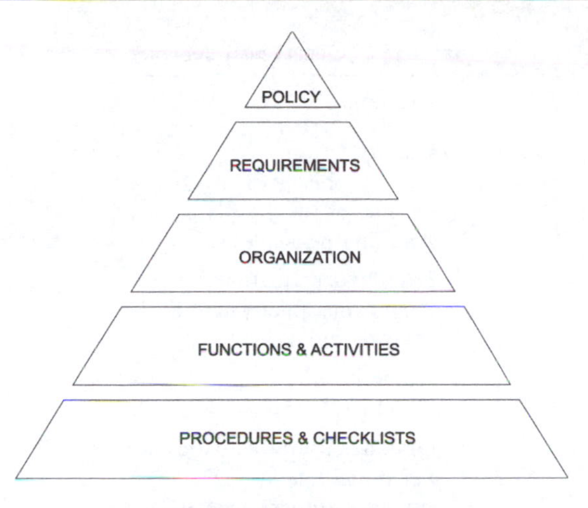

Figure 15.4. The quality plan.

24. Relations with external parties
25. Product stewardship
26. Agency permits, compliance reports and inspections
27. Off-site waste management

15.2.3 Quality Plan Structure

The quality plan, referring to the quality planning element of the 'Juran Triology' in Figure 15.1, is the underlying basis for all quality and safety management. The key elements of a quality plan are depicted in Figure 15.4. The starting points are the quality policy and the requirements placed upon the organization. The plan itself consists of careful descriptions of organizational, functional and procedural plan elements. Examples of such quality plan elements are given in Table 15.6.

15.2.4 Quality Programme Structure

It is well known from experience that most organizations are able to define a quality plan and establish the desirable standards that they want to achieve, but have greater difficulty in living up to these standards as a part of the daily routine. In order to succeed in improving quality (e.g. safety), a company needs not only a plan that defines the relevant activities and quality objectives but also certain measures that will transform the organization into one that lives by its new policies and standards. This requires a more rigorous approach in terms of a quality programme. The quality programme structure for safety outlined in Figure 15.5 is more process-oriented than the quality plan previously

Table 15.6. Quality plan elements

Quality plan elements	Examples	
Policy	Quality vision Quality objectives Outline of quality plan Auditing principles	
Requirements	Laws and regulations Applicable quality standards Own requirements	
Organization	Responsibility Authority	
Functions and activities	Organization of activities Personnel management Information management Handbooks, procedures Material management Follow-up of non-conformities Quality auditing Experience feedback	
Procedures and checklists	Accident reporting Agents Auditing Chartering Contracts Contingency Document control Economy Experience feedback General	HSE Maintenance Marketing Modifications Operations Organization Planning Pollution prevention Project management Purchasing

shown. This safety programme structure contains elements that focus more directly on how to obtain a working safety plan.

15.2.5 Standardization

One of the key objectives of international maritime safety regulations is to attain an acceptable level of risk in shipping. As has been discussed earlier in this chapter, traditional prescriptive rules and requirements have several distinct shortcomings. It is therefore increasingly argued that, in order to more dynamically achieve improved safety, companies must integrate the notion of total quality into their business strategy, and in particular see health, safety and environment (HSE) as a part of the total quality concept. A promising approach to achieving higher quality in production and service is the introduction of international quality standards. The International Standards Organization (ISO)

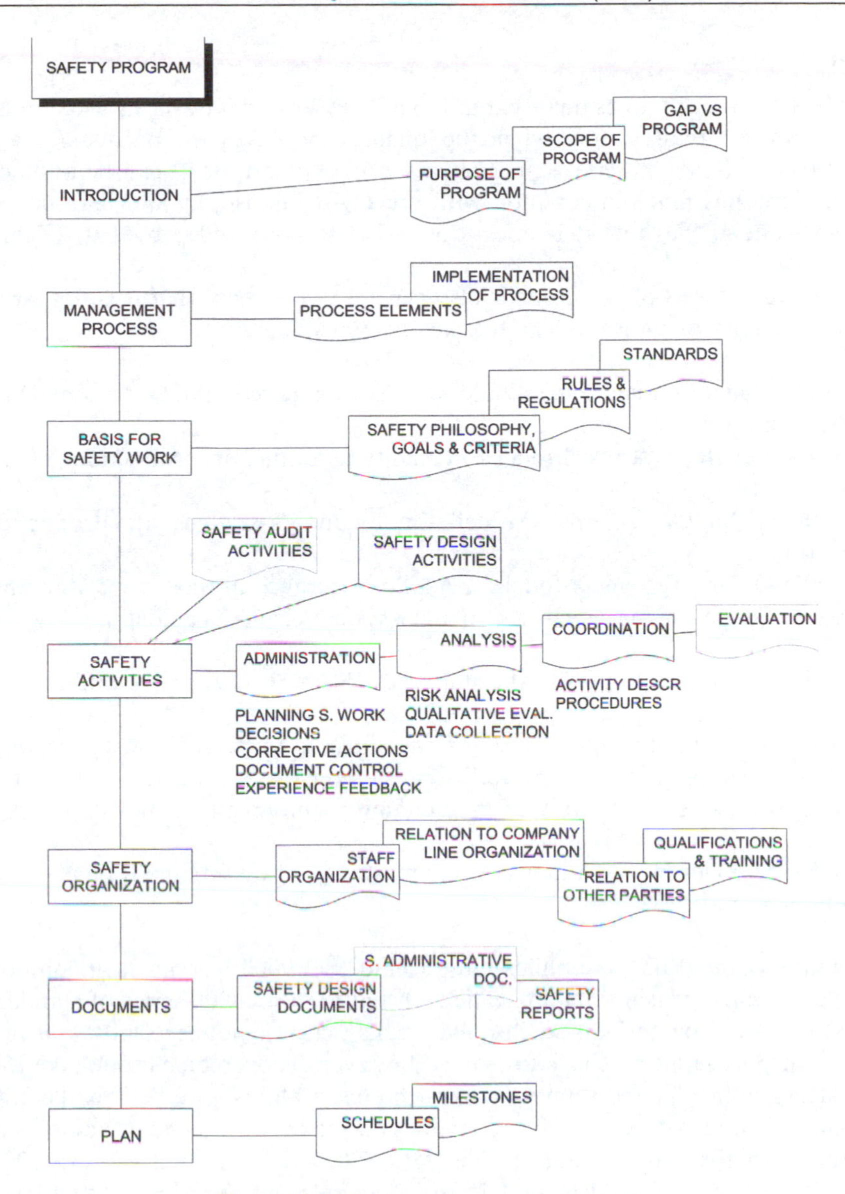

Figure 15.5. A safety programme structure.

is playing a prominent role in the development of such standards, and in particular two TQM-based ISO standard series have had a profound influence on a wide range of different industries and activities. These standard series are the ISO 9000 series, particularly addressing quality assurance, and the ISO 14000 series focusing on environmental management.

ISO 9000

The ISO 9000 series addresses quality, but in another sense than what is usual in industry standards. Rather than stating what the quality should be in particular cases, the standards in the ISO 9000 series state how a specified quality level can be attained. This way of thinking has much in common with the TQM philosophy and may be termed a consistency system: 'Says what it does; does what it says; and proves it. (Zahren and Duncan, 1994.)'

The ISO 9000 series of standards is also generic in the sense that it is not written for any specific activity or industry. The five key documents are:

- ISO 9000: Quality management and quality assurance standards: Guidelines for selection and use.
- ISO 9001: Quality systems: Model for quality assurance in production, installation, and servicing.
- ISO 9002: Quality systems: Model for quality assurance in production and installation.
- ISO 9003: Quality systems: Model for quality assurance in final inspection and test.
- ISO 9004: Quality management and quality system elements guidelines.

Zharen and Duncan point out that the standards ISO 9001–9003 have different scopes:

- ISO 9003: Supplier must demonstrate a capability to inspect and test a product.
- ISO 9002: Supplier must, in addition, demonstrate that the relevant manufacturing processes are capable of maintaining requirements as per design specifications.
- ISO 9001: Supplier must demonstrate the mastering of quality in all phases of design, development and servicing.

As already pointed out, the philosophy behind ISO 9000 has much in common with current quality management thinking. The fact that the ISO 9000 series of standards was developed with land-based industry in mind means that it is not especially well suited for maritime transport services. This hurdle may, however, be overcome if one take ISO 9004 as the starting point. This document covers the basic philosophy behind the ISO 9000 series standards and makes it easier to apply and make the necessary adjustments and modifications to the other standards (i.e. ISO 9001–9003). Part 2 of ISO 9004 gives guidelines for supply of services and is therefore relevant for shipping and maritime management services.

Adherence to the whole ISO 9000 series means that the following functions have to be addressed:

1. Contract review
2. Design control
3. Document control

4. Purchasing, including assessment of subcontractors and suppliers
5. Product identification and traceability
6. Process control
7. Inspection and testing
8. Inspection, measuring and test equipment
9. Control of non-conforming product
10. Corrective action
11. Handling, storage, packaging and delivery
12. Quality records
13. Internal quality audits
14. Training
15. Services
16. Statistical techniques

ISO 9000:2000

In order to reflect modern management approaches and to improve organizational practices, it was found useful and necessary to introduce structural changes to the ISO 9000 standards while maintaining the essential requirements of the current standards. The following paragraphs are taken from the ISO homepage on the web (ISO, 2000) and summarize the present status of the revision of ISO 9000.

The current ISO 9000 family contains some 27 standards and documents. This proliferation of standards has been a particular concern of ISO 9000 users and customers. To respond to this concern, ISO TC 176 (i.e. Technical Committee 176) has agreed that from the year 2000 the ISO 9000 family will consist of four primary standards supported by a considerably reduced number of other documents (i.e. guidance standards, brochures, technical reports, technical specifications). To the extent possible, the key points in the current 27 documents will be integrated into the four primary standards, and sector needs will be addressed while maintaining the generic nature of the standards. The four primary standards will be:

- ISO 9000: Quality management systems – Fundamentals and vocabulary
- ISO 9001: Quality management systems – Requirements
- ISO 9004: Quality management systems – Guidance for performance improvement
- ISO 19011: Guidelines on quality and environmental auditing

The revised ISO 9001 and ISO 9004 standards are being developed as a 'consistent pair' of standards. Whereas the revised ISO 9001 will more clearly address the quality management system requirements for an organization to demonstrate its capability to meet customer needs, the revised ISO 9004 is intended to lead beyond ISO 9001 towards the development of a comprehensive quality management system, designed to address the needs of all interested parties.

Both standards will use a common vocabulary as defined in ISO 9000:2000, which also describes the underlying fundamentals. A logical, systematic approach has been adopted in formulating the definitions used in ISO 9000:2000, with the intention of generating a more consistent terminology that is 'user-friendly'.

The current ISO 9001, ISO 9002 and ISO 9003 standards will be consolidated into the single revised ISO 9001:2000 standard. Clause 1.2 'Application' will permit the exclusion of some clauses of ISO 9001:2000 where the related processes are not performed by the organization, and these requirements do not affect the organization's ability to provide products that meet customer and applicable statutory or regulatory requirements.

ISO 14000

ISO 14000 is a series of voluntary international environmental management standards (ISO, 1995, 2000). The 14000 series of standards addresses the following aspects of environmental management:

- Environmental Management Systems (EMS)
- Environmental Auditing and Related Investigations (EA&RI)
- Environmental Labels and Declarations (EL)
- Environmental Performance Evaluation (EPE)
- Life Cycle Assessment (LCA)
- Terms and Definitions (T&D)

The ISO 14000 series of standards effectively addresses the needs of organizations world-wide by providing a common framework for managing environmental issues. The standards promise to have the effect of a broadly based improvement in environmental management, which in turn can facilitate trade and improve environmental performance world-wide. The ISO 14000 standards are being developed with the following key principles in mind:

- They must result in better environmental management.
- They must be applicable in all nations.
- They should promote the broad interests of the public and the users of the standards.
- They should be cost-effective, non-prescriptive and flexible to allow them to meet the differing needs of organizations of any size world-wide.
- As part of their flexibility, they should be suitable for internal or external verification.
- They should be scientifically based.
- And above all, they should be practical, useful and usable.

The benefits of an Environmental Management System (EMS) have been stated by ISO as follows:

- Assuring customers of commitment to provable environmental management
- Maintaining good public/community relations

- Satisfying investor criteria and improving access to capital
- Obtaining insurance at reasonable cost
- Enhancing image and market share
- Meeting vendor certification criteria
- Improving cost control
- Reducing incidents that result in liability
- Demonstrating reasonable care
- Conserving input materials and energy
- Facilitating the attainment of permits and authorizations
- Fostering development and the sharing of environmental solutions
- Improving industry–government relations

The standards in the ISO 14000 series fall into two major groups: organization-oriented standards and product-oriented standards. The organization-oriented standards provide comprehensive guidance for establishing, maintaining and evaluating an Environmental Management System (EMS). They are also concerned with other organization-wide environmental systems and functions. The product-oriented standards are concerned with determining the environmental impacts of products and services over their life-cycles, and with environmental labels and declarations. These standards will help an organization gather the information it needs in order to support its planning and decision processes, and to communicate specific environmental information to consumers and other interested parties.

15.3 THE ISM CODE

15.3.1 Background

The task that should face all shipping companies is to minimize the risk of poor human and organizational decisions that could have negative effects on operational safety, which may eventually lead to accidents. Human and organizational factors can have both direct and indirect effects on safety. One aim should be to ensure that staff are properly informed and equipped to fulfil their operational responsibilities safely. Decisions taken ashore can be as important as those taken at sea, and there is a need to ensure that every action affecting safety, taken at any level within the organization, is based on a sound understanding of its consequences. The adoption by the International Maritime Organization (IMO) of the 'International Code for the Safe Operation of Ships and for Pollution Prevention', normally referred to as the International Safety Management (ISM) Code, is the reflection of this objective on the part of governments (IMO 1994, 1995). The ISM Code establishes an international standard for the safe management and operation of ships by setting requirements for the organization of company management in relation to safety and pollution prevention, and for the implementation of a safety management system (SMS). The ISM Code addresses the very important issues relating to human factors, and some argue that it is one of the most significant documents to be produced by the IMO.

The ISM Code is a creation of the so-called 'culture of self-regulation' (see earlier discussion) in which regulations go beyond the setting of externally imposed compliance criteria, often known as the prescriptive regime. The ISM Code concentrates on internal management and organization for safety, which means that safety is organized by those who are directly affected by the implications of failure.

15.3.2 Implementation of the ISM Code

The Assembly of the IMO has adopted a series of resolutions dealing with guidelines on management procedures to ensure the safest possible operation of ships and the maximum attainable prevention of marine pollution. These resolutions culminated in the ISM Code, which was adopted in November 1993 by resolution A.741(18). In May 1994 a SOLAS (i.e. the International Convention for Safety of Life at Sea) Conference decided on a new Chapter IX of SOLAS which makes the ISM Code mandatory for ships, regardless of the date of construction. For most ship types the Code was implemented on 1 July 1998, but from 1 July 2002 the Code has been mandatory for all ships, including mobile offshore drilling units. For passenger ships and high-speed passenger craft there is no lower limit in terms of vessel size, but for all other ship types there is a lower limit of 500 gross register tons (grt).

15.3.3 The Substance of the Code

The ISM Code is based on a set of general principles and objectives to be achieved. It comprises 13 sections over 9 pages, and is therefore a fairly short document. The main purpose of the ISM Code is to demand that individual ship operators create a safety management system that actually works (ICS and ISF, 1996). The Code does not describe in detail how the company should undertake this task, but just states that some areas of measures have to be addressed. The underlying philosophy behind the Code is commitment from the top, verification of competence, clear placement of responsibility, and quality control of work.

The IMO has stated the following objectives for the adoption of a safety management system (SMS):

- To provide for safe practices in ship operation and a safe working environment.
- To establish safeguards against all identified risks.
- To continuously improve the safety management skills of personnel ashore and aboard.
- Preparing for emergencies related both to safety and environmental protection.

These objectives clearly show that the ISM Code has relations to existing or traditional safety management approaches such as technical solutions, training, emergency preparedness and risk analysis. The Code has 13 chapters, which are listed in Table 15.7.

Table 15.7. The ISM Code: table of contents

1. General
2. Safety and environmental protection policy
3. Company responsibility and authority
4. Designated person(s)
5. Master's responsibility and authority
6. Resources and personnel
7. Development of plans for shipboard operations
8. Emergency preparedness
9. Reports and analysis of non-conformities, accidents and hazardous occurrences
10. Maintenance of the ship and equipment
11. Documentation
12. Company verification, review and evaluation
13. Certification, verification and control

When discussing the effect of ISM on safety, there are two basic aspects: the content of the Code's regulations and what is an acceptable compliance with the Code. The content of the ISM Code is described in detail in Table 15.8. The ISM Code specifies certain requirements for the safety management system (SMS) of the operating company. In order for the SMS to work properly, certain distinct elements and functions have to be in place. The different chapters in the ISM Code cover these elements or functions of the system. Chapter 11 of ISM Code states that the SMS may be seen as the product of the establishment of controls that are defined in terms of:

- Responsibility and authority
- Supply of resources and support
- Procedures for the checking of competence and operational readiness, training, shipboard operations, etc.
- Minimum standards of the maintenance system

Chapter 11 of the Code also states that the safety management system shall be adequately documented. Another key feature of the ISM Code is the definition and introduction of a monitoring function based on audits and the reporting of events. Auditing shall ensure that errors and shortcoming in the SMS are corrected and that the system is updated in view of new requirements and conditions. Auditing and event reporting shall also address hazards and system errors directly, and this may lead to corrective actions in terms of modified systems and improved human competence. The interactions between the different elements of the ISM Code are outlined in Figure 15.6.

Chapter 13 mainly states that the company should have a certificate of approval, known as a Document of Compliance (DOC), which states that its SMS is in accordance with the intention and specific requirements of the ISM Code. In addition every vessel should have a Safety Management Certificate (SMC). The Code indicates several

Table 15.8. The content of the ISM Code

Chapter	Organizational tasks
2. Safety and environmental protection policy	Establish a policy that is in accordance with the objectives of the ISM Code. Ensure that the policy is implemented and maintained at all levels onboard and ashore.
3. Company responsibilities and authority	Identification of the company responsible for the operation. Specify responsibility, authority and interrelation of key personnel. Ensure adequate resources and shore-based support.
4. Designated persons	Identity/assign person(s) serving as a link between vessel and company. Monitor safety and secure resources and support.
5. Master's responsibility and authority	Statement of Master's authority for the safety onboard, i.e. superior responsibility for the implementation of the SMS. The Master's tasks include: motivate crew, issue orders, verify adherence, review SMS and report events.
6. Resources and personnel	The company should ensure that: The Master is qualified and conversant with the SMS. The ship is manned with qualified, certificated (i.e. STCW) and medically fit seafarers. Procedures for familiarization of personnel with new duties exist. Rules, regulations, codes and guidelines are understood. Procedures for identification and provision of training exist. Procedures exist in a working language for provision of information about the SMS. The ship's personnel are able to communicate effectively.
7. Development of plans for shipboard operations	Develop procedures for plans and preparation of instructions for key shipboard operations. Definition of tasks and assignment of personnel.
8. Emergency preparedness	Develop procedures to identify, describe and respond to emergencies. Establish programmes for drills and exercises. Ensure that the company organization can respond.
9. Reports and analysis of non-conformities, accidents and hazardous occurrences	Develop procedures for the reporting of events. Ensure that procedures exist for the implementation of corrective actions.
10. Maintenance of the ship and equipment	Maintenance in accordance with rules and company-based requirements. Ensure inspections, non-conformity reporting, corrective action and record-keeping. Identify safety-critical systems. Ensure reliable operation and testing.

(*continued*)

Table 15.8. Continued

Chapter	Organizational tasks
11. Documentation	Develop procedures for establishing and maintaining documentation of the SMS. Establish procedures for the review and approval of changes to documents. Ensure that the documentation is available.
12. Company verification, review and evaluation	Carry out internal safety audits, evaluate efficiency of SMS (auditing personnel should have an independent role). Feedback of findings to involved personnel and responsible management.
13. Certification, verification and control	The company should have a Document of Compliance (DOC), and the vessels a Safety Management Certificate (SMC).

supplementary approaches in terms of assessing compliance with the ISM regulations. These approaches include:

- Documentation on how the ISM Code will be implemented
- External verification and certification by an independent party
- Logging/reporting of the safety management processes
- Internal auditing and verification

Apart from this, the *Guidelines on Implementation of the ISM Code by Administrations* (IMO, 1995) are fairly vague on how to verify that a safety management system conforms with the Code. This document admits that certain criteria for assessment are necessary, but also warns against prescriptive requirements and solutions prepared by external consultants. The obvious philosophy behind this attitude is that the SMS should be an integral part of the management thinking in the company and hence should be a product of the company.

It is important to recognize that the ISM Code and its requirements to company safety management systems must be seen in the context of already existing international safety regulations. The main safety conventions in this respect are:

- SOLAS: Safety of Life at Sea (1974) and SOLAS Protocol (1978, 1988).
- STCW: Standards for Training, Certification and Watchkeeping for Seafarers.
- MARPOL: The International Convention for the Prevention of Pollution from Ships (1973), and its 1978 Protocol.
- COLREG: Convention on the International Regulations for Preventing Collisions at Sea (1972).
- ILLC: The International Convention on Load Lines (1966).

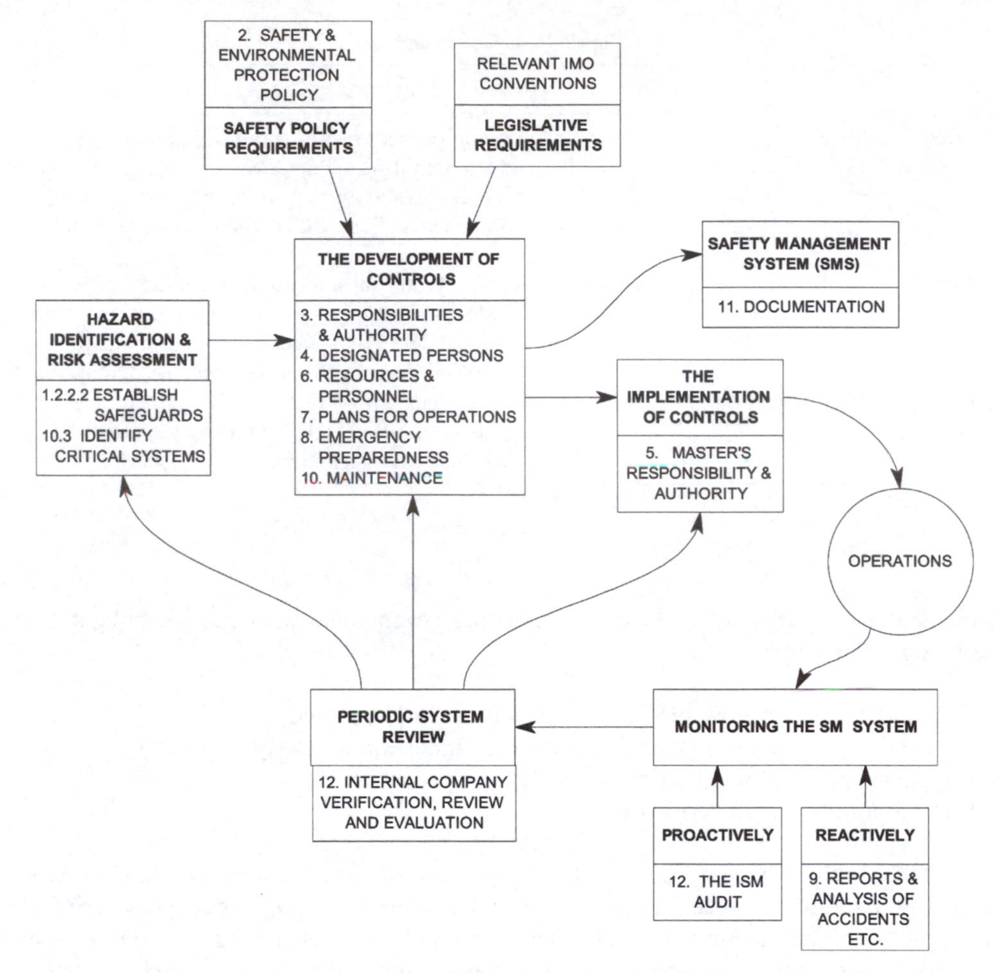

Figure 15.6. The functional interrelationships between the different elements of the ISM Code.

ISM does not address any of the specific requirements in these conventions, but just assumes that the management system should ensure that they are adhered to.

15.3.4 The ISM Code in Practice

The ISM Code may at first glance look relatively complex, but as Bromby (1995) has pointed out, the very idea behind the Code is a fairly straightforward learning process which may be illustrated by Figure 15.7. Based on risk assessments, a set of controls is developed and implemented. The safety management system will be monitored and reviewed on a regular basis in order to ensure that it functions adequately and properly.

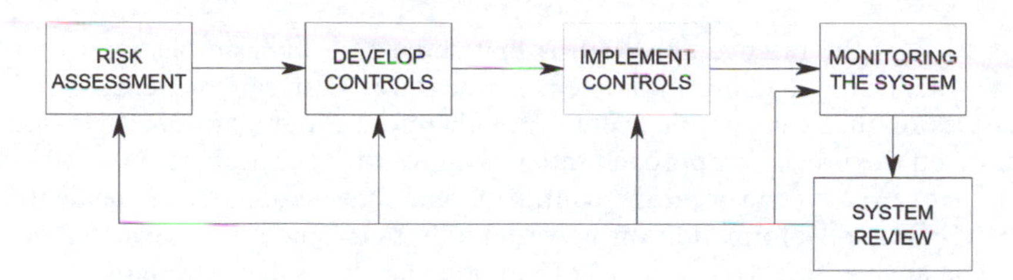

Figure 15.7. The safety management model (adapted from Bromby, 1995).

The process of risk assessment has been discussed in detail in earlier chapters and is not commented on any further here. The concept of controls has been described earlier in this chapter. The monitoring of the SMS in terms of the adequacy of the controls can be undertaken using different approaches. It is common to make a distinction between proactive and reactive methods. Proactive methods focus on how operations are performed relative to the safety objectives, whereas reactive methods are based on what we can learn from non-conformities and accidents. The following proactive methods can be applied:

- Inspections
- Safety tours
- Surveys
- Audits

Inspections normally focus on the condition of equipment and systems, and have for a long time been an established approach in safety work. Typical questions asked in inspection include: Is the equipment/system functioning? Is the equipment damaged in any way? Should the equipment be maintained?

A safety tour concentrates on how people perform their jobs/tasks relative to established procedures. A key question is whether personnel are safety-conscious, which can be ascertained through observation of work and conversation with individuals.

Surveys are directed towards assessing/checking the condition of safety-related systems and equipment, such as evacuations systems, life-saving equipment and personal protection equipment.

Audits are closely linked to the introduction of systematic safety management. It is not oriented towards hardware or people but rather the management system itself. Auditing will be discussed in greater detail in the following section.

The reactive approach is, on the other hand, based on information about:

- Accidents
- Incidents or near-accidents
- Non-conformities or deviations

The drawback of the reactive approach is that something undesirable has to take place before we can learn anything or implement measures that can prevent or reduce the consequences of such undesirable events. It is therefore a popular view that one should concentrate on the proactive approach rather than wait for the accident with all its misery and costs. However, one may on the contrary argue that accidents and incidents are the only direct source of information on how and why these undesirable events/phenomena happen. The answer seems to be that both approaches have their strengths.

15.4 AUDITING

15.4.1 What to Audit

The basis for implementing a safety management system (SMS) is the definition of policies, an assessment of the requirements placed upon the organization, a clear view of necessary functions and activities, the development of management procedures/routines, as well as appropriate training. In order to assess whether the SMS is functioning as expected, different approaches may be used:

- Inspection of vessels
- Work observation
- Vetting of non-conformities
- Incident reporting
- Accident investigation
- Auditing

Apart from auditing, most of these approaches/methods are explained and studied in other chapters of this book. Auditing differs from the other approaches because it is so closely related to the introduction of formal safety managements systems (SMS). While the other and more traditional approaches assess external parameters such as accidents, fatalities, pollution, system conditions, and the real competence of employees, audits focus on the elements of the SMS. In auditing, typical questions asked would include: 'Are people well trained?', 'Are system reviews performed regularly?' and 'Are accidents followed up with analysis and the implementation of reactive measures?', and so on. Saunders (1992) defines the main objects of an audit as follows:

- Policies
- Procedures
- Practices
- Programmes
- Organization

A detailed outline of these objects is given in Table 15.9.

The process of auditing consists of a number of distinct steps which are outlined in Figure 15.8. In the initiating preparations phase the objective of the audit is defined,

Table 15.9. The main objects of an audit

Area	Description	
Policies	Legal requirements	Budgetary provision
	Statement of objectives	Staffing: correct people
	Mission, strategy	Organizational structure
Procedures	Administrative structure	Recruitment
	Technical procedures	Safety training
	Communication	Supervision
	Time management	Discipline
	Internal/public relations	
Practices	Costing of accidents	Hazard assessment
	Accident investigation	Accident analysis
	Data collection	Equipment inspections
	Medical examination	HSE codes of practice
	Welfare	
Safety programme	Enforcement	Working environment
	Engineering	Education
Organization	Organizational structure	Fundamental activities
	Necessary functions	Emergency response teams

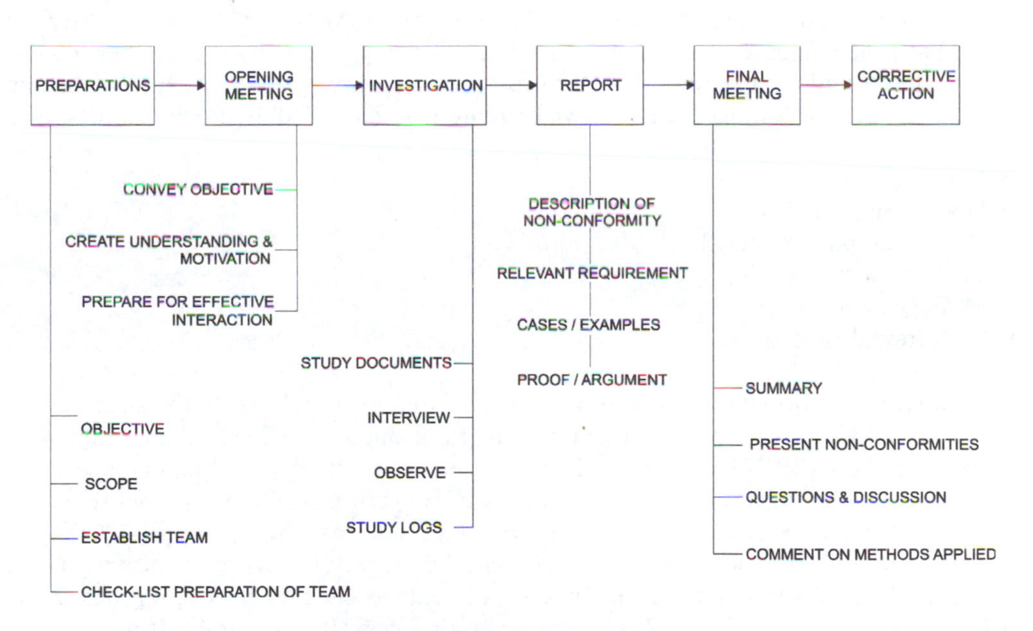

Figure 15.8. The auditing process.

and the scope (i.e. extent) of the audit is specified. The audit may cover the whole company or alternatively only specific departments or functions. An important task in the preparations phase is to establish and prepare the team to perform the audit. The opening meeting will establish the first contact between the audit team and the department/function that will be audited. The main purpose of this meeting is to explain the objectives of the auditing process and how the parties will interact during that process.

The core activity of the auditing process is the investigation phase. The starting point of this phase will normally be to study all relevant safety management system (SMS) documents. Through the study of logs, interviews, and observation of key personnel it is established whether the SMS is functioning as intended. The investigation results in a report where all non-conformities are documented. It is important that all deviations are supported with relevant argumentation and documentation, and that relevant requirements are clearly stated. The preliminary report is then circulated for comments from all the involved departments and individuals. The final task of the audit team is to arrange a final meeting where the findings and assessments are presented and discussed. The audit report will then serve as a basis for taking corrective action by the company.

15.4.2 Audit Findings

Gray and Sims (1997) have made an analysis of non-conformities found in audits on different safety management systems, including ISO- and ISM-based systems. A statistical summary of their findings is given in Figure 15.9. The diagram shows the number of non-conformities found relative to given checklists for each company function. It is interesting to note that the six dominating areas, which represent 66% of all non-conformities, are:

- Maintenance
- Documentation
- Resources and personnel
- Emergency preparedness
- Management system
- Operational procedures

An alternative format is to present the degree of attainment relative to the target values set by the company. This may be more meaningful if one has defined a long-range plan to improve the SMS. Figure 15.10 shows the audit results for a shipping operation within Exxon. The functions with the greatest gap between target and achieved score are listed from the top in the diagram. It is interesting to note that the largest gap was found for 'Personnel' and 'Reporting, investigation and analysis [of non-conformities]'. These are typical 'soft' aspects of the SMS that are difficult to change. Illustrating this is the finding that the quality plan itself is 100% in accordance with the target. It is a general experience with safety management systems that the easiest part of quality management is to produce plans and documentation.

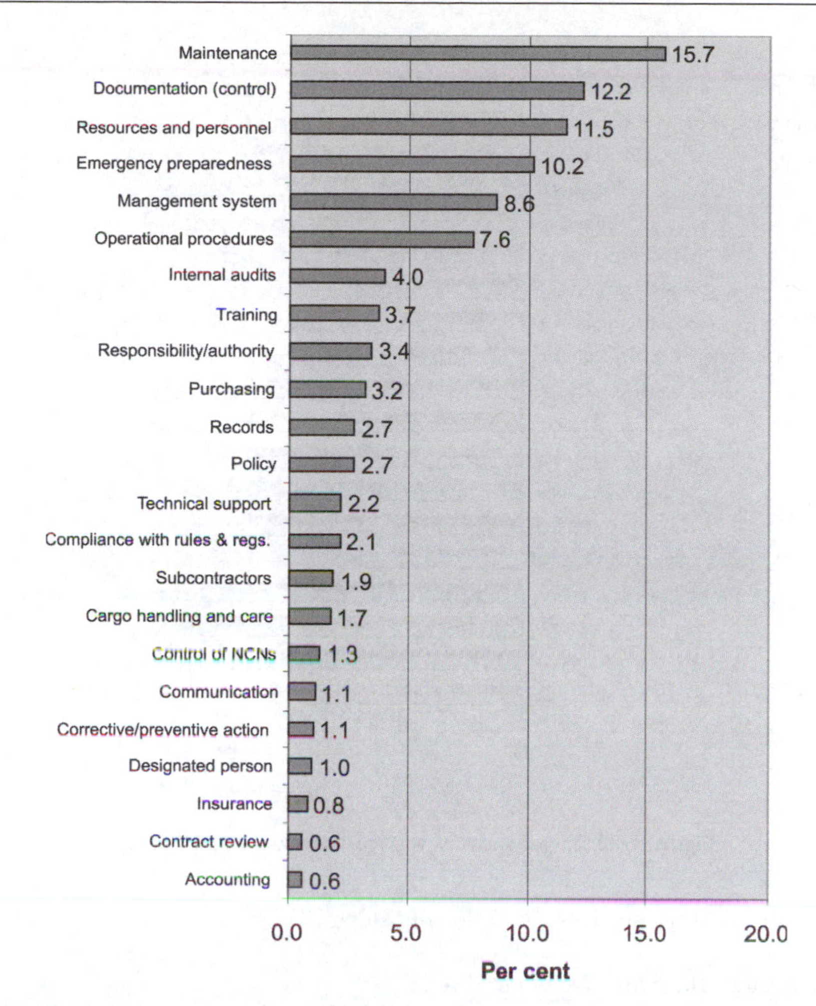

Figure 15.9. Non-conformities found in audits on different safety management systems.

15.4.3 Auditing in the Context of the ISM Code

The guidelines for the auditing process of the ISM Code are specified by IMO (1995). Auditing will be performed both before certification of ISM compliance and at regular intervals after this certification. The intention of the audit is to:

- Determine whether the SMS conforms with the requirement of ISM.
- Assess the effectiveness of the company's SMS.
- Determine that the ships' SMS are in compliance with rules and regulations.
- Assess the effectiveness of SMS in ensuring that other rules and regulations are adhered to.
- Assess whether the 'safe practices' recommended have been taken into account.

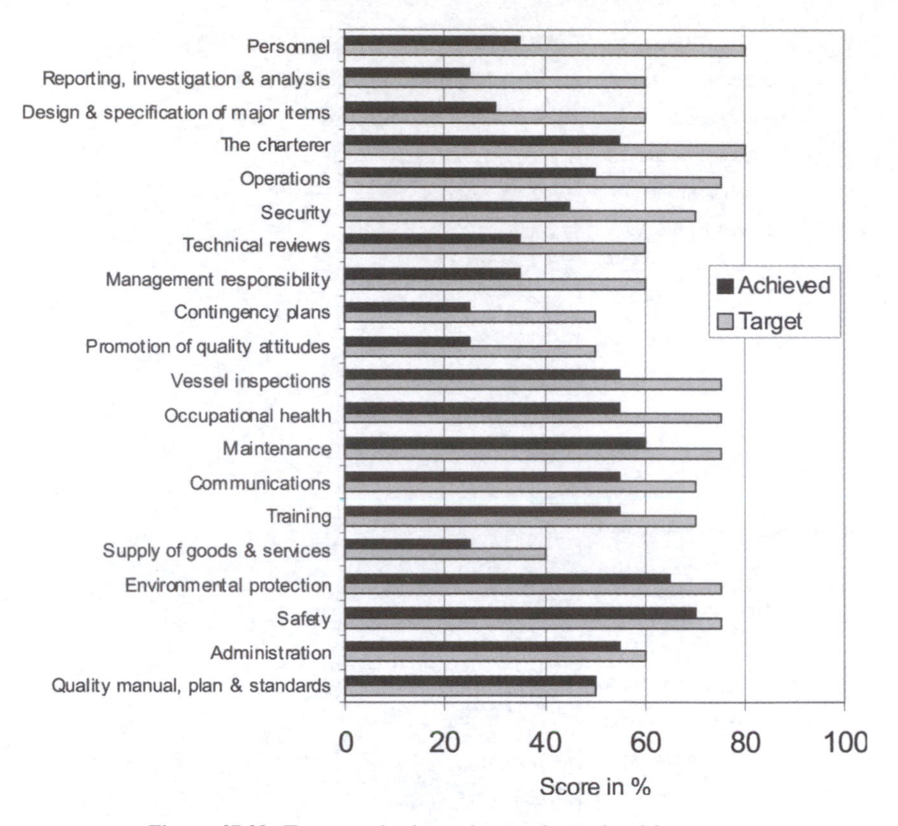

Figure 15.10. Target and achieved score for audited functions.

The process of auditing involves the following steps:

- Going through the SMS documentation.
- Checking logs and other documentation that reflects the practice of the SMS.
- Interviews with key personnel in the company.
- Assessing the competence of key personnel.
- Summarizing and identifying non-conformities in the SMS.
- Assessing corrections of non-conformities.
- Granting a Document of Compliance (DOC).

One potential problem in terms of auditing compliance with the ISM Code is that the Code itself and the SMS documents, which establish the basis for the audit, are fairly general and open to subjective evaluation. Problems related to auditing are discussed by Anderson (1998) and Sagen (1999), and some of these are as follows:

- General formulations in the regulations are open to interpretation.
- Difficulty in giving precise formulations of non-conformities.

- How to assess the seriousness of non-conformities.
- Lack of factual evidence.
- How to pin-point lack of real competence.
- The efficiency aspect of SMS is vague.

15.5 CORPORATE SAFETY CULTURE

15.5.1 What is Safety Culture?

So far in this chapter we have focused on how systematic management can contribute to quality and safety. There is, however, an emerging realization that the performance in different aspects/functions within an organization is rooted in the culture of that organization, as illustrated in Figure 15.11 (Krause et al., 1990).

Some hold the belief that the true basis for systematic effort and genuine concern for safety must be rooted in a safety culture of organizations. It is largely accepted that organizational culture determines the behaviour and performance of its individual members. Organizational culture may be defined as a common set of norms and values within an organization that override differing subunit orientations. The culture concept is inherited from organizational theory and became a key topic in works such as *In Search of Excellence* by Peters and Waterman (1982) and *Corporate Cultures: The Rites and Rituals of Corporate Life* by Deal and Kennedy (1982). Culture is interdisciplinary in nature and the understanding of the concept differs considerably between academic disciplines. The governing view is that there are two main perspectives of organizational culture: the socio-anthropological perspective and the organizational psychology perspective (Wiegmann et al., 2002).

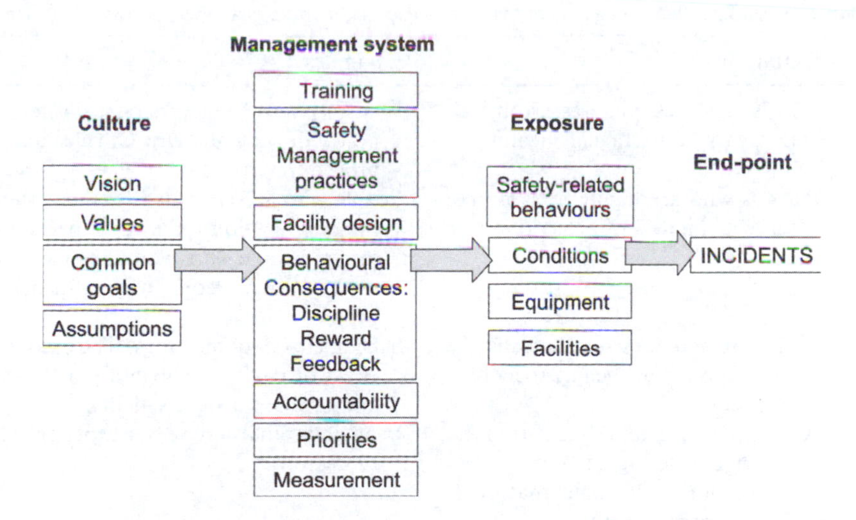

Figure 15.11. Causation of incidents. (Source: Krause et al., 1990.)

The socio-anthropological perspective focuses on the underlying structure of symbols, heroes and rituals, manifested in shared values tied together as practices or as visible manifestations. This perspective can be illustrated by Figure 15.12, and the meaning of the underlying concepts is briefly outlined in Table 15.10. The deeper structure of culture is not immediately observable by outsiders (i.e. those external to the organization), and may even in some cases be difficult to articulate precisely for those within the organization because the concepts are only subtly embedded in 'the way we do things around here'. The global chemical giant Du Pont is an example of a company that is seen as a pioneer in understanding and implementing a safety culture (Koenig, 1993).

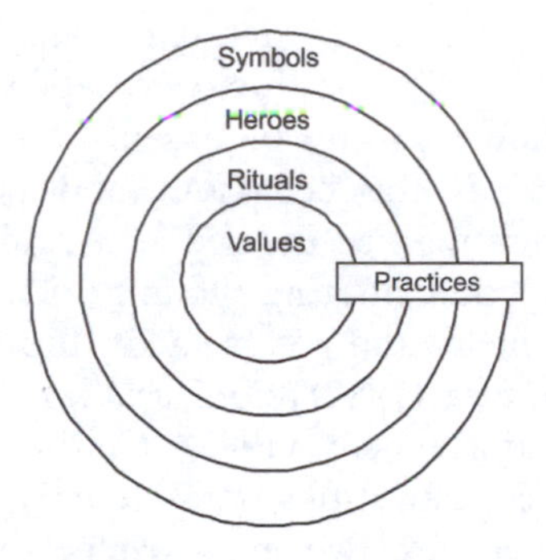

Figure 15.12. The socio-anthropological perspective of culture.

Table 15.10. The underlying concepts of the socio-anthropological perspective of culture

Concept	Definition	Examples
Symbols	Words, gestures, pictures or objects that carry a particular meaning	Warning signs, posters, slogans, safety awards, policy documents, written rules and procedures
Heroes	Persons who are highly praised and serve as models of behaviour	People who are rewarded by their peers and the organization for their effort towards safety. People who 'walk the talk' or practise as they teach. People who influences other people.
Rituals	Collective activities – technically superfluous but seen as socially essential	Scheduled safety meetings. The secondary effect of paying compensation to widows and orphans after accidents.
Values	Attitudes within an organization that override differing subunit orientations. The 'glue' that holds the company together.	Behaviours that have been experienced to be successful

The organizational psychology perspective of (organizational) culture also focuses on shared values and beliefs manifested through symbols, rituals and specialized language. But organizational psychologists hold the view that culture has functional aspects that can be manipulated and thereby contribute to improved productivity. In this view, relevant aspects of the culture concept are organizational commitment, social stability and motivation. The perspective is also more rational in the sense that the culture concept can be broken down analytically and manipulated.

Wiegmann et al. (2002) made a comprehensive literature review on studies of the safety culture concept, and this review clearly showed that people understand the term safety culture differently. Despite this the authors were able to pin-point some common or related definitions:

- Shared values by a group or organization
- Emphasizes the contribution by everyone at every level
- Impact of the behaviour of each member at work
- Reflects the contingency between reward systems and safety performance
- Willingness to learn from errors, incidents and accidents
- Is relatively enduring, stable and resistant to change
- High value put on worker and public safety
- Shared values, beliefs and norms
- Sub-facet of organizational culture
- A joint belief in the importance of safety

The understanding of the term 'safety culture' is not made easier by the fact that the term 'safety climate' also is commonly used. According to Wiegmann et al. (2002), a possible interpretation of the latter term may be that the safety climate is a temporal measure of the safety culture, or perceived state of the culture, at a particular place and time. The following definitions of safety climate are offered in the literature:

- A psychological phenomenon: perception of a safety state
- Intangible issues such as situational and environmental factors
- A 'snapshot' of the safety culture – unstable and subject to change
- Procedures and rules
- The surface features of the safety culture
- Perceptions of safety systems, as well as job and individual factors
- Perceptions about the relative importance of safe conduct

15.5.2 Measuring Safety Culture

Wiegmann et al. (2002) propose a set of organizational indicators that may give a more operationally oriented view on safety culture. These indicators are presented and explained in Table 15.11, and may be used in terms of measuring safety culture.

Similarly, one may assess safety cultures using the following benchmarking criteria, which are more concrete than the organizational indicators for safety culture

Table 15.11. Organizational indicators for safety culture

Indicator	Comment
Organizational commitment	The top management of the organization identify safety as a core value and demonstrate an enduring positive attitude towards improving safety even in times of economic/fiscal austerity
Management involvement	Management participation in day-to-day operations influences the degree to which employees comply with operating rules
Employee empowerment	Employees are given substantial power of influence and responsibility in safety decisions
Reporting systems	The willingness and ability of the organization to learn, i.e. the degree of free and uninhibited reporting of safety issues.
Reward systems	The manner in which behaviour is evaluated and rewards and penalties are doled out.

Source: Wiegmann et al. (2002).

presented in Table 15.11:

- Importance put on safety training
- Degree of job satisfaction
- Labour turnover (in percent)
- Workplace conditions
- Workplace risk, number of occupational accidents
- Status of the safety committee
- Status of safety officers
- Effect of safe conduct on promotion
- Effect of safe conduct on social status

Itoh and Boje Andersen (1999) studied the relation between accident rate and cultural factors with relevance to the terms of safety for a Japanese company responsible for railway track maintenance. They carried out a questionnaire-based survey that focused on the train drivers, who have to operate under fairly stressful conditions. The company in question consists of five different branches and, as Figure 15.13 shows, the workers' morale and motivation, as well as attitudes on management, operating procedures and organizational issues, differed markedly between the branches.

The safety culture factors shown in Figure 15.13 were then correlated with accident frequencies for each branch of the company as shown in Figure 15.14. Accident frequency was expressed by weighting the sum of big and small accidents, considering five small accidents as equal to one big accident (i.e. the scale is 5:1). The upper graph in Figure 15.14 shows accident frequency against the motivation score. The graph gives a clear indication of a correlation between these factors. The branch with the lowest measured motivation (i.e. branch B) had the highest accident rate. The lesson learned with

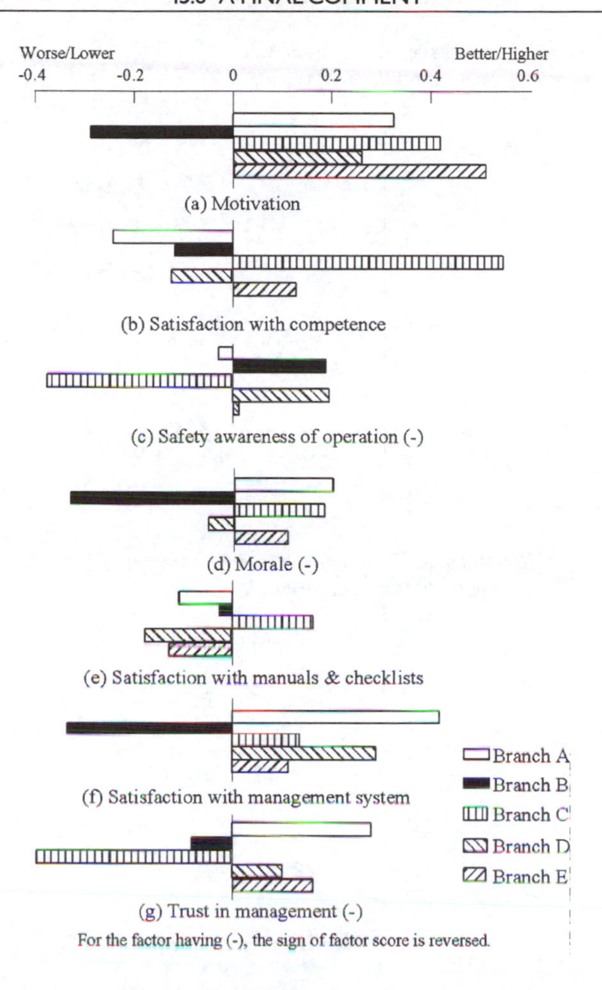

Figure 15.13. Measurement of safety culture factors within branches of a railway track maintenance company. (Source: Itoh and Boje Andersen, 1999.)

respect to the safety culture factor of morale is basically the same, as shown in the lower diagram in Figure 15.14.

15.6 A FINAL COMMENT

This chapter has discussed the importance and role of safety management in achieving high levels of safety. It has been shown how systematic management approaches such as Total Quality Management (TQM) and the International Safety Management (ISM) Code have brought the safety performance of maritime organizations forward. It should also be kept in mind that important progress has been made through a greater focus on education and training. It is also essential to remember that in the end everything boils down to

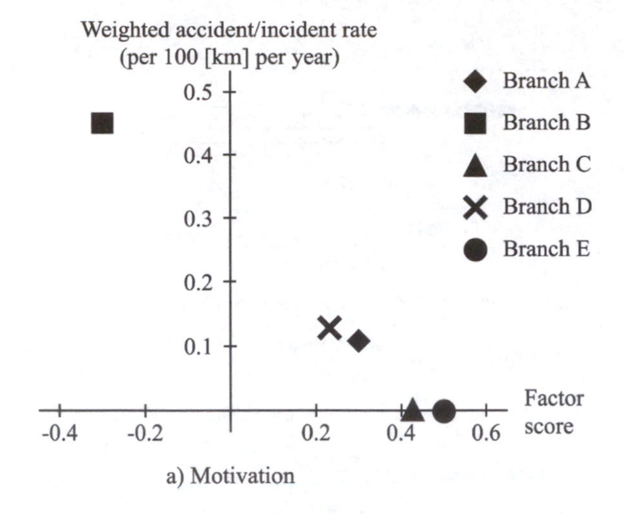

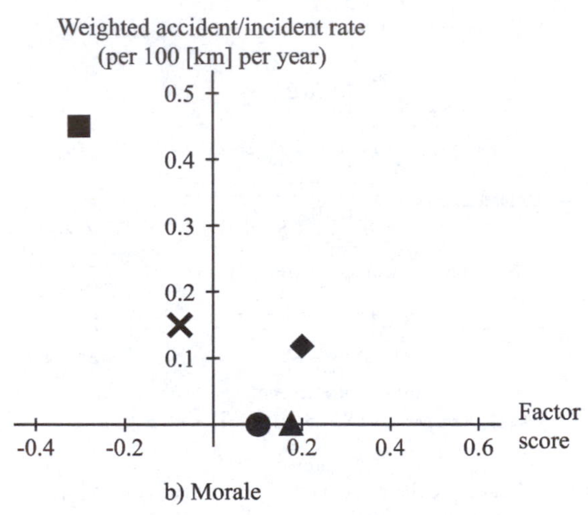

Figure 15.14. The correlation between accident frequency and motivation and morale, respectively, for train drivers. (Source: Itoh and Boje Andersen, 1999.)

improvement of human performance in all parts of the companies and organizations. However, despite the extensive focus put on education, and planning and drilling of safety-critical operations, it should be remembered that humans have a natural tendency to resist change both with respect to values and behaviour in day-to-day operations. Krause et al. (1990) offer the following explanations for this tendency:

- Protection of habits
- Entails additional work
- Means facing the unknown

Table 15.12. Strategies to resist change of behaviour

Strategy	Example
Emotional	Meet with anger, aggression
Cognitive	Change subject. Failing to understand. Make you lose track of the issue.
Social	'We're buddies, why get down on me.' Alter the nature of the relationship.
Behavioural	Fighting: saying OK, but not really complying.
Psycho-physiological	Being too tired. Getting sick.

- Feeling of insecurity and anxiety
- Giving up what seems right
- Feeling that the way things are currently done is adequate

The authors also outline different strategies that are applied in order to resist the pressure for change. These strategies are outlined in Table 15.12. The obvious conclusion from this is that we have to show perseverance and consistency in our efforts to improve or maintain an acceptable safety level.

REFERENCES

Aitken, J. D. et al., 1996, Committed HSE management vs. TQM: is there any difference?, International Conference on Health, Safety and Environment, New Orleans, 9–12 June. Society of Petroleum Engineers Inc., Paper no. SPE 35760.

Anderson, P., 1998, *ISM Code – A Practical Guide to the Legal and Insurance Implications*. LLP Reference Publishing, London.

Bromby, M., 1995, Ensuring compliance with the IMO's Code and its place within quality management systems. Conference on Quality Management Systems in Shipping, 27–28 March, arranged by Institute for International Research Ltd., London.

Clement, D. L. et al., 1996, Business integration of safety, health and environmental management. International Conference on Health, Safety and Environment, New Orleans, 9–12 June. Society of Petroleum Engineers Inc., Paper no. SPE 35852.

Costin, H. I. (ed.), 1998, *Strategies for Quality Improvement: TQM, Reengineering, and ISO 9000*. Dryden Press, Fort Worth, TX.

Deal, T. E. and Kennedy, A. A., 1982, *Corporate Culture: The Rites and Rituals of Corporate Life*. Addison-Wesley, Menlo Park, CA.

Gray, J. R. and Sims, M. D., 1997, Management system audits for ship operators – an auditor's experience, *Transactions of the Institute of Marine Engineers,* Vol. 109(3), 233–255.

ICS and ISF, 1996, *Guidelines on the Application of the IMO International Safety Management (ISM) Code*, 3rd ed. International Chamber of Shipping and International Shipping Federation.

ILCI, 1995, *Texaco EHS Management Systems Audit*. International Loss Control Institute Inc.

IMO, 1994, *International Safety Management Code (ISM Code)*. ISBN 92801 1311 9, International Maritime Organization, London.

IMO, 1995, *Guidelines on Implementation of the ISM Code by Administrations*, resolution A.788(19) adopted on 23 November 1995. International Maritime Organization, London.

ISO, 2000, *ISO 9000 and 14000*. http://www.iso.ch/9000e/9k14ke.htm

ISO, 1995, *ISO 14001, 14004, and 14010 – Draft Standards on Environmental Management*. International Organization for Standardization.

Itoh, K. and Boje Andersen, H., 1999, Motivation and morale of night train drivers correlated with accident rates. Proceedings of CAES '99: International Conference on Computer-Aided Ergonomics and Safety, 19–21 May, Barcelona.

Juran, J. M., 1998, The quality trilogy: a universal approach to managing for quality. In Costin, H. I. (ed.), *Strategies for Quality Improvement: TQM, Reengineering, and ISO 9000*. Dryden Press, Fort Worth, TX.

Koenig, D. W., 1993, The corporate safety culture: its workings and its importance. Offshore Technology Conference, Houston, 3–6 May, Paper no. OTC 7096.

Krause, T. R. et al., 1990, *The Behaviour-Based Safety Process: Managing Involvement for an Injury-Free Culture*. Van Nostrand Reinhold, New York.

Mitroff, I. I. and Pauchant, T. C., 1990, *We're So Big and Powerful Nothing Bad Can Happen to Us: An Investigation of America's Crisis Prone Corporations*. Birch Lane Press, New York.

Morone, J. G. and Woodhouse, E. J., 1986, *Averting Catastrophe: Strategies for Regulating Risky Technologies*. University of California Press, Berkeley.

Peters, T. J. and Waterman, R. H., 1982, *In Search of Excellence: Lessons from America's Best-Run Companies*. Harper & Row, New York.

Sagen, A., 1999, *The ISM Code – In Practice*. Tano Aschehoug, Oslo.

Saunders, R., 1992, *The Safety Audit: Designing Effective Strategies*. Financial Times/Pitman Publishing, London.

Wiegmann, D. A. et al., 2002, *A Synthesis of Safety Culture and Safety Climate Research*. Technical report ARL-02-3/FAA-02-2, Aviation Research Lab, University of Illinois at Urbana-Champaign.

Zharen, W. M. von and Duncan, W., 1994, Environmental risk assessment and management in the maritime industry: the interaction with ISO 9000, ISM and ISM management system. *SNAME Transactions*, Vol. 102, 137–164.

INDEX